IET ENERGY ENGINEERING SERIES 215

Industrial Demand Response

Other volumes in this series:

Volume 1 **Power Circuit Breaker Theory and Design** C.H. Flurscheim (Editor)
Volume 4 **Industrial Microwave Heating** A.C. Metaxas and R.J. Meredith
Volume 7 **Insulators for High Voltages** J.S.T. Looms
Volume 8 **Variable Frequency AC Motor Drive Systems** D. Finney
Volume 10 **SF_6 Switchgear** H.M. Ryan and G.R. Jones
Volume 11 **Conduction and Induction Heating** E.J. Davies
Volume 13 **Statistical Techniques for High Voltage Engineering** W. Hauschild and W. Mosch
Volume 14 **Uninterruptible Power Supplies** J. Platts and J.D. St Aubyn (Editors)
Volume 15 **Digital Protection for Power Systems** A.T. Johns and S.K. Salman
Volume 16 **Electricity Economics and Planning** T.W. Berrie
Volume 18 **Vacuum Switchgear** A. Greenwood
Volume 19 **Electrical Safety: a guide to causes and prevention of hazards** J. Maxwell Adams
Volume 21 **Electricity Distribution Network Design, 2nd Edition** E. Lakervi and E.J. Holmes
Volume 22 **Artificial Intelligence Techniques in Power Systems** K. Warwick, A.O. Ekwue and R. Aggarwal (Editors)
Volume 24 **Power System Commissioning and Maintenance Practice** K. Harker
Volume 25 **Engineers' Handbook of Industrial Microwave Heating** R.J. Meredith
Volume 26 **Small Electric Motors** H. Moczala *et al.*
Volume 27 **AC-DC Power System Analysis** J. Arrillaga and B.C. Smith
Volume 29 **High Voltage Direct Current Transmission, 2nd Edition** J. Arrillaga
Volume 30 **Flexible AC Transmission Systems (FACTS)** Y-H. Song (Editor)
Volume 31 **Embedded generation** N. Jenkins *et al.*
Volume 32 **High Voltage Engineering and Testing, 2nd Edition** H.M. Ryan (Editor)
Volume 33 **Overvoltage Protection of Low-Voltage Systems, Revised Edition** P. Hasse
Volume 36 **Voltage Quality in Electrical Power Systems** J. Schlabbach *et al.*
Volume 37 **Electrical Steels for Rotating Machines** P. Beckley
Volume 38 **The Electric Car: Development and future of battery, hybrid and fuel-cell cars** M. Westbrook
Volume 39 **Power Systems Electromagnetic Transients Simulation** J. Arrillaga and N. Watson
Volume 40 **Advances in High Voltage Engineering** M. Haddad and D. Warne
Volume 41 **Electrical Operation of Electrostatic Precipitators** K. Parker
Volume 43 **Thermal Power Plant Simulation and Control** D. Flynn
Volume 44 **Economic Evaluation of Projects in the Electricity Supply Industry** H. Khatib
Volume 45 **Propulsion Systems for Hybrid Vehicles** J. Miller
Volume 46 **Distribution Switchgear** S. Stewart
Volume 47 **Protection of Electricity Distribution Networks, 2nd Edition** J. Gers and E. Holmes
Volume 48 **Wood Pole Overhead Lines** B. Wareing
Volume 49 **Electric Fuses, 3rd Edition** A. Wright and G. Newbery
Volume 50 **Wind Power Integration: Connection and system operational aspects** B. Fox *et al.*
Volume 51 **Short Circuit Currents** J. Schlabbach
Volume 52 **Nuclear Power** J. Wood
Volume 53 **Condition Assessment of High Voltage Insulation in Power System Equipment** R.E. James and Q. Su
Volume 55 **Local Energy: Distributed generation of heat and power** J. Wood
Volume 56 **Condition Monitoring of Rotating Electrical Machines** P. Tavner, L. Ran, J. Penman and H. Sedding
Volume 57 **The Control Techniques Drives and Controls Handbook, 2nd Edition** B. Drury
Volume 58 **Lightning Protection** V. Cooray (Editor)
Volume 59 **Ultracapacitor Applications** J.M. Miller
Volume 62 **Lightning Electromagnetics** V. Cooray
Volume 63 **Energy Storage for Power Systems, 2nd Edition** A. Ter-Gazarian
Volume 65 **Protection of Electricity Distribution Networks, 3rd Edition** J. Gers
Volume 66 **High Voltage Engineering Testing, 3rd Edition** H. Ryan (Editor)
Volume 67 **Multicore Simulation of Power System Transients** F.M. Uriate
Volume 68 **Distribution System Analysis and Automation** J. Gers
Volume 69 **The Lightening Flash, 2nd Edition** V. Cooray (Editor)
Volume 70 **Economic Evaluation of Projects in the Electricity Supply Industry, 3rd Edition** H. Khatib
Volume 72 **Control Circuits in Power Electronics: Practical issues in design and implementation** M. Castilla (Editor)
Volume 73 **Wide Area Monitoring, Protection and Control Systems: The enabler for Smarter Grids** A. Vaccaro and A. Zobaa (Editors)
Volume 74 **Power Electronic Converters and Systems: Frontiers and applications** A. M. Trzynadlowski (Editor)
Volume 75 **Power Distribution Automation** B. Das (Editor)
Volume 76 **Power System Stability: Modelling, analysis and control** A.A. Sallam and B. Om P. Malik
Volume 78 **Numerical Analysis of Power System Transients and Dynamics** A. Ametani (Editor)
Volume 79 **Vehicle-to-Grid: Linking electric vehicles to the smart grid** J. Lu and J. Hossain (Editors)

Volume 81	**Cyber-Physical-Social Systems and Constructs in Electric Power Engineering** S. Suryanarayanan, R. Roche and T.M. Hansen (Editors)
Volume 82	**Periodic Control of Power Electronic Converters** F. Blaabjerg, K.Zhou, D. Wang and Y. Yang
Volume 86	**Advances in Power System Modelling, Control and Stability Analysis** F. Milano (Editor)
Volume 87	**Cogeneration: Technologies, Optimisation and Implentation** C. A. Frangopoulos (Editor)
Volume 88	**Smarter Energy: from Smart Metering to the Smart Grid** H. Sun, N. Hatziargyriou, H. V. Poor, L. Carpanini and M. A. Sánchez Fornié (Editors)
Volume 89	**Hydrogen Production, Separation and Purification for Energy** A. Basile, F. Dalena, J. Tong and T.N. Veziroğlu (Editors)
Volume 90	**Clean Energy Microgrids** S. Obara and J. Morel (Editors)
Volume 91	**Fuzzy Logic Control in Energy Systems with Design Applications in Matlab/Simulink®** İ H. Altaş
Volume 92	**Power Quality in Future Electrical Power Systems** A. F. Zobaa and S. H. E. A. Aleem (Editors)
Volume 93	**Cogeneration and District Energy Systems: Modelling, Analysis and Optimization** M. A. Rosen and S. Koohi-Fayegh
Volume 94	**Introduction to the Smart Grid: Concepts, technologies and evolution** S.K. Salman
Volume 95	**Communication, Control and Security Challenges for the Smart Grid** S.M. Muyeen and S. Rahman (Editors)
Volume 96	**Industrial Power Systems with Distributed and Embedded Generation** R Belu
Volume 97	**Synchronized Phasor Measurements for Smart Grids** M.J.B. Reddy and D.K. Mohanta (Editors)
Volume 98	**Large Scale Grid Integration of Renewable Energy Sources** A. Moreno-Munoz (Editor)
Volume 100	**Modeling and Dynamic Behaviour of Hydropower Plants** N. Kishor and J. Fraile-Ardanuy (Editors)
Volume 101	**Methane and Hydrogen for Energy Storage** R. Carriveau and D. S-K. Ting
Volume 104	**Power Transformer Condition Monitoring and Diagnosis** A. Abu-Siada (Editor)
Volume 106	**Surface Passivation of Industrial Crystalline Silicon Solar Cells** J. John (Editor)
Volume 107	**Bifacial Photovoltaics: Technology, applications and economics** J. Libal and R. Kopecek (Editors)
Volume 108	**Fault Diagnosis of Induction Motors** J. Faiz, V. Ghorbanian and G. Joksimović
Volume 109	**Cooling of Rotating Electrical Machines:** Fundamentals, modelling, testing and design D. Staton, E. Chong, S. Pickering and A. Boglietti
Volume 110	**High Voltage Power Network Construction** K. Harker
Volume 111	**Energy Storage at Different Voltage Levels: Technology, integration, and market aspects** A.F. Zobaa, P.F. Ribeiro, S.H.A. Aleem and S.N. Afifi (Editors)
Volume 112	**Wireless Power Transfer: Theory, Technology and Application** N.Shinohara
Volume 114	**Lightning-Induced Effects in Electrical and Telecommunication Systems** Y. Baba and V. A. Rakov
Volume 115	**DC Distribution Systems and Microgrids** T. Dragičević, F. Blaabjerg and P. Wheeler
Volume 116	**Modelling and Simulation of HVDC Transmission** M. Han (Editor)
Volume 117	**Structural Control and Fault Detection of Wind Turbine Systems** H.R. Karimi
Volume 119	**Thermal Power Plant Control and Instrumentation: The control of boilers and HRSGs, 2nd Edition** D. Lindsley, J. Grist and D. Parker
Volume 120	**Fault Diagnosis for Robust Inverter Power Drives** A. Ginart (Editor)
Volume 121	**Monitoring and Control using Synchrophasors in Power Systems with Renewables** I. Kamwa and C. Lu (Editors)
Volume 123	**Power Systems Electromagnetic Transients Simulation, 2nd Edition** N. Watson and J. Arrillaga
Volume 124	**Power Market Transformation** B. Murray
Volume 125	**Wind Energy Modeling and Simulation Volume 1: Atmosphere and plant** P. Veers (Editor)
Volume 126	**Diagnosis and Fault Tolerance of Electrical Machines, Power Electronics and Drives** A.J. M. Cardoso
Volume 128	**Characterization of Wide Bandgap Power Semiconductor Devices** F. Wang, Z. Zhang and E.A. Jones
Volume 129	**Renewable Energy from the Oceans: From wave, tidal and gradient systems to offshore wind and solar** D. Coiro and T. Sant (Editors)
Volume 130	**Wind and Solar Based Energy Systems for Communities** R. Carriveau and D. S-K. Ting (Editors)
Volume 131	**Metaheuristic Optimization in Power Engineering** J. Radosavljević
Volume 132	**Power Line Communication Systems for Smart Grids** I.R.S Casella and A. Anpalagan
Volume 134	**Hydrogen Passivation and Laser Doping for Silicon Solar Cells** B. Hallam and C. Chan (Editors)

Volume 139	**Variability, Scalability and Stability of Microgrids** S. M. Muyeen, S. M. Islam and F. Blaabjerg (Editors)
Volume 143	**Medium Voltage DC System Architectures** B. Grainger and R. D. Doncker (Editors)
Volume 145	**Condition Monitoring of Rotating Electrical Machines** P. Tavner, L. Ran, C. Crabtree
Volume 146	**Energy Storage for Power Systems, 3rd Edition** A.G. Ter-Gazarian
Volume 147	**Distribution Systems Analysis and Automation 2nd Edition** J. Gers
Volume 151	**SiC Power Module Design: Performance, robustness and reliability** A. Castellazzi and A. Irace (Editors)
Volume 152	**Power Electronic Devices: Applications, failure mechanisms and reliability** F Iannuzzo (Editor)
Volume 153	**Signal Processing for Fault Detection and Diagnosis in Electric Machines and Systems** M. Benbouzid (Editor)
Volume 155	**Energy Generation and Efficiency Technologies for Green Residential Buildings** D. Ting and R. Carriveau (Editors)
Volume 156	**Lithium-ion Batteries Enabled by Silicon Anodes** C. Ban and K. Xu (Editors)
Volume 157	**Electrical Steels, 2 Volumes** A. Moses, K. Jenkins, Philip Anderson and H. Stanbury
Volume 158	**Advanced Dielectric Materials for Electrostatic Capacitors** Q Li (Editor)
Volume 159	**Transforming the Grid Towards Fully Renewable Energy** O. Probst, S. Castellanos and R. Palacios (Editors)
Volume 160	**Microgrids for Rural Areas: Research and case studies** R.K. Chauhan, K. Chauhan and S.N. Singh (Editors)
Volume 161	**Artificial Intelligence for Smarter Power Systems: Fuzzy Logic and Neural Networks** M. G. Simoes
Volume 166	**Advanced Characterization of Thin Film Solar Cells** N. Haegel and M Al-Jassim (Editors)
Volume 167	**Power Grids with Renewable Energy** Storage, integration and digitalization A. A. Sallam and B. OM P. Malik
Volume 169	**Small Wind and Hydrokinetic Turbines** P. Clausen, J. Whale and D. Wood (Editors)
Volume 170	**Reliability of Power Electronics Converters for Solar Photovoltaic Applications** F. Blaabjerg, A.l Haque, H. Wang, Z. Abdin Jaffery and Y. Yang (Editors)
Volume 171	**Utility-scale Wind Turbines and Wind Farms** A. Vasel-Be-Hagh and D. S.-K. Ting
Volume 172	**Lighting interaction with Power Systems, 2 volumes** A. Piantini (Editor)
Volume 174	**Silicon Solar Cell Metallization and Module Technology** T. Dullweber (Editor)
Volume 180	**Protection of Electricity Distribution Networks, 4th Edition** J. Gers and E. Holmes
Volume 182	**Surge Protection for Low Voltage Systems** A. Rousseau (Editor)
Volume 184	**Compressed Air Energy Storage: Types, systems and applications** D. Ting and J. Stagner
Volume 186	**Synchronous Reluctance Machines: Analysis, optimization and applications** N. Bianchi
Volume 193	**Overhead Electric Power Lines: Theory and practice** S. Chattopadhyay and A. Das
Volume 194	**Offshore Wind Power Reliability, availability and maintenance, 2nd edition** P. Tavner
Volume 198	**Battery Management Systems and Inductive Balancing** A. Van den Bossche and A. Farzan Moghaddam
Volume 199	**Model Predictive Control for Microgrids: From power electronic converters to energy management** J. Hu, J. M. Guerrero and S. Islam
Volume 204	**Electromagnetic Transients in Large HV Cable Networks: Modeling and calculations** Ametani, Xue, Ohno and Khalilnezhad
Volume 208	**Nanogrids and Picogrids and their Integration with Electric Vehicles** S. Chattopadhyay
Volume 212	**Battery State Estimation: Methods and Models** S. Wang
Volume 225	**Fusion-Fission Hybrid Nuclear Reactors: For enhanced nuclear fuel utilization and radioactive waste reduction** W. M. Stacey
Volume 905	**Power system protection, 4 volumes**

Industrial Demand Response

Methods, best practices, case studies, and applications

Edited by
Hassan Haes Alhelou, Antonio Moreno-Muñoz and Pierluigi Siano

The Institution of Engineering and Technology

Published by The Institution of Engineering and Technology, London, United Kingdom

The Institution of Engineering and Technology is registered as a Charity in England & Wales (no. 211014) and Scotland (no. SC038698).

First published 2022

The Institution of Engineering and Technology
Futures Place
Kings Way, Stevenage
Hertfordshire, SG1 2UA, United Kingdom

www.theiet.org

British Library Cataloguing in Publication Data
A catalogue record for this product is available from the British Library

ISBN 978-1-83953-561-1 (hardback)
ISBN 978-1-83953-562-8 (PDF)

Typeset in India by MPS

Contents

About the editors

Hassan Haes Alhelou is a Senior Member of IEEE. He is with the Department of Electrical and Computer Systems Engineering, Monash University, Clayton, Australia. At the same time, He is a Professor and faculty member at Tishreen University in Syria, and a consultant with Sultan Qaboos University (SQU) in Oman. Previously, He was with the School of Electrical and Electronic Engineering, University College Dublin (UCD), Dublin 4, Ireland between 2020–21, and with Isfahan University of Technology (IUT), Iran. He completed his B.Sc. from Tishreen University in 2011, M.Sc. and Ph.D. from Isfahan University of Technology, Iran all with honors. He was included in the 2018 and 2019 Publons and Web of Science (WoS) list of the top 1% best reviewer and researchers in the field of engineering and cross-fields over the world. He was the recipient of the Outstanding Reviewer Award from many journals, e.g., Energy Conversion and Management (ECM), ISA Transactions, and Applied Energy. He was the recipient of the best young researcher in the Arab Student Forum Creative among 61 researchers from 16 countries at Alexandria University, Egypt, 2011. He also received the Excellent Paper Award 2021/2022 from IEEE CSEE Journal of Power and Energy Systems (SCI IF: 3.938; Q1). He has published more than 200 research papers in high-quality peer-reviewed journals and international conferences. His research papers received 2850 citations with H-index of 26 and i-index of 56. He authored/edited 15 books published in reputed publishers such as Springer, IET, Wiley, Elsevier, and Taylor & Francis. He serves as an editor in a number of prestigious journals such as *IEEE Systems Journal*, *Computers and Electrical Engineering* (CAEE-Elsevier), *IET Journal of Engineering, and Smart Cities*. He has also performed more than 800 reviews for high prestigious journals, including *IEEE Transactions on Power Systems*, *IEEE Transactions on Smart Grid*, *IEEE Transactions on Industrial Informatics*, *IEEE Transactions on Industrial Electronics*, *Energy Conversion and Management*, *Applied Energy*, *International Journal of Electrical Power & Energy Systems*. He has participated in more than 15 international industrial projects over the globe. His major research interests are Renewable energy systems, Power systems, Power System Security, Power system dynamics, Power System Cybersecurity, Power system operation, control, Dynamic state estimation, Frequency control, Smart grids, Micro-grids, Demand response, and Load shedding.

Antonio Moreno-Muñoz is with the Department of Electronics and Computer Engineering, University of Cordoba, Cordoba, Spain. He is a Professor at the University of Córdoba, Spain, where he is Chair of the Industrial Electronics and Instrumentation R&D Group. Besides his Senior Membership with the IEEE Technical Committee on Smart Grids and extensive experience with the Spanish rail company RENFE, he is member of various related committees and boards. His research focuses on industrial electronics for smart grids and renewable energy systems and he has published extensively in this area.

Pierluigi Siano (M'09–SM'14) received the M.Sc. degree in electronic engineering and the Ph.D. degree in information and electrical engineering from the University of Salerno, Salerno, Italy, in 2001 and 2006, respectively. He is a Professor and Scientific Director of the Smart Grids and Smart Cities Laboratory with the Department of Management & Innovation Systems, University of Salerno. Since 2021 he has been a Distinguished Visiting Professor in the Department of Electrical & Electronic Engineering Science, University of Johannesburg. His research activities are centered on demand response, energy management, the integration of distributed energy resources in smart grids, electricity markets, and planning and management of power systems. In these research fields, he has co-authored more than 660 articles including more than 390 international journal that received in Scopus more than 13,600 citations with an H-index equal to 57. In 2019, 2020, and 2021 he has been awarded as Highly cited Researcher in Engineering by Web of Science Group. He has been the Chair of the IES TC on Smart Grids. He is Editor for the Power & Energy Society Section of *IEEE Access*, *IEEE Transactions on Power Systems*, *IEEE Transactions on Industrial Informatics*, *IEEE Transactions on Industrial Electronics*, and *IEEE Systems*.

Foreword

Power systems are the direct reflection of their countries' economy, security, and advancement. They should be operated in safe and stable modes in order to avoid negative consequences on their nation's security and economy. For decades, the traditional power systems were operated safely and securely with no problems related to its stability and security. This view of power systems has been changed during the last decade due to environmental challenges and energy security risks which enabled the movement toward the feasible implementation of smart grid concept. In the last decade, the penetration level of renewable energy sources (RESs) has been highly increased. Although RESs reduce the environmental concerns, but have negative impacts on the stability and security of existing power systems. For instance, the high share of renewables in modern power systems has reduced the total rotating inertia and as consequences the frequency variable has been affected.

In conventional power systems, the frequency which is a global variable was controlled by well-operating and managing primary and secondary reserves came from generation side. However, this is not yet valid due to change in the power system situation and high reduction of the rotating inertia, therefore the operators need faster reserve that cannot be available from generation side. It has been found that the best source of such reserves is the demand side. Later, new topics has been initialized which are demand side management and demand response for controlling the intelligent appliances that can provide some of their capacities as source for the necessary reserves. As a consequence, it has been suggested to take advantages of demand response for providing ancillary services in power systems.

There are specific types of loads and smart appliances that can provide ancillary services to power systems. These loads should not affect the conformable of the consumers and in the same time the management of these appliances should be done based on reserve and energy markets. For instance, there are great research activities these days on building a practical aggregators of electrical vehicles so that they can participate in ancillary services market by well-controlling their charging and stage of charge situation during a specific period of time. Likewise, researchers have suggested models of air conditioners, refrigerators, water heaters and other thermostat appliances for considering them as good storage aggregators that can provide some services to power systems.

It is clear that there is a need for comprehensive book on demand response topic that can cover the latest methods, best practices, case studies, and applications for encouraging researchers and governments for supporting the movement toward smart grid concept.

It is obvious that the penetration level of renewable energy sources is highly increasing over the world. Solar and wind energy are among the most percentage shares of renewables in modern power systems. By nature of the photovoltaic cells, they provide zero inertia to power systems. This means that increasing the power generation from solar power plants would at least reduce the inertia with the same percentage of their active power generation share increase. As aforementioned, this high reduction in inertia would bring new challenges and technical issues to the operators of modern power systems where the main problems related to stability and security of energy systems. On the other hand, different types of wind turbines provide neglectable inertia (almost zero) to modern power systems bringing the same problems that would be arisen from solar energy systems. Therefore, there is a serious need for new sources to keep the balance in power system operation especially in the view of providing ancillary services. With the low inertia in future power systems, the traditional and conventional reserve sources would not act accurately and properly in the aim of maintaining the power system stability, therefore, it is highlighted that the demand side can be considered as a good source of such services based on demand response programs. The main advantage of demand response over energy and reserve sources in the generation sides is its flexibility which is the most important feature to the operators. Different types of demand side loads can provide demand response services to power systems, where this book will focus on industrial demand response and the types of demands that have the feasibility of implementation.

Introduction

The book is principally focused on industry demand responses (DRs) and their roles in modern power systems. The book is sorted out and organized in 17 chapters. Each chapter begins with the fundamental structure of the problem required for a rudimentary understanding of the methods described. The book starts with chapters that give comprehensive review and discussions on industry DR and their security. The next set of chapters discusses the aggregation, optimization, and technical issues related to industry DR. Finally, the last set of chapters presents best practices and case studies. Brief descriptions of the book chapters are given below.

Chapter 1: One of the key points marking the transition from traditional toward smart energy grids is the provision of flexibility services from the demand side. Power flexibility facilitates the integration of renewable energy resources (RESs), while the balance between the supply and the demand side is maintained. DR techniques are providing the opportunity for high exploitation of the flexibility potential, since they enable reducing, increasing or shifting a portion of the electrical demand, for a specific time period. The industrial sector is expected to have a more significant contribution compared to the residential and the commercial sector, mainly because of the high-consuming equipment, the scheduled operations and the already installed metering equipment on the facilities. The present work explores the state-of-the-art DR applications implemented in the industrial sector. The individual characteristics of each type of industry are analyzed, as they play a major role in the identification of DR potential, since industrial processes may involve critical loads, being highly correlated, that must follow strict operational constraints. The current level of participation in industrial DR programs is being assessed, identifying possible technical or regulatory limitations that prevent further adoption.

Chapter 2: Cybersecurity in industry DR is crucial for modern power systems due to their high digitization. The open information and communication technology, that ICT is being used for the operation of such systems, is highly vulnerable to cyber threats. The adopted smart grid concept around the globe enables the utilization of demand-side for providing ancillary services based on well-known DR programs. These programs aggregate smart appliances in homes and electric vehicles (EVs) for providing vital services such as frequency regulation a voltage support. Since the aggregation is based on cyber layer, any cyber threat could affect the ancillary services that are being delivered from the aggregators, which might lead to stability and security issues resulting in brownout or massive blackouts. This chapter discusses the cybersecurity in DR program and shows its importance for modern and future smart power systems due to their stability and security margins. Furthermore, the cyberattack case study is implemented in a power system with demand side program responsible for providing primary frequency support ancillary service, where the results confirm the high vulnerability of modern power systems to cyber threats on DR-active power reserve providers. Moreover, technical suggestions are provided for enhancing the cybersecurity in DR programs in power systems with high-power share from renewable energy sources (RESs).

Chapter 3: Electric load forecasting is a fundamental technique to understand end-user behavior and therefore a crucial factor in the design of DR programs. Load forecasting will also identify the appropriate design of DR programs. In this chapter, a range of different machine learning applications are covered to represent the influential factors for electrical load demand forecast in a DR context, with a variety of different data scenarios, temporal and technical scenario. This chapter explores and compares the load prediction analysis through basic recurrent neural networks (RNNs); vanilla RNN, gated recurrent units (GRU), and long short-term memory (LSTM), using principal component analysis (PCA). It is found that PCA can be used to reduce the number of principal components for vanilla RNN, GRU, and LSTM networks. Reducing the number of principal components using PCA is one of the techniques that are used in dimensionality reduction. Reduction in dimensionality will relieve the computational burden. In this work, the dimensionality reduction improves the predictive output. It is observed that for electric load demand forecasting, the preferred technique is GRUs, trained with principal components. The performance is evaluated through mean absolute percentage error (MAPE), which is relatively lower than other techniques.

Chapter 4: DR, which is an important feature of smart grid, can play avital role by making the demand side more responsive to the varying gap between demand and supply. DR is utilized by power utilities to maintain system reliability, security and stability while customers utilize it to reduce the electricity cost by increasing or decreasing the load during valley or peak demand periods. Industries consume huge amount of electricity; therefore, DR strategies are required to be implemented by industrial customers to enhance the saving. Further, industrial customers can provide DR by employing many different technologies or strategies to achieve shifts in demand in the following ways: (i) reducing or interrupting consumption temporarily with no change in consumption in other periods, (ii) shifting consumption to other time periods, and (iii) temporarily utilizing onsite generation in place of energy from the grid.

Chapter 5: Smart grid enables active participation of consumers' daily operation of the grid through DR. DR refers to the actions initiated from contracted customers by changing their demand in response to price signals, incentives, or directions from grid operators. In this chapter, industrial DR suitable for frequency regulation is discussed. For this, a mathematical model of price-based DR from thermostatically controlled loads (TCL) for controlling the temperature of the chillers in large academic complex environment is presented. A probabilistic model of the density function of aggregated TCL loads is discussed. The variation of the thermostat set point demand temperature an increase in the price is presented. In order to match the power demand and power supply, a new method for dynamic demand control (DDC) with automatic generation control (AGC) in smart grid environment is proposed. A load frequency control using DDC was modeled in this study. The load frequency control model was simulated for a step load change of 0.01. The frequency deviation was compared with the frequency deviation obtained when generation control, using PI controller, alone was implemented for frequency control. Thus, DDC alone is required to maintain the system frequency, during small load variations. DDC will play a major role in reducing these losses caused to the GENCOs under a smart grid environment.

Chapter 6: In 2020, the sales volume of EVs in China reached 1.367 million. A rapid growth trend was witnessed by the huge increment of electric EVs in past several years. By the end of 2020, China has nearly 5 million new energy vehicles. Meanwhile, China's charging infrastructure has reached 1,681,000 units. It is expected that the global penetration rate of new energy vehicles will exceed 30% in 2030. At that time, the number of EVs in China is prospected to be 80–100 million. As the largest EV market in the world, China has unique conditions to develop and study the interactive application of EVs and power grid. The power system can have a chance of promoting comprehensive innovation thanks to the booming of the EV industry. A smart energy transportation network that can participate in the grid DR) timely, would possibly consist of massive EVs, the power grid, renewable energy network and transportation network. Because of the inherent mobile energy storage characteristics of EVs, flexible large-scale EV clusters have great potential in power load regulation, renewable energy consumption, power quality improvement, etc. Thus EVs can be used to participate in auxiliary services such as peak shifting and valley filling, frequency regulation, emergency support so as to interaction friendly with the grid. In recent years, many cities in China have tried to include EVs in the pilot, and made positive exploration in vehicle network interaction.

Chapter 7: DR programs are defined as the ability of customers to change their consumption pattern in response to market/system signals. Nowadays, DR programs are interested worldwide as an essential part of the future power system and also considered as virtual generation resources. However, an accurate measurement and verification (M&V) approach is needed to implement these programs successfully. Indeed, the evaluation of the real potential of a DR program that is enabled during a DR event is depended on an evaluation method that should be employed to estimate the consumption behavior of the customers if they have not participated in DR. In this regard, customer base-load (CBL) estimation is defined as the approach to estimate the customers' load levels if they have not received DR calls. Then, by computing the difference among the estimated baseline and measured load data, the real potential of DR would be calculated. So, the determination of the real potential of DR is dependent on the difference between the estimated baseline and measured load data. Since various factors (such as load type, weather condition, and day of a week) could affect the CBL, it is a challenging and complex task to provide an accurate estimation of the CBLs.

Chapter 8: In a smart grid paradigm, the concepts of DR and transactive energy (TE) are used to optimize the consumption and generation in the power networks. In this chapter, two models for DR are analyzed based on the well-known Cobb–Douglas utility function. Both models maximize their utility, subject to different constraints. A time-of-use price-based DR program is employed. Restructuring in the electricity sector, with an increase in RERs and distributed energy management technologies, offers the potential for significant improvement in the efficiency of power systems through the TE framework. In a TE framework, prosumers of all sizes can participate in the double auction electricity markets via automated home energy management systems. Heating, ventilation, and air conditioning (HVAC) and energy storage devices are the two important loads in residential buildings that account for a large proportion of building energy consumption. A two-way exchange of energy and information is possible with the current advent of communication systems and net metering. In this work, we consider the case of solar photovoltaics (PV), HVAC, and energy storage devices (EVs) and battery energy storage systems (ESSs) of prosumers participating in the retail real-time double auction market. The problem is formulated as maximization of social welfare subject to power balance and network constraints. Simulation studies and results are presented for the modified IEEE 13 node distribution system.

Chapter 9: Supply demand balance is imperative for reliability of power system. Inability to maintain this balance results in frequency deviation and system failure. The recent integration of RESs such as wind and solar has reduced inertia and variable output which leaves the power system at risk to disturbance while also reducing controllability of generators. However, latter day DR is coming across as an economical and effective way of adding to the reliability and security of power system by managing electricity demand of customers at times of severe power imbalance. This chapter carries out a detailed literature review of centralized and decentralized demand control approaches. As well as presents a novel demand control approach for providing frequency regulation by using domestic refrigerators as control loads. This chapter also carries out a detailed study of large-scale appliance level interval meter consumption data from Australia's largest network provider Ausgrid. Appliance level data is used in combination with household level data to study the contribution of air-conditioners in summer peak demand. Clustering is performed on air-conditioner data to identify various air-conditioner load profile patterns. These patterns are then used with demand control strategies to study the possible load reductions from residential air-conditioner control across the Australian State of New South Wales.

Chapter 10: Current environmental trends such as the rapid penetration of RES and decommissioning of controllable but polluting generators are putting stress on the reliable operation of electricity systems. This reduction of flexibility and increase in volatility in the supply side calls for compensation from other sources in the grid. Although the development of ESSs is creating an opportunity to relax the energy balance constraints in the grid, it is currently not sufficient to solve the constantly growing need for flexibility.

Chapter 11: DR is one of the pillars of the modern distribution system, where consumers would voluntarily reduce consumption in response to financial incentives. While for residential consumers, demand curtailment is mainly a matter of inconvenience, for industrial customers, reduction in electric demand can lead to severe operational ramifications such as halt in production, pile-up of inventory, or wasted labor. These challenges have caused industrial DR to remain less explored compared to residential demand side management. Although the industrial sector may be small by numbers, its energy consumption is the dominant load on most distribution systems. This further underlines the potential benefits gained by involving industrial loads in DR events. One way to motivate industrial DR is to improve opportunities for indirect cost savings as a result of participation in a DR event. For instance, it has been shown in the literature that demand curtailment can be performed in conjunction with inventory management in order to help the plant operate at or near just-in-time. Another option can be to coordinate DR with asset management, i.e., by shutting down lines and workstations under stress or those that are due for maintenance. This way, the direct financial incentives from participating in DR can be augmented with long-term benefits of optimal asset utilization. Providing one such solution is the goal of the current chapter. A multi-objective optimization framework is proposed here that allows plant operators to balance and optimize different financial, operational, and resource objectives while taking advantage of DR to alleviate operational stress on assets. The approach can further incentivize plant managers to participate in DR events, while allowing electric utilities to employ this significant untapped potential.

Chapter 12: Considerable interest is now being vested in low-carbon energy sources in other to meet the world's ever-growing energy demand without causing damage to the environment, which has given rise to the increasing contributions of RESs to the energy grid. This development is not without its challenges to modern electric power systems. Due to the intermittent nature of RERs, its increase has resulted in energy demand-supply mismatch, grid imbalance, or grid instability. A reliable and cost-effective approach is required to address this energy trade imbalance caused by the influx of RESs. DR is a concept that aims at achieving energy balance in the grid by controlling and adjusting flexible loads. Industrial DR has the potential for a significant contribution to the operational flexibility of power systems. Since the industrial sector is one of the major electricity consumers in the world, as many industrial loads consume much electrical energy, therefore, a proper industrial DR regime will help in ensuring a safe and secured grid, improve energy balance, promote decarbonization, more grid reliability and cost reduction for customers.

Chapter 13: Industrial DR (IDR) has been used for regulation and balance purposes for many years. Large energy intensity manufactures have responded to external signals, generally from system operator (SO) to shift or shutdown loads according with emergence or unexpected situations in power system. Historically, such ancillary services were very constrained and only remunerated by means of payments according with an individual contract agreement between SO and industrial customer. Nowadays, new IDR programs have been investigated, which have received a great impulse because of the development of new control and communication systems, distributed energy resources (DER) based on RES, energy storage capacity, alongside market liberalization and pricing diversification. Such new IDR programs tend to focus on controllable, deferrable and interruptible loads aggregation, as well as on DER and ESS aggregation through virtual power plants (VPP) or the energy market participation of an aggregator on behalf of the prosumers.

Chapter 14: DR is a basic tool to achieve power systems flexibility in the short and medium terms. The effective deployment of DR and the engagement of new resources need both knowledge about how DR performs and how to evaluate their flexibility to give a correct economic feedback to customers and aggregators. DR verification requires a reference in absence of control: the customer baseline load (CBL). The aim of this chapter is to describe several baselines that provide an acceptable evaluation of load response as well as the use of different adjustment methods to improve the CBL. Some of these adjustment factors can be justified through the simulation of physical-based load models (PBLM), which are also used in DR for planning and operational tasks. The chapter discusses some issues reported by grid operators: detection of abnormal responses (before and after DR) that can be due to gaming or are reactions to maintain load service such as preheating, precooling or the change of tasks timeline. All these approaches have been illustrated using real data of an industrial customer. Results show that the adjustment of CBLs can improve several conventional approaches described in the literature.

Chapter 15: Despite its comparative advantages with respect to residential and commercial DR, industrial DR (IDR) in general, and modeling of different types of flexible industrial processes in particular, has received relatively limited research attention, with previous work having only explored limited and industry-sector-specific subsets of such processes. This chapter adopts an alternative, sector-agnostic modeling approach and develops generic models of all conceivable types of flexible industrial processes, with the aim to shed light on their key operating differences and assist industrial consumers interested in IDR schemes to identify and assess the types that are more relevant to their systems. In this context, this chapter identifies and discusses seven different types: (1) uninterruptible processes with fixed power, (2) interruptible processes with fixed power, (3) uninterruptible processes with discretely adjustable power, (4) interruptible processes with discretely adjustable power, (5) uninterruptible processes with continuously adjustable power, (6) interruptible processes with continuously adjustable power, and (7) material storage buffers.

Chapter 16: DR has proven to be a crucial mechanism in the process of flexibility exploitation on the demand side. Throughout the years, it has evolved and expanded, reaching more and more previously untapped potential sources. In that process, residential users have provided a significant buffering capacity for balancing energy production and demand, but this came with a few challenges. With more and more households transitioning from being purely energy users to smart homes and energy prosumers with distributed renewable energy generation, new possibilities have opened up for integrated optimization approaches that make the best use of both locally generated and grid-supplied energy as well as energy storage systems.

Chapter 17: Despite all achievements, and advances in energy markets, microgrids, and smart grids within the world, issues such as power distribution, consumption, or optimization are among the important and significant areas within the industry and technology. As industrialization and technology improve, these subjects become more important. Most of the experts attempt to have far better control on power consumption/distribution, and technologies like combined heat, and power (CHP), or gas-electricity, or demand forecasting, especially in smart sustainable cities (SSCs). Using artificial intelligence (AI) and neural networks (NNs) can have an important role in performing, and optimization that will lead to lowering the issues in future power systems. An NN-LSTM-based model can help the experts to control, predict, and optimize the facility consumption, and power distribution. Conceptually, in industrial and smart sustainable cities, more they develop, more the quantity of data is going to be generated that a simple and practical tool to research about and analyze these big data is AI. Regarding an outsized amount of data, the training and predicting process of AI is going to be far more accurate, due to the low root mean square error (RMSE). Accordingly, the result is going to be near the actual and help the SSCs to possess controlled power consumption, distribution, and CHPs. Also, the combination of quantum technology with smart grids, and NNs are analyzed. Accordingly, the mentioned technologies cause preventing power loss and promoting a way to a smarter, technology-based, and sustainable world with high ability of DR.

The editors
Hassan Haes Alhelou, Antonio Moreno-Muñoz, Pierluigi Siano

Chapter 1

A comprehensive review on industrial demand response strategies and applications

Christos Timplalexis,[1] Georgios-Fotios Angelis,[1] Stylianos Zikos,[1] Stelios Krinidis,[1,2] Dimosthenis Ioannidis[1] and Dimitrios Tzovaras[1]

One of the key points marking the transition from traditional toward smart energy grids is the provision of flexibility services from the demand side. Power flexibility facilitates the integration of Renewable Energy Resources (RESs), while the balance between the supply and the demand side is maintained. Demand Response (DR) techniques are providing the opportunity for high exploitation of the flexibility potential, since they enable reducing, increasing or shifting a portion of the electrical demand, for a specific time period. The industrial sector is expected to have a more significant contribution compared to the residential and the commercial sector, mainly because of the high-consuming equipment, the scheduled operations and the already installed metering equipment on the facilities. The present work explores the state-of-the-art DR applications implemented in the industrial sector. The individual characteristics of each type of industry are analyzed, as they play a major role in the identification of DR potential, since industrial processes may involve critical loads, being highly correlated, that must follow strict operational constraints. The current level of participation in industrial DR programs is being assessed, identifying possible technical or regulatory limitations that prevent further adoption.

1.1 Introduction

As the energy landscape is rapidly evolving into a digital and more dynamic era, future power systems are also changing, trying to adapt to this challenging transition. At the same time, governments and other public stakeholders have also provided for

[1]Information Technologies Institute, Centre for Research and Technology Hellas, Thessaloniki, Greece
[2]Department of Management Science and Technology, International Hellenic University (IHU), School of Economics and Business Administration, Thessaloniki, Greece

extremely ambitious future plans regarding significant reduction targets in greenhouse gas (GHG) emissions. In European Union (EU), for example, the most recent Climate Target Plan [1] commits for a reduction of GHG emissions by at least 55% by 2030 and aims toward achieving climate neutrality by 2050. The energy sector is making an effort to contribute to those targets by introducing the Smart Grid (SG) paradigm. SGs are based on the widespread implementation of Advanced Metering Infrastructure (AMI) and on the huge capabilities that derive from the development of Artificial Intelligence (AI) solutions in the field of electrical energy monitoring and control.

The operational model on which traditional power systems were built considers the demand side as non-flexible. The supply side is therefore responsible for making all the necessary adaptations in order to provide power to the grid in a reliable manner. This model mainly depends on fossil fuel-based power stations, whose output is determined by the amount of fuel that they are supplied, consequently they can be considered controllable to some extent. However, the high penetration of Renewable Energy Resources (RES) has introduced a level of uncertainty on the supply side, since solar or wind power output for example depends on stochastic complex natural phenomena. The concept of Demand Side Management (DSM) was first introduced by the Electric Power Research Institute (EPRI) in the 1980s as a series of activities that utilities undertake to change their load shape and/or energy consumption pattern for benefit maximization, investment delay, and reliability enhancement [2]. DSM activities can be classified either as "Energy Efficiency (EE)" or as "Demand Response (DR)." DR denotes a power consumption shift made by a utility customer (residential, commercial or industrial), as a response to a price signal or an incentive-based reward. It inherently tries to mitigate some of the challenges that are deriving from the SG transition, such as the intermittent and stochastic nature of RESs and Electric Vehicles (EVs) or the high cost and flexibility of the Electrical Storage Systems (ESS). DR consists of a highly dynamic interaction between demand and supply, capable of efficiently handling energy equilibrium toward several goals such as greater RES penetration, adjusting the demand according to the generation available, increasing the overall reliability and stability of the power grid.

According to the US Energy Information Administration [3], the industrial sector uses more delivered energy than any other end-use sector, consuming about 54% of the world's total delivered energy. However, research interest and the provision of SG services are mainly focused on the residential and commercial sector, even though they have a much smaller consumption footprint and environmental impact. Optimal energy management in the industrial sector could provide the necessary flexibility, avoiding the use of expensive storage units and peaking power plants [4]. In order to unlock the full flexibility potential of industrial applications, operation scheduling needs to be done considering the complicated processes taking place in each type of industry, which may set some limitations to the extent that their loads can be automatically controlled.

More specifically, industrial DR can help industries gain significant economic benefits without compromising the smooth progress of the production process, while giving the aggregators a large margin of flexibility that can help them manage the grid in a more efficient way, minimizing the inconsistencies between generation and consumption.

Implementation of DR programs can be more challenging for industrial facilities compared to residential customers, as reliability issues are usually vital for industries [5]. The violation of the operational constraints could lead to the interruption or even the stoppage of production. Moreover, two-way communication between the system operator and the participating industry is required. Most industrial sites already have metering infrastructure installed, so their participation is facilitated. However, a low participation rate is observed, mainly due to lack of knowledge, technical constraints, and complexity issues [6]. Review of the literature implementations suggests that cement, steel, and aluminum plants (which mainly belong to the construction industry) have a large flexibility potential which has been extensively studied over the years. Electrochemical manufacturing is also a sector in which multiple DR applications have been developed. Food industry participates with refrigerator warehouses, which are excellent candidates for DR implementations due to their thermal inertia. Finally, the IT industry and more specifically data centers are continuously growing in size and in energy needs, so an increasing number of studies have been studying their flexibility potential. The rest of this paper is organized as follows. Section 1.2 analyzes DSM techniques and ancillary services in the context of smart grid. In Section 1.3, state-of-the-art implementations are presented, elaborating on the special characteristics of industrial sectors with higher participation in DR programs. Section 1.4 investigates the barriers and the limitations identified, preventing the wider adoption of industrial DR solutions. Finally, conclusions are drawn in Section 1.5.

1.2 Demand side management and ancillary services in smart grid

1.2.1 Smart grid

Three main categories at the top level for the customer section of smart grids are commonly recognized: residential, commercial, and industrial. Even though public attention on smart grids had been focused mostly on the residential sector during the previous decade, commercial and industrial sectors have larger consumption footprint and peak load contribution [7]. In particular for the industrial sector, it has several common individual facilities to represent large loads, due to high use of electricity. Therefore, there is potential to achieve economic, environmental, and other benefits. A short overview of main smart grid technologies with link to the industrial sector is presented next.

Energy efficiency: It is a main objective for energy management, especially in places where energy costs play a crucial role and need to be reduced. Applying control actions and utilizing intelligent systems for energy conservation in recent years have been made possible due to the advances in Internet-of-Things (e.g. low-cost wireless sensors, interconnected networks) and asset management software. Modern automation systems that are used in industry provide meaningful information to facility managers via dashboards, allowing them to assess the status on the facility operations. These tools provide to facility managers all needed information to respond to signals that are received from the electricity grid in case of grid problems.

Direct load control: It refers to the case when a utility or service provider directly controls the loads by sending control signals to assets at a facility. Direct load control is more common in the residential section, where incentives are offered to the owners to install the necessary equipment for controlling HVAC units and electric water heaters. On the contrary, direct load control in industry is more complex, especially in production environments, as further context is required on the type of the load and its role in the production process. This is because it must be ensured that in all cases, any safety rules are not violated and the operation of affected production lines is not at risk.

Storage: The use of storage allows to decouple to an extent the purchase of electricity with the operation of the facility. This is especially useful in cases of low availability, fluctuating quality, or high price of electricity supply. The two main types of storage are the "Electrical storage" and the "Thermal storage." The former usually relies on batteries, and the common scenario is to apply it when electricity is offered at low price in order to consume the stored energy later when needed. Stationary batteries come in various capacities and may be of different technologies. For example, the capacity of utility-scale storage batteries ranges from several megawatt-hours to hundreds, and lithium-ion batteries are the most prevalent type [8]. Moreover, using the batteries of electric vehicles by applying smart charging, V2H, and V2G modes, has attracted much attention due to the increasing availability of such vehicles and related EV charging equipment.

Power generation: Industrial facilities may generate power to cover part of their electricity needs or even to provide to the grid. Biomass and fossil fuels were the main source of cogenerated power in the past, however, distributed generation using renewable energy sources (e.g. PVs and wind turbines) is the current trend. Cost is an important factor, and given the variability of production from renewable energy sources, high level of automation and efficient control strategies are needed. Therefore, the facility managers will be able to plan the production process accordingly based on information about electricity cost and predicted onsite power generation.

Microgrids: A microgrid is defined as a group of distributed energy resources, including renewable energy sources and energy storage systems, plus loads that operate locally as a single controllable entity [9], having monitoring, control, and optimization functionalities. Facilities with these characteristics that can form a microgrid can operate off the grid completely or partially, allowing them to enable uninterrupted operation in cases where grid supply is not reliable. According to [10], industrial microgrids consist of factories with distributed energy resources that rely on combined heat and power (CHP) systems for energy generation, while renewables and plug-in electric vehicles may also be included. Moreover, as indicated in [11] the industrial microgrid can be optimized, by managing the storage and generation resources and implementing appropriate control actions for load curtailment.

1.2.2 Demand response automation schemes

Demand side activities that are coordinated directly with supply system operators' requirements and utility communications, mainly related to grid power quality, availability, and price, are referred to as Demand Response. For example, such demand side activities include daily peak load reduction during high price time intervals and participation in electricity transactions with various response times.

The term "Demand Response" encompasses electric load-reduction strategies that are both manually initiated and automatically initiated. There are three main approaches to demand response, which are the following [12]:

- Manual—A person performs operation (e.g. turns off) on the equipment. Human intervention does not guarantee persistent and low response times, making the manual approach not suitable for participation in all markets.
- Semi-automated—A person launches an energy management system (EMS), which determines and performs all the necessary actions on the equipment. Therefore, the reliability of the response depends on the availability of the initiator.
- Automated—The EMS in this case automatically initiates the execution of the control strategy, upon request from the grid operator without any human intervention. This is achieved through communication signals that are received by the automation systems from the grid. The advantages of the automated Demand Response are high speed and reliability, provided that the automated control actions have been properly designated.

Automated demand response can be implemented at several levels with regard to where the logic resides. In large-scale facilities or buildings that are equipped with EMS, it is common to have the EMS handle the Demand Response logic. Alternatively, some local energy resources may directly implement the Demand Response logic rules. A third option that is common for small-scale facilities is to allow the Demand Response logic to be handled by a service provider or the utility. As far as the implementation of demand response is concerned, the Open Automated Demand Response (OpenADR) standard [13] has been adopted and widely used. OpenADR enables the exchange of various related information, such as measurement reports, forecasts, schedules, baselines, price data, and other. It has been identified as a key standard for demand response. The current version of the standard, OpenADR 2.0b, has received full approval as International Electrotechnical Commission (IEC) Standard IEC 62746-10-1. The main advantages of OpenADR 2.0b are the following: (1) open architecture and publicly available; (2) support of hierarchical architecture; (3) dynamically configurable message exchange intervals, and transmission of past and future flexibility data [14]. OpenADR is already in use at several installations to provide demand response, as well as ancillary services and DER management.

1.2.3 Ancillary services in the industrial sector

Ancillary services programs are of major importance, as they have the potential to increase the amount of flexibility that is provided to the smart grid and improve power quality and grid resilience. The increasing share of renewable power generation is having an impact on the grid that could be balanced through the provision of ancillary services by DR programs. The involvement of industrial consumers could bring profit to the industries and at the same time increase the overall power system efficiency.

Ancillary services include all services and actions that are needed in order to support a power system, which must be secure and reliable [4]. The advancements that have been made in real-time communication technologies and automated control especially during the last 15 years have allowed the engagement of small loads that are suitable for fast DR [15]. Two main differences of DR for ancillary services compared to typical DR applications are the reduced notification time and more advanced technical requirements related to measurements (speed and accuracy). Moreover, ancillary services may be needed and requested any time during the day and not just during peak hours. There are special types of loads with characteristics, such as the ability to be shortly interrupted without negative impact, that allow them to provide ancillary services. These loads include refrigerated warehouses, electric water heaters, dual-fuel boilers, water pumping, etc. [4].

According to the report presented in [16], ancillary services are classified as (a) frequency ancillary services (mainly for balancing); (b) services for congestion management; and (c) non-frequency ancillary services, such as voltage control and grid restoration. A high-level description of the most common types of ancillary services provided by conventional units in EU is provided below [17].

- *Frequency control*: This service restores the frequency in the nominal operating value after possible deviation due to imbalance between generation and demand. The main ancillary services for frequency restoration are frequency containment reserves (FCR), frequency restoration reserves (FRRs), and replacement reserves (RRs).
- *Voltage control and reactive power supply*: This service controls the voltage to maintain it to the desired acceptable limits. To this end, reactive power is required to be injected at specific locations of the grid network, close to the voltage deviation point, using controllable devices. According to the activation time, there are three voltage control types: primary voltage control, secondary voltage control, and tertiary voltage control.
- *Black-start capability*: This service is provided by generating units that can inject energy into the system for restoring it after a general or partial interruption of its operation.

1.3 Industrial DR case study implementations

The energy requirements the equipment and the special characteristics of each industry will be analyzed in this section. Case studies of DR implementations found in the literature for each industry will also be investigated.

1.3.1 Manufacturing processes

The largest part of the industrial sector is consisted of procedures that are related to manufacturing. The term manufacturing implies the processing of raw materials, in order to physically, mechanically or chemically transform them into finished goods or more complex items. DR scheduling in this case may face some severe limitations, since there are multiple machines and complex processes involved in the production procedure. Furthermore, some of those processes are executed in parallel, so the implementation of DSM actions may interrupt the production procedure. Thus, the available DR potential should be studied separately at each case, taking into consideration the level of criticality of the performed operations.

1.3.1.1 Iron and steel

The iron and steel industry is the second largest energy user in the global industrial sector [18], being responsible for a large percentage of carbon dioxide emissions [19] (up to 1.6 billion tons of CO_2 annually). There are two methods that are used predominantly in the steel production process [20]: (1) Blast Furnace—Basic Oxygen Furnace (BF-BOF) is the most widely used method at a percentage of 75% of the global production. The main steps followed in this process include coking, sintering, pelletizing, ironmaking, primary and secondary steelmaking, casting and hot rolling [21]. The final product can be delivered in various forms, such as coils, plates, sections or bars. (2) Electric arc furnace (EAF) is an alternative steel production method used in 25% of global production. In this case, the main raw material is recycled steel scrap and electricity is the main source of energy. EAF route is considered to be more environmentally friendly, since it has lower CO_2 intensity. However, the global percentage of EAF steelmaking is smaller compared to BF-BOF, mainly because of shortness of raw materials and scrap resources [22,23]. A graphical explanation of the two methods can be seen in Figure 1.1.

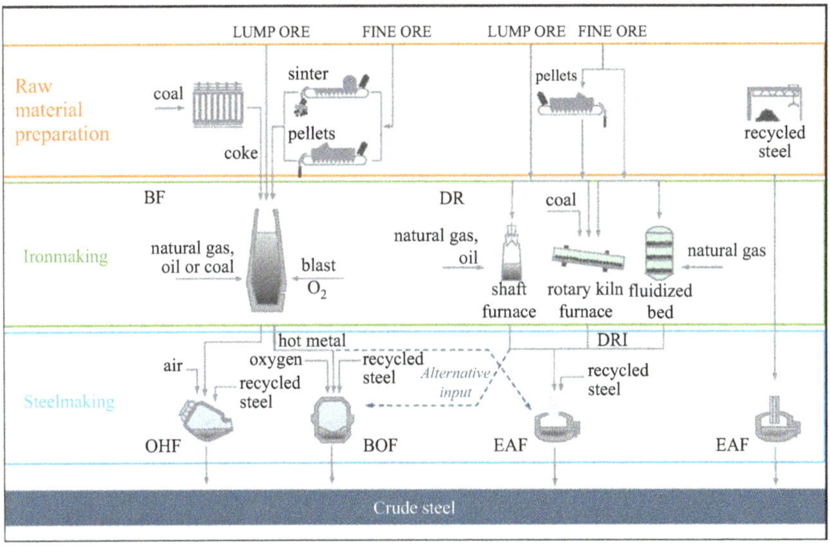

Figure 1.1 BF-BOF and EAF iron and steel production methods [20]

The authors of [24] are considering the steel production process, utilizing weighted directed acyclic graphs. They define representations corresponding to events, procedures and time spent on procedures, while a notional process is simulated using MATLAB software. Certain processes of the graph are considered as critical, which means that time adjustment is not possible at these routes. Other processes, such as steel rolling, are found to have a considerable DR participation time, improving the adjustment ability of the steelmaking process to DR programs. The work presented in [25] highlights the complexity of the processes taking place in the steel industry, focusing on reducing the computational cost of load scheduling on DR events. The production process requires fine grained computations, in terms of time resolution, but that results into long computation time which is not acceptable in the dynamic environment of this specific industry. This issue is addressed by applying two different techniques, whose effectiveness is numerically validated. At first, additional constraints are added, in an effort to reduce the search space of the MIP problem. Then, the commercial solvers which are typically used for the solution of optimization problems are substituted by a tailored method which utilizes the special characteristics of the steel manufacturing industry. EAF steel production method is considered in [26], where the optimal day-ahead scheduling is attempted, for participation in the electricity market. Similar to [27], the steel plant is modeled as a RTN, so the whole production process is described by interactions between the resources (e.g. equipment) and operational or transfer tasks. Optimal scheduling is based on the study analyzed in [27], however, it has been extended to include spinning reserve provision. Four different scheduling scenarios are simulated with results indicating that at all cases spinning reserve provision slightly increases electric energy cost, but significantly reduces the overall operational cost of the steel plant. A larger optimization horizon is implemented in [28], where load scheduling is running within a 5-day rolling window and maintenance planning is considered for 30-days ahead. Actual data from a steel plant in Taiwan are used for the current study. The schedulable parts of the equipment include two hot strip mills (HSM) and picking and cold reduction (PCR). Results indicate that the maintenance planning maximizes the DR potential, while energy scheduling with DR minimizes the operational cost of the plant.

1.3.1.2 Aluminum

Aluminum production is a less complex and less energy intensive procedure, compared to steelmaking. Yet, it constitutes a large share of the total global energy usage, reaching up to 1.53% [29]. The dominant electricity consumption process occurring at an aluminum plant is smelting, which utilizes energy for an electrolysis process that involves the chemical reduction of aluminum oxide into aluminum. Smelters are estimated to account for 85% of the total energy needs of an aluminum plant [29]. Moreover, these systems have been designed for a stable operation at constant power levels, as their energy consumption is directly related to the production of the plant. Hence, maintaining a stable load profile with minimum variations ensures the seamless operation of the production process. Minor rapid variations may be implemented for the provision of regulation services to the power system, without significantly reducing the quantity of the aluminum produced [30]. It becomes clear that the DR potential is restricted due to the plant operational constraints, however a strong advantage compared to other industries is that the smelting process is able to change its power consumption very quickly and with high accuracy, controlling the DC voltage of the smelting pots. The industrial aluminum production process can be seen in Figure 1.2.

A DR strategy candidate for implementation on aluminum smelters is the participation to the day-ahead energy and spinning reserve bidding markets. This case is presented in [32], where the optimal bidding strategy is investigated. Flexibility provision is considered only by controlling rectifiers, while it is assumed that the smelter can inject energy back to the grid. A stochastic optimization approach is followed, taking into consideration all the possible scenarios for the day-ahead electricity price forecast. All of the price scenarios are modeled with the same probability. The solution of the MILP problem suggests that energy bidding is higher when electricity prices are high. As a result, it is preferable to sell energy during those hours, so there is less space for providing spinning reserve. An approach trying to maximize the aluminum plant profits is presented in [33], where the optimal combination of aluminum production and regulation of the electric power grid is investigated. The automatic generation control (AGC) signal, which is used for the balancing between supply and demand, is linearly simplified. Then, various scenarios are created and given as an input to a stochastic optimization problem. Various simulations are conducted utilizing different price and cost parameters, as both hourly and multi-hour regulation provision are examined. According to the results, when the profit price is higher, the electricity consumption as well as the aluminum production are higher, while the regulation participation is lower. The authors of [34] follow a slightly different approach, formulating the problem as a mixed integer quadratic programming problem. A two-level optimization strategy is implemented, where the system operational cost is minimized and the profit of the aluminum plant is maximized, taking into account the profit from the DR participation. The maximization of profit is a convex and continuous problem, so it is replaced by the Karush–Kuhn–Tucker (KKT) optimality conditions. The

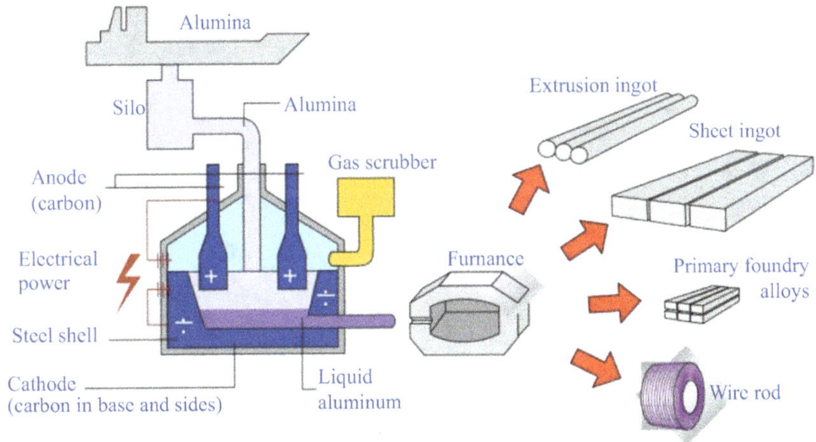

Figure 1.2 The industrial aluminum production process [31]

IEEE-39 bus system is used as a case study, at a time resolution of 15 minutes. Results indicate that implementing DR is a great solution for electrolytic aluminum plants, since it can be a source of profit even at periods when raw material prices fluctuations reduce product sales profits. In [35], electrolytic aluminum advantages for DR participation are analyzed. The large the installed capacity, the large thermal inertia and the small peak valley difference between day and night make aluminum plants ideal candidates for DR participation. A separation is made between aluminum enterprises with and without their own power plants, since in the first case, there seems to be a large potential in peak shaving, while in the latter, time shift DR is more suitable. At any case, most of the applications on the field use price-based methods for DR participation. The authors support that future research should move toward the direction of combining price-based and incentive-based schemes for optimal results.

1.3.1.3 Cement

One of the most energy intensive industries in the world, being responsible for more than 8% of China's total industrial energy consumption, is the cement industry. The largest part of the energy used comes from fossil fuel, while electrical energy consumption in cement power plants is broken down as follows: 35% at the cement mill, 30% in clinker production, 25% in kiln preparation, and 10% to other destinations [36]. The energy consumption breakdown of a cement plant can be seen in Figure 1.3. The cement mill, which is the machine with the highest electrical consumption, is preparing the raw materials for the kiln by utilizing wet or dry grinding. From the power consumption perspective, the wet grinding process was widely used due to lower energy requirements, but advances in technology have now made dry grinding the most efficient method [37]. Energy savings actions in cement plants are usually focused on upgrades of the existing equipment or savings in fuel. DR potential exists in those processes of the production process that are not continuous. For example, since the kiln is operating continuously, its interruption would cause stoppage of production or other damages. On the contrary, raw mix grinding, fuel grinding, and clinker grinding are the processes on which DR implementations are mostly built on.

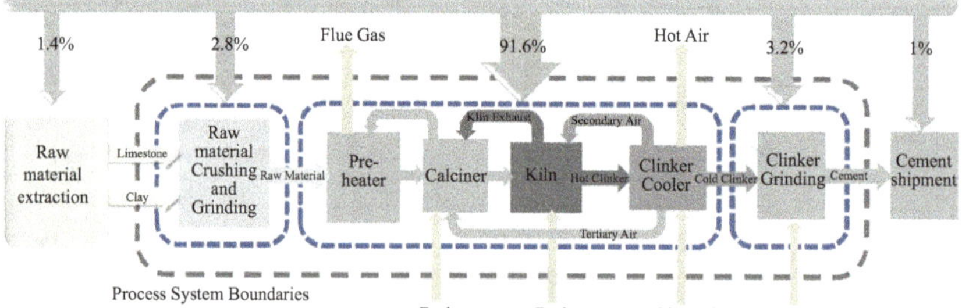

Figure 1.3 Cement plant energy consumption breakdown [38]

One of the first studies that considered DR implementation in industrial cement plants is presented in [39]. The production process was modeled by using start probabilities and constraints for all of the sub-processes, the objective function was set to minimize electricity tariff and finally the optimization problem was solved by using particle swarm optimization (PSO) technique. The simulated case study involved a dry process cement plant and a ToU DR scheme was considered. It is concluded that raw materials storage gives a larger flexibility potential to the plant. Optimal scheduling was achieved by mostly modifying the grinding operations, while crushing and homogenization were also shifted. The work presented in [40] investigates the potential for ancillary services provision through continuous power changes. This scenario is not easily implemented in cement plants, since only discrete power changes are made. A model predictive control (MPC) framework is deployed with hourly regulation signal forecasts given by an ARMA model. Cement plant parameters are considered and MILP optimization is solved, estimating the benefit from the installation of an ESS system. Three scheduling options are compared, namely real-time MPC coordination, hourly regulation cost and day-ahead optimal scheduling. Results indicate that loads are able to support the power system operation by providing regulation, while the authors suggest that the proposed framework to be extended to include more system components such as commercial building and electric vehicles. A data-driven approach for flexibility estimation is presented in [41]. Data from two Korean factories were used for the evaluation of the proposed solution. Load curves were decomposed to three components: seasonal, trend and residual. DR events were implemented and the level of demand reduction was analyzed from the collected data. According to the results, the usual power consumption could be cut down to its half, if the potential score on ramping down the demand for cement loads is 0.27. The score scale varies from 0 to 1, with 1 being the highest possibility. An interesting implementation, aiming toward the provision of aggregated DR flexibility from cement plants is introduced in [42]. The aggregator controls the operation of 8 plants, separating their production line into 4 discrete sub-processes and making day-ahead (24 h) planning, by taking into consideration the constraints set by the industries. Numerical results simulate the coordinated operation of the plants for a hot summer day, with the network experiencing serious power shortage. The problem is mathematically formulated as a dual robust mixed-integer linear programming (R-MIQP) problem. Results suggest that the storage capacity of the siloes is the key to unlock the maximum flexibility potential. The authors argue that the aggregator could possibly integrate various diverse heavy industries, optimally coordinating their operation. As indicated in [38], energy saving actions in energy intensive industries can be achieved not only with DSM techniques but also by performing energy assessment and maintenance on the existing facilities. In the presented case study of an Iranian cement plant, the electrical cost was reduced by 10% by implementing the following actions: reducing moisture on the raw materials, replacing typical fans with variable speed fans on the kiln air flow system, leveling the rotational speed of the kilns, insulating parts of the high temperature area after visual inspection. This indicates the importance of implementing simple maintenance solutions for electrical cost reduction, before resorting to more complex DSM solutions.

1.3.1.4 Other manufacturing sectors

Within the industrial sector, there are multiple other energy intensive industries, such as chemicals, textile, automotive, aerospace, wood, plastics and rubber, which have manufacturing processes as their main activity. However, DR implementations are found in the literature only for those industries that are considered appropriate from an operational point of view. For example, safety critical operations may not be interrupted for the provision of flexibility services, as a production failure could have more significant effects, compared to the benefits obtained from DR implementation.

According to [43], the average electricity consumption for the vehicle assembly process is 80 kWh/vehicle. Thus, several studies have implemented DR schemes to the automobile manufacturing industry. In [44], a MILP approach attempts to optimize the metal stamping process, which is the mechanism that creates consistently sized and shaped car body parts, by pressing sheets of metal. The objective function minimizes the total manufacturing cost, while the constraints are customized for the processes taking place in car manufacturing. Simulation results indicate significant cost reduction because of the use of DR, which is further cut down when DR is combined with ESS interaction. A similar case, where an automotive assembly line is simulated, considering as equipment seven machines and five buffers, is presented in [45]. Real-time DR scenarios are considered, since non-real time strategies fail to satisfy the requested DR events, without compromising system throughput. Real-time control manages to maintain the production target unaffected, while it is highlighted that more sophisticated solutions are required for long lasting DR events. The work presented in [46], also focuses on real-time, event-driven DR requests. The power control model is built utilizing a Markov decision process (MDP) framework, capturing both deterministic and stochastic properties of the model. Dynamic programming is used as a tool for the solution of the formulated MDP problem, while forward recursion is preferred over backward recursion, due to the curse of dimensionality. A five machine and four buffer manufacturing line is considered as a case study, utilizing real data from an actual automotive industrial facility. According to the results, during the DR event, electrical power consumption can be reduced by up to 22%, while the system throughput remains unchanged.

The chemical industry is another energy-intensive sector which is responsible for a significant amount of industrial carbon emissions [47]. The work presented in [48] investigates the DR potential in a chemical manufacturing facility. Real-time optimization (RTO) and economic model predictive control (EMPC) are compared on different optimization horizons. It is observed that EMPC profit is depending on the horizon size, while computational concerns are raised for realistic examples that could create substantially larger models. Operating profits are achieved, however it is highlighted that the cost of the installation of new equipment needs to be taken into account in the profits calculation. Electrolysis-based chlor-alkali manufacturing is an energy-intensive technology that can be optimized to provide significant demand response services. In [49], the operation schedule of the plant is modified every 15 min, following the fluctuations of the electricity prices. Significant reduction of 29% in power demand was achieved, exploiting the inherent inertia of the process, which allows short-term scheduling. The authors highlight that operation constraints, such as the cell temperature, should be constantly monitored during DR events. In the current work, electricity prices were considered to be accurate, however in real-life scenarios, they derive from a forecasting process that may introduce errors that will affect the method efficiency. Chlor-alkali manufacturing DR is also studied in [50], where day-ahead and real-time optimization scenarios are implemented. In this case study, the plant is powered by a grid-connected hybrid renewable energy system, while energy forecasts are used for the prediction of the output from the generation units. The optimization problem determines the optimal dispatch of the available resources for a specific time horizon. Various DR schemes were considered, with the results suggesting that the electricity contract and incentive based scheme are the most beneficial both for the industrial customer and the operator.

The case study presented in [51] is simulating processes of the ceramic manufacturing industry. A data-driven approach is selected, although the cost of implementing a real-time data collecting and monitoring framework is highlighted. DR implementation is focused on the ball mill machine, which is a type of grinder that blends materials of ceramics. The suggested solution is described in three different layers, namely data capturing, monitoring and energy efficiency optimization. Data analysis reveals the most efficient ball mill of the plant, while the authors suggest that energy and cost savings can be achieved by extending this behavior to the rest of the equipment. Pulp and paper production also require a significant amount of energy for the execution of the related manufacturing processes. Wood chips can produce pulp with chemical or mechanical pulping. The latter process, which consumes the most electric power, is examined in [52]. The problem constraints are defined, including steam demand, capacity and cost of alternative steam sources, future production plan, maximum production of each line and minimum feasible production of each line. The production plan is used to generate forecasts for the entire process. Those forecasted data are used for the optimal scheduling of the operations avoiding peak price hours.

1.3.2 Refrigerator warehouses

According to [53], refrigerator warehouses are responsible for approximately 16% of the total energy use of the food industry. Taking into consideration that their peak demand coincides with extreme hot ambient temperatures, when the grid is stressed the most, the significance of implementing DSM techniques is highlighted. Their high DR potential is also underlined by the fact that they can show high tolerance to brief interruptions of their power supply. Since they are closed spaces, well insulated and usually full of stored products, they are able to maintain high thermal inertia. A key advantage compared to other industrial facilities is that they are usually already equipped with sophisticated metering infrastructure used for monitoring and control of the warehouse conditions, in order to ensure that the products maintain their quality.

Refrigerator warehouses are excellent candidates for DR programs since the processes taking place are much simpler, compared to other industries. Both load shifting and load shedding techniques can be used as shown in Figure 1.4. In load shifting, power demand is moved from high demand to off-peak periods, ensuring lower prices and less stress for the grid. The thermal capacitance of the stored products plays

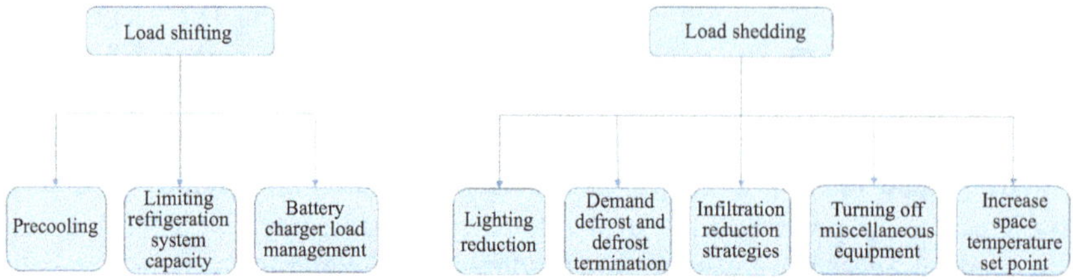

Figure 1.4 DR operations on refrigerator warehouses [12]

an important role for the warehouse's ability to perform this type of operation. Load shedding reduces power consumption on peak periods, but only affects operations that do not need to be recovered in the future. DR methods that are affecting the indoor temperature of a warehouse should be cautiously implemented, since the product quality could be reduced, resulting into additional costs [54].

One of the first studies that considered the implementation of DSM techniques in refrigerated warehouses is presented in [53]. In two different case studies where Auto-DR was implemented, an average load shed of 36% and 25%, respectively, was achieved. In nine cases where manual DR was adopted, the analysis showed that the results were less consistent. Furthermore, it is observed that the actual total DR potential obtained from California's refrigerated warehouses in 2008 is much less than the estimated potential due to low participation. A more realistic approach is made in [54], where only a part of the industrial consumers is expected to comply to the DR set points. A cooperative demand response (CDR) scheme for refrigerated warehouses is proposed, incorporating a punishment mechanism that tries to avoid selfish behavior from the consumers. The problem is mathematically formulated making assumptions for the warehouses consumption, the cost coefficients, and the electricity prices. A constrained optimization approach is followed, deriving the optimal strategy for each consumer. Results indicate that CDR strategy reduces electricity costs, especially when a small number of warehouses (~60) are participating in the program. The work presented in [55] tries to define all the important parameters that could affect the DR potential of refrigerated warehouses. Various constructional and operational characteristics such as the walls, the pallets, and the refrigerator machines are taken into account. Moreover, external weather conditions are also considered. A model is finally built trying to predict the temperature of a warehouse during DR events. Temperature measurements of an actual warehouse validate the accuracy of the simulation model. Various DR scenarios are then implemented, coming to the conclusion that season, the percentage of loaded products and the DR event duration are the most important parameters. As expected, optimal results are obtained on winter, for short period DR events and a large percentage of loaded products that ensure high thermal inertia. A similar work, studying load, temperature (outside and set points), and DR duration effects, is presented in [56]. The cold room parameters such as heat transfer coefficient and coefficient of performance are experimentally measured on site.

An important finding is that after the completion of each DR event, even though the product surface temperature started to decrease (heat transfer by convection), the core temperature continued to increase for a while (heat transfer by conduction) and this should be taken into consideration in order to ensure the quality of the stored products. Furthermore, a cumulative effect was observed on the core temperature of the products after three consecutive days of implementation of DR events. The authors of [57] highlight the need to make sure that DR events respect food temperature regulation ensuring the safety of the products. They propose four different deep learning architectures, making an effort to predict the effects of the implementation of a DR event. Three experimental datasets were used for the study, while it was found that power demand prediction is more accurate than the temperature prediction, since temperature is related to more complex phenomena. It is concluded that data-driven predictive models can assist toward more informed decisions for DR applications in cold storage.

1.3.3 IT industry/data centers

We are in the midst of a digital transformation that has been called the fourth industrial revolution that aims toward the digitalization and automation of the entire value chain process. In the distant past, industries embraced IT infrastructure in manufacturing, however the recent advances in networking, data storage, AI, edge devices, enable the potential of a fully automated production process. Definitely, this ongoing digital transformation has a vast impact on the power and electricity demand. A few years earlier data centers could be categorized as part of the commercial sector, but nowadays there are multiple data centers (DCs) that are industrial-scale operations using huge amounts of energy. Moreover, traditional industrial sectors are significantly contributing toward the increase of the electricity demand of data centers, since the automation of their processes requires the implementation of IT support services with high storage and computational requirements. According to [58], the IT sector will contribute 50% to the total consumed electricity power, approaching 3,200 TWh over the next few years. Considering this along with the great potential that data centers have to turn into key players for flexibility provision to the grid, we can conclude that they are proper candidates for demand response. DC facilities can be divided in two main categories, IT-network and site infrastructure. The first category consists of servers, storage, networking equipment, while the second is composed of equipment that is supporting to IT (fans, cooling, etc.).

Over the last years, several studies have been published that considered DR implementation, while there is a limited number of review publications [59–62]. In [63], the authors proposed two algorithms to achieve energy savings and also performed simulations to evaluate numerically their performance. The first algorithm aims at minimizing expenditures of a DC estimating from historical data when electricity demand will be the highest and cautions will occur. The second algorithm presents another DC optimization method that emphasizes minimizing the worst-case cost. They claim that there is always a limited number of warnings that will occur, in a period, and knowing the exact workload demand and renewable generation, they solve a convex optimization formula. The main difference between these two approaches is that the first algorithm takes for granted that it has the exact knowledge of the distribution of coincident peak warnings, while the second uses an upper limit. The experimental results show that their methods manage to reduce energy expenditures by 40% compared to baseline algorithms. The authors of [64] presented a study aiming at reducing energy costs in small and medium HPC clusters. They utilized SLURM workload manager to perform job scheduling and manage queue(s) of work and they combined it with EDEALS, to assess the magnitude of energy savings. Their approach reduced the power consumption utilized and they claimed that if they apply the same procedure in HPC clusters it would save 7% of data center consumption. In [65], the authors modeled a two stage interaction between the smart grid and the data centers. First, the utility company estimates pricing procedure in order to stabilize the consumed power and then the data centers perform all the necessary actions in order to reduce their cost. The formulate this method as a bi-level quadratic program, rejecting optimization techniques. Finally they aim to attain decent results in electric load index (ELI) showing a win–win situation for both utilities and DC. This publication explored the topic of emergency demand response (EDR) in data centers and it introduced an environmentally friendlier technique that was able to extract load reduction from tenants at an EDR period. They proposed coloEDR and validated it through simulation tests. The study presented in [66] provided two novel methods that performed simultaneously as a holistic DR for the DC. The first one is a workload scheduling method with time-shifting and the second one uses a UPS for energy storage. By enabling these two methods, the DC can participate in the DR since high-frequency variance can be smoothed and the low frequency can be reshaped by the UPS storage. In [67], the authors implemented a control strategy for DC that maintained all its key activities. Based on this, they have implemented a system model that was based on the physical attributes of all the DC components in order to adjust the electrical, thermal and IT subsystems power consumption. This method ensures the DR participation. The approach introduced in [68] suggests a real-time pricing framework and modeled a decision support system for monitoring utility company choices and workload scheduling. According to [69], the implemented two-way DR method can reduce the energy expenses of the DC. The first stage aimed at utilizing the interaction between the DC and the SG, by allowing the second to send DR signal based on the price of electricity. In the second stage, an auction method was proposed, where the user had to submit bids when the DC sent the DR signals. A recent study [70] introduced MOEA, a multi-objective algorithm for calculating Pareto efficiency for workload and energy scheduling values. This method ensures escalation and optimization for IDC networks and QoS respectively. The numerical evaluation on Google and IBM internet data centers shows that the suggested algorithm outperforms all existing approaches.

1.4 Barriers and limitations

In a rapidly evolving environment based on the transformation of the traditional power systems, the successful implementation of DSM techniques requires the coordinated operation of several mechanisms. Industries have to cover the knowledge gap that probably exists, making their participation to DR programs technically feasible. Moreover, they have to constantly monitor and understand the complex structure of the novel market schemes that are emerging in the electrical energy sector. Regulatory authorities should respond in a timely manner to the challenges that may arise, setting the rules for the involved stakeholders by adopting regulatory measures toward the full liberalization of the electricity market.

1.4.1 Financial

The participation in DR programs has as a prerequisite an initial investment that has to be made for the installation of equipment that enables the execution of DR requests. The level of participation of the industrial customer to DR schemes has to ensure that future energy cost savings compensate for the cost of this initial investment. The return of investment (ROI) calculation is a key factor that a DR participant would take into consideration as a strong financial incentive, in order to accept even a slight disturbance on the production process. Another fundamental issue that has a combined financial and regulatory nature is that ToU and dynamic pricing are not widely adopted yet. Thus, participation is discouraged because load shifting is not able to provide the expected profit for the industries. Moreover, as indicated in the previous sections, each industry has a certain flexibility potential, so each request should be individually assessed by the industry. Implementing the requested setpoints could possibly risk the quality of production, causing a larger financial damage, compared to the profits obtained for implementing DR.

1.4.2 Behavioral/social

Widespread adoption of the DR programs among the industry's employees is of utmost importance. Shutting down or even minimizing energy usage may sometimes jeopardize the accomplishment of the operational targets. That is why planning for the DR strategies should align with the KPIs and the targets of the employees of the business. This way, employees stay motivated and the interruption of operation of the equipment does not affect their productivity. Moreover, communicating the benefits (financial, environmental, etc.) of the DR concept and its benefits to the whole team is important. When employees are educated, they are able to understand the value that DR adds to the business. An important point is also raised in [71] where it is highlighted that there should be a trustworthy relationship among the interacting stakeholders. This ensures that the exchanged information between the DR provider and the industries is correct as expected.

1.4.3 Regulatory

Regulation and policies, regarding DSM framework, are different across countries [72]. Pricing strategies on the wholesale and the retail electricity market may significantly vary. It is also noticed that, being affected from the operation of the traditional power systems, the regulations usually focus only on the supply side, since power generation was considered only on this side of the grid [4]. The most important barrier from regulatory aspect seems to be the lack of a framework defining the market participants and their respective roles. A stable regulatory framework would ensure that industries are able to estimate the expected profits, coming from the provision of DSM services. Moreover, when regulations fail to keep up with the technical progress, the energy efficiency and GHG emissions reduction targets also fall behind.

1.4.4 Technological

Implementation of DR programs requires the installation of the necessary IoT infrastructure, which is used for communication and metering purposes. The installed equipment should successfully exchange large volume of data, in real-time and at high sampling frequencies. According to [71], IoT for DR systems has a high level of complexity, hence experts with high technological skills should design and assemble these technologies. No common approach is followed in the data exchange process by the involved stakeholders and as a result this interaction is a very slow process. But even if DR signals can be technically handled with success, industries should properly assess the impact that the adoption of DR strategies might have on the production process. Risk should be minimized by defining the maximum number of events or the length of the events that they could respond to, under certain circumstances. DR impact is obviously different for each type of industry and this mainly depends on the equipment that is being used. Hence, analysis of the load characteristics should ensure that the flexibility potential is maximized, the energy cost is minimized and at the same time the seamless operation of the production process is ensured.

1.5 Conclusions

Industrial sector accounts for a large part of the total electrical energy globally consumed in modern power systems. Given the progress that is being made on electricity market mechanisms and monitoring and control infrastructure, there is a large flexibility potential that can be exploited, providing benefits both for the industries and the grid operators. In this study, a comprehensive review was made, analyzing the most energy-intensive industries, while state-of-the-art DSM case studies were presented for each sector. Most of the studies still remain at a simulation level, without on-site implementation on industrial facilities, highlighting the fact that more effort should be given toward this direction. The analysis of the barriers reveals the factors that should be considered in order to accelerate the wider adoption of industrial DR. Smart grid transition is expected to increase the percentage of RES penetration to the grid, introducing severe challenges such as power quality issues and frequency instability. DR holds the key that can provide the necessary flexibility to the grid, with the industrial sector being a major contributor toward this effort.

References

[1] Commission E. 2030 Climate Target Plan; 2020. [Online; accessed 10-09-2021]. https://ec.europa.eu/clima/policies/eu-climate-action/.

[2] Associates CR. Primer on Demand-Side Management; 2005. [Online; accessed 10-09-2021]. https://silo.tips/download/primer-on-demand-side-management.

[3] Administration UEI. Industrial sector energy consumption; 2016. [Online; accessed 10-09-2021]. https://www.eia.gov/outlooks/ieo/pdf/industrial.pdf.

[4] Shoreh MH, Siano P, Shafie-khah M, *et al.* A survey of industrial applications of Demand Response. Electric Power Systems Research. 2016;141:31–49.

[5] Shafie-khah M, Siano P, Aghaei J, *et al.* Comprehensive review of the recent advances in industrial and commercial DR. IEEE Transactions on Industrial Informatics. 2019;15(7):3757–3771.

[6] Siddiquee SMS, Howard B, Bruton K, *et al.* Progress in demand response and it's industrial applications. Frontiers in Energy Research. 2021;9:330. Available from: https://www.frontiersin.org/article/10.3389/fenrg.2021.673176.

[7] Samad T, Kiliccote S. Smart grid technologies and applications for the industrial sector. Computers & Chemical Engineering. 2012;47:76–84.

[8] BRIEF IL. Utility Scale Batteries; 2019.

[9] Olivares DE, Mehrizi-Sani A, Etemadi AH, *et al.* Trends in microgrid control. IEEE Transactions on Smart Grid. 2014;5(4):1905–1919.

[10] Derakhshandeh S, Masoum AS, Deilami S, *et al.* Coordination of generation scheduling with PEVs charging in industrial microgrids. IEEE Transactions on Power Systems. 2013;28(3):3451–3461.

[11] Li H, Eseye AT, Zhang J, *et al.* Optimal energy management for industrial microgrids with high-penetration renewables. Protection and Control of Modern Power Systems. 2017;2(1):1–14.

[12] Scott D, Castillo R, Larson K, *et al.* Refrigerated Warehouse Demand Response Strategy Guide; 2015.

[13] Alliance O. OpenADR 2.0b Profile Specification Distributed Energy Resources (DER). 2019.

[14] Kolenc M, Ihle N, Gutschi C, *et al.* Virtual power plant architecture using OpenADR 2.0 b for dynamic charging of automated guided vehicles. International Journal of Electrical Power & Energy Systems. 2019;104:370–382.

[15] Heffner G. Loads Providing Ancillary Services: Review of International Experience; 2008.

[16] CEDEC, ENTSO-E, GEODE, E.DSO and EURELECTRIC. An integrated approach to active system management with the focus to TSO-DSO coordination in congestion management and balancing; 2019. https://docstore.entsoe.eu/Documents/Publications/Position%20papers%20and%20reports/TSO-DSO_ASM_2019_190416.pdf.

[17] Oureilidis K, Malamaki KN, Gallos K, *et al.* Ancillary services market design in distribution networks: review and identification of barriers. Energies. 2020;13(4):917.

[18] Johansson MT. Effects on global CO_2 emissions when substituting LPG with bio-SNG as fuel in steel industry reheating furnaces — the impact of different perspectives on CO_2 assessment. Energy Efficiency. 2016;9(6): 1437–1445.

[19] Zhang F, Zhou Y, Sun W, *et al.* CO_2 capture from reheating furnace based on the sensible heat of continuous casting slabs. International Journal of Energy Research. 2018;42(6):2273–2283.
[20] Sun W, Wang Q, Zhou Y, *et al.* Material and energy flows of the iron and steel industry: status quo, challenges and perspectives. Applied Energy. 2020;268:114946.
[21] Wen Z, Wang Y, Li H, *et al.* Quantitative analysis of the precise energy conservation and emission reduction path in China's iron and steel industry. Journal of Environmental Management. 2019;246:717–729.
[22] Zhang J, Wang G. Energy saving technologies and productive efficiency in the Chinese iron and steel sector. Energy. 2008;33(4):525–537.
[23] Cavalier P. *Clean Ironmaking and Steelmaking Processes: Efficient Technologies for Greenhouse Emissions Abatement.* Cham: Springer; 2016.
[24] Gong F, Han N, Li D, *et al.* A method for calculating steel demand response potential based on AOE network. In: 2020 IEEE 9th Joint International Information Technology and Artificial Intelligence Conference (ITAIC). vol. 9. IEEE; 2020. pp. 890–894.
[25] Zhang X, Hug G, Kolter Z, *et al.* Computational approaches for efficient scheduling of steel plants as demand response resource. In: 2016 Power Systems Computation Conference (PSCC). IEEE; 2016. pp. 1–7.
[26] Zhang X, Hug G, Kolter Z, *et al.* Industrial demand response by steel plants with spinning reserve provision. In: 2015 North American Power Symposium (NAPS). IEEE; 2015. pp. 1–6.
[27] Castro PM, Sun L, Harjunkoski I. Resource—task network formulations for industrial demand side management of a steel plant. Industrial & Engineering Chemistry Research. 2013;52(36):13046–13058.
[28] Tang W, Liou LL, Yang HT. Mid-short term industrial demand response strategy—case study for a steel mill. In: 2020 IEEE Power & Energy Society General Meeting (PESGM). IEEE; 2020. pp. 1–5.
[29] Gupta R. Energy resources, its role and use in metallurgical industries. In: Treatise on Process Metallurgy. Elsevier; 2014. pp. 1425–1458.
[30] Todd D, Caufield M, Helms B, *et al.* Providing reliability services through demand response: a preliminary evaluation of the demand response capabilities of Alcoa Inc. ORNL/TM. 2008. p. 233.
[31] Kvande H, Drabløs PA. The aluminum smelting process and innovative alternative technologies. Journal of Occupational and Environmental Medicine. 2014;56(5 Suppl):S23.
[32] Zhang X, Hug G. Bidding strategy in energy and spinning reserve markets for aluminum smelters' demand response. In: 2015 IEEE Power & Energy Society Innovative Smart Grid Technologies Conference (ISGT). IEEE; 2015. pp. 1–5.
[33] Zhang X, Hug G. Optimal regulation provision by aluminum smelters. In: 2014 IEEE PES General Meeting — Conference & Exposition. IEEE; 2014. pp. 1–5.
[34] Zeng K, Wang H, Liu J, *et al.* A bi-level programming guiding electrolytic aluminum load for demand response. In: 2020 IEEE/IAS Industrial and Commercial Power System Asia (I&CPS Asia). IEEE; 2020. pp. 426–430.

[35] Gong F, Ren K, Zhang A, *et al.* Review of electrolytic aluminum load participating in demand response to absorb new energy potential and methods. In: 2021 IEEE 2nd International Conference on Big Data, Artificial Intelligence and Internet of Things Engineering (ICBAIE). IEEE; 2021. pp. 1015–1019.

[36] Worrell E, Galitsky C, Price L. Energy efficiency improvement and cost saving opportunities for cement making. LBNL-54036-Revision Ernest Orlando Lawrence Berkeley National Laboratory, University of California, March 2008.

[37] Olsen D. *Opportunities for Energy Efficiency and Demand Response in the California Cement Industry*. Lawrence Berkeley National Laboratory; 2011. https://escholarship.org/uc/item/7856f8vn.

[38] Afkhami B, Akbarian B, Beheshti N, *et al.* Energy consumption assessment in a cement production plant. Sustainable Energy Technologies and Assessments. 2015;10:84–89.

[39] Zhao X, He B, Xu FY, *et al.* A model of demand response scheduling for cement plant. In: 2014 IEEE International Conference on Systems, Man, and Cybernetics (SMC). IEEE; 2014. pp. 3042–3047.

[40] Zhang X, Hug G, Kolter JZ, *et al.* Demand response of ancillary service from industrial loads coordinated with energy storage. IEEE Transactions on Power Systems. 2017;33(1):951–961.

[41] Lee E, Baek K, Kim J. Evaluation of demand response potential flexibility in the industry based on a data-driven approach. Energies. 2020; 13(23):6355.

[42] Golmohamadi H, Keypour R, Bak-Jensen B, *et al.* Robust self-scheduling of operational processes for industrial demand response aggregators. IEEE Transactions on Industrial Electronics. 2019;67(2): 1387–1395.

[43] Galitsky C, Galitsky C, Worrell E. Energy efficiency improvement and cost saving opportunities for the vehicle assembly industry: an energy star guide for energy and plant managers. Lawrence Berkeley National Lab. (LBNL), Berkeley, CA; 2008.

[44] Luo Z, Hong SH, Kim JB. A price-based demand response scheme for discrete manufacturing in smart grids. Energies. 2016;9(8):650.

[45] Li L, Sun Z, Tang Z. Real time electricity demand response for sustainable manufacturing systems: challenges and a case study. In: 2012 IEEE International Conference on Automation Science and Engineering (CASE). IEEE; 2012. pp. 353–357.

[46] Sun Z, Li L. Potential capability estimation for real time electricity demand response of sustainable manufacturing systems using Markov Decision Process. Journal of Cleaner Production. 2014;65:184–193.

[47] Worrell E, Bernstein L, Roy J, *et al.* Industrial energy efficiency and climate change mitigation. Energy Efficiency. 2009;2(2):109.

[48] Mendoza-Serrano DI, Chmielewski DJ. Demand response for chemical manufacturing using economic MPC. In: 2013 American Control Conference. IEEE; 2013. pp. 6655–6660.

[49] Otashu JI, Baldea M. Demand response-oriented dynamic modeling and operational optimization of membrane-based chlor-alkali plants. Computers & Chemical Engineering. 2019;121:396–408.

[50] Wang X, El-Farra NH, Palazoglu A. Optimal scheduling of demand responsive industrial production with hybrid renewable energy systems. Renewable Energy. 2017;100:53–64.

[51] Ma S, Zhang Y, Liu Y, *et al.* Data-driven sustainable intelligent manufacturing based on demand response for energy-intensive industries. Journal of Cleaner Production. 2020;274:123155.

[52] Merkert L, Harjunkoski I, Isaksson A, *et al.* Scheduling and energy—industrial challenges and opportunities. Computers & Chemical Engineering. 2015;72:183–198.

[53] Goli S, McKane A, Olsen D. Demand response opportunities in industrial refrigerated warehouses in california. Lawrence Berkeley National Lab. (LBNL), Berkeley, CA; 2011.

[54] Ma K, Hu G, Spanos CJ. A cooperative demand response scheme using punishment mechanism and application to industrial refrigerated warehouses. IEEE Transactions on Industrial Informatics. 2015;11(6):1520–1531.

[55] Akerma M, Hoang HM, Leducq D, *et al.* Demand response in refrigerated warehouse. In: 2018 IEEE International Smart Cities Conference (ISC2). IEEE; 2018. pp. 1–5.

[56] Akerma M, Hoang HM, Leducq D, *et al.* Experimental characterization of demand response in a refrigerated cold room. International Journal of Refrigeration. 2020;113:256–265.

[57] Hoang HM, Akerma M, Mellouli N, *et al.* Development of deep learning artificial neural networks models to predict temperature and power demand variation for demand response application in cold storage. International Journal of Refrigeration. 2021;131:857–873

[58] Institute SESR. Digital Economy and Climate Impact—A bottom-up forecast of the IT sector energy consumption and carbon footprint to 2030; 2021.

[59] Ghatikar G. Demand response opportunities and enabling technologies for data centers: findings from field studies. 2012. https://escholarship.org/content/qt7bh6n6kt/qt7bh6n6kt.pdf.

[60] Wierman A, Liu Z, Liu I, *et al.* Opportunities and challenges for data center demand response. In: International Green Computing Conference. IEEE; 2014. pp. 1–10.

[61] Vasques TL, Moura P, de Almeida A. A review on energy efficiency and demand response with focus on small and medium data centers. Energy Efficiency. 2019;12(5):1399–1428.

[62] Chen M, Gao C, Song M, *et al.* Internet data centers participating in demand response: a comprehensive review. Renewable and Sustainable Energy Reviews. 2020;117:109466.

[63] Liu Z, Wierman A, Chen Y, *et al.* Data center demand response: avoiding the coincident peak via workload shifting and local generation. Performance Evaluation. 2013;70(10):770–791.

[64] McFadden W, Nikolich A, Parpart R, *et al.* Saving on Data Center Energy Bills with {EDEALS}: Electricity Demand-response Easy Adjusted Load Shifting. In: {USENIX} Workshop on Cool Topics on Sustainable Data Centers (CoolDC 16); 2016.

[65] Wang H, Huang J, Lin X, *et al.* Proactive demand response for data centers: a win–win solution. IEEE Transactions on Smart Grid. 2015;7(3):1584–1596.

[66] Yang T, Zhao Y, Pen H, *et al.* Data center holistic demand response algorithm to smooth microgrid tie-line power fluctuation. Applied Energy. 2018;231: 277–287.

[67] Cupelli L, Schütz T, Jahangiri P, *et al.* Data center control strategy for participation in demand response programs. IEEE Transactions on Industrial Informatics. 2018;14(11):5087–5099.

[68] Bahrami S, Wong VW, Huang J. Data center demand response in deregulated electricity markets. IEEE Transactions on Smart Grid. 2018;10(3):2820–2832.

[69] Li Y, Wang X, Luo P, *et al.* A two-stage demand response strategy for datacenters in the smart grid environment. In: Journal of Physics: Conference Series. vol. 1585. IOP Publishing; 2020. p. 012025.

[70] Chiu WY, Hsieh WK, Chen CM, *et al.* Multiobjective demand response for Internet data centers. IEEE Transactions on Emerging Topics in Computational Intelligence. 2022;6(2):365–376.

[71] Shafie-khah M, Siano P, Aghaei J, *et al.* Comprehensive review of the recent advances in industrial and commercial DR. IEEE Transactions on Industrial Informatics. 2019;15(7):3757–3771.

[72] Xenias D, Axon CJ, Whitmarsh L, *et al.* UK smart grid development: an expert assessment of the benefits, pitfalls and functions. Renewable Energy. 2015;81:89–102.

Chapter 2

Demand response cybersecurity for power systems with high renewable power share

Hassan Haes Alhelou[1] and Behrooz Bahrani[1]

Cybersecurity is crucial for modern power systems due to their high digitization. The open information and communication technology, which is (ICT) being used for the operation of such systems, is highly vulnerable to cyber threats. The adopted smart grid concept around the globe enables the utilization of demand-side for providing ancillary services based on well-known demand response (DR) programs. These programs aggregate smart appliances in homes and electric vehicles (EVs) for providing vital services such as frequency regulation and voltage support. Since the aggregation is based on the cyber layer, any cyber threat could affect the ancillary services that are being delivered from the aggregators, which might lead to stability and security issues resulting in brownout or massive blackouts. This chapter discusses the cybersecurity in DR program and shows its importance for modern and future smart power systems due to their stability and security margins. Furthermore, the cyberattack case study is implemented in a power system with a demand side program responsible for providing primary frequency support ancillary service, where the results confirm the high vulnerability of modern power systems to cyber threats on DR-active power reserve providers. Moreover, technical suggestions are provided for enhancing the cybersecurity in DR programs in power systems with high power share from renewable energy sources.

2.1 Introduction

Energy systems are among the most complex systems created by humans during history. Their stability, security, and growth are direct reflections of the advances and prosperity of the countries in which they are being operated. For decades, the traditional power systems were operated safely and securely with no problems related to their stability and security. This view of power systems has been changed during the last few decades due to environmental challenges and energy security risks

[1]Department of Electrical and Computer Systems Engineering, Monash University, Clayton, Australia

which enforced the upgrade resulted in new version called smart grid concept, where most countries around the globe are working on the adoption of such concept in their energy systems. In the last decade, the penetration level of renewable energy sources (RESs) has been highly increased. Although RESs reduce the environmental concerns, but have negative impacts on the stability and security of existing power systems. For instance, the high share of renewables in modern power systems reduces the total rotating inertia and as consequences, the power system frequency stability and security has been affected. In conventional power systems, the frequency which is a global variable was controlled by well-operating and managing of primary and secondary reserves came from the generation side. However, this is not yet valid due to changes in the power system situation and high reduction of the rotating inertia, therefore the operators need a faster reserve that cannot be available from generation side. It has been found that the best source of such active power reserves for supporting the frequency is the demand-side. Later, new topics have been initialized which are demand-side management and demand response (DR) for controlling the intelligent appliances that can provide some of their capacities as source for the necessary reserves. As a consequence, it has been suggested to take advantage of DR for providing ancillary services in power systems. There are specific types of loads and smart appliances that can provide ancillary services to power systems. These loads should not affect the conformable of the consumers and at the same time, the management of these appliances should be done based on reserve and energy markets. For instance, there are great research activities these days on building practical aggregators of electrical vehicles so that they can participate in the ancillary services market by well-controlling their charging, discharging and state of charge situation during a specific period of time. Likewise, the participation of air conditioners, refrigerators, water heaters, and other thermostat appliances have been suggested for as considerable and good storage aggregators that can provide some services to power systems instead of traditional battery energy storage systems. It is clear that there is a need for a comprehensive look on the security and stability issues related to DR programs.

It is obvious that the penetration level of RESs is highly increasing over the world. Solar and wind energies are among the most percentage shares of renewables in modern power systems. By nature of the photovoltaic cells, they provide zero inertia to power systems. This means that increasing the power generation from solar power plants would at least reduce the inertia with the same percentage of their active power generation share increase. As aforementioned, this high reduction in inertia would bring new challenges and technical issues to the operators of modern power systems where the main problems related to stability and security of energy systems. On the other hand, different types of wind turbines provide neglectable inertia, almost zero, to modern power systems bringing the same problems that would be arisen from solar energy systems. Therefore, there is a serious need for new sources to keep the balance in power system operation especially in the view of providing ancillary services. With the low inertia in future power systems, the traditional and conventional reserve sources would not act accurately and properly in the aim of maintaining the power system stability, therefore, it is highlighted that the demand side can be considered as a good source of such services based on DR programs. The main advantage of DR over energy and reserve sources in the generation sides is its high flexibility, which is the most important feature to the operators. Different types of demand-side loads can provide DR services to power systems, where more discussion on them can be found on the other chapters in this book.

Since demand–response (DR) programs will provide a considerable reserve for supporting the stability of both voltage and frequency in future power electronic dominated-power systems, the security in such programs will be crucial due to their direct impact on the overall stability and security in the energy systems. The DR providers are built based on aggregators, which are a cyber-layer. Therefore, the biggest threat to such systems would be cyber threats, including false-data injection cyberattack and other malicious malware cyberattacks. This gives an overview on the importance of cybersecurity in virtual power plants based on DR, electrical vehicles (EV)-based DR, and other DR programs. Therefore, this chapter focuses on DR cybersecurity for future smart power systems with high-power share from renewable energy sources.

This chapter is organized as follows. Section 2.2 presents a literature survey on DR, where the review of DR cybersecurity is presented in Section 2.3. Section 2.4 models the EVs-based DR for supporting frequency in power systems with high renewable power share. Section 2.5 presents a new method for detecting cyberattacks on measurements used in DR aggregators. The results are presented in Section 2.6, while Section 2.7 concludes.

2.2 An overview of DR and EV-based DR

Traditional power systems consider the demand side as non-flexible, where the flexibility comes from generation side by providing load following service and other ancillary services such as frequency control and regulation for keeping the grid operated in a stable and secure mode. These services depend on fossil fuel-based power stations, whose output is determined by the amount of fuel that they are supplied, and consequently, they can be considered controllable to some extent. However, the high-power share from RES has introduced a level of uncertainty on the generation side and reduced the total rotating inertia resulting in new challenges in providing high-speed services in such new systems with high penetration of RES. Therefore, the demand side has been introduced as a solution for providing flexibility in modern power grid instead of generation side. The concept of demand side management (DSM) was first introduced by the Electric Power Research Institute (EPRI) in the 1980s as a series of activities that utilities undertake to change their load shape and/or energy consumption pattern for benefit maximization, investment delay, and reliability enhancement [1]. DSM activities can be classified either as "Energy Efficiency (EE)" or as "Demand Response (DR)." DR denotes a power consumption shift made by a utility customer, as a response to a price signal or an incentive-based reward.

It inherently tries to mitigate some of the challenges that are deriving from the smart grid transition, such as the intermittent and stochastic nature of RES and EVs or the high cost and flexibility of the electrical storage systems. DR consists of a highly dynamic interaction between demand and supply, capable of efficiently handling energy equilibrium toward several goals such as greater RES penetration, adjusting the demand according to the generation available, increasing the overall reliability and stability of the power grid. DR can be categorized into three main branches, namely industry DR, commercial DR, and residential DR. The industry DR is based on heavy industries and manufacturers, such as iron and steel, aluminum, and cement industries, while the residential one is based on smart appliances such as AC air conditioners, refrigerators, and water heaters. It is worth mentioning that the transportation system can be devised into either residential or industry DR based on the type and operation of its individual units. Usually, the common private EVs-based aggregation method can be considered as a residential DR, while public E-buses, trams, and E-trains-based aggregation methods can be considered as industry DR providers.

The industry DR can provide a considerable reserve amount for providing the required flexibility in future power systems. According to the US Energy Information Administration [2], the industrial sector uses more delivered energy than any other end-use sector, consuming about 54% of the world's total delivered energy. However, the research activities mainly focused on providing the ancillary services and flexibility from residential DR instead of industry one, which show a research gap that should be discussed. It has been shown in [3] that optimal energy management in the industrial sector could provide considerable flexibility, avoiding the use of expensive storage units and peaking power plants. However, the implementation of DR programs can be more challenging for industrial facilities compared to residential customers, as reliability issues are usually vital for industries [4]. The violation of the operational constraints could lead to the interruption or even the stoppage of production. Moreover, two-way communication between the system operator and the participating industry is required. Most industrial sites already have metering infrastructure installed, so their participation is facilitated. However, a low participation rate is observed, mainly due to lack of knowledge, technical constraints, and complexity issues [5]. For more information on industry and residential DR programs, readers are referred to Chapter 1, where a comprehensive discussion is presented. In what follows, the electric vehicles-based DR methods for supporting the frequency in power systems are reviewed and discussed.

The environmental concern and energy during last few decades have led to several changes in energy and transportation systems. The main change in energy system was the replacement of fossil fuel power plants with renewable energy sources which introduce stability issues to the operation of the grid. On the other hand, the main change in the transportation system was its electrification leading to high deployment of EVs over the globe, where the connection of EVs to distribution system would change the power consumption pattern and introduce new uncertainties to the operation of the power grid. However, the studies showed that the problems and technical challenges due to the high penetration of RES can be solved by DR program based on EVs, which is also known as EV-based virtual power plants (VPPs). The management and real-time control of the charging and discharge power of EVs can provide a considerable virtual generator that can highly help in maintaining the stability, especially the frequency stability in the modern grid.

Due to its fast response, EVs have been aggregated for providing primary frequency support in modern power systems with high power share from renewable energy sources. In [6], an aggregated dynamic model of plug-in EVs has been proposed for primary frequency control. The model allows the power system operators to consider EVs as DR program for providing the primary active power reserve used to intercept the frequency decline before triggering under frequency protection relays. Alhelou *et al.* have developed a hierarchical plug-in EV control method in [7] that considers EV for supporting the primary frequency in interconnected power systems with low rotating inertia due to the high-power share from renewables. In the same context, a multi-agent system is developed by Alhelou *et al.* in [8] for aggregating EVs for providing sufficient frequency support to microgrids. Similarly, several methods have been developed in [9,10] for controlling the charging and discharging power for providing considerable ancillary reserve services for regulating and controlling the frequency in smart grids. In addition to the primary frequency control, EVs have been widely considered for providing secondary reserve services used in secondary frequency control in both bulk power systems and microgrids. A dynamic model of EVs for secondary frequency reserve support has been proposed by Alhelou *et al.* in [11,12], which provides a sophisticated tool based on DR of EVs that can be used in wide-area monitoring systems for online controlling the frequency variations due to generated power fluctuations from renewable energy sources. The model builds an aggregator of EVs with dynamic behavior similar to sophisticated conventional power plants with high ramp rates by modeling the governor response for an aggregated EV model. In [13], EVs have been considered as an industry DR that can support frequency in two-area power systems. The proposed model is also updated for wide-area power systems in [14].

The recent research activities try to build VPP based on the aggregation of EVs that could provide a considerable vehicle-to-grid services (V2G) [15]. A closed-loop V2G control method has been suggested in [16] for load frequency control, to meet the charging demand and control the frequency in weak power systems. This work has also considered the modeling of time-delay in the EV model and analyzed it, to study its impact on the DR of EVs. Likewise, a wide-area interconnected power system with high integration of EVs that contribute to load frequency control has been investigated in [17]. Generally, the results in the aforementioned study need to be analyzed for practical usage in light of the presence of nonlinearties in modeling EVs. A similar method has been developed in [15,18] for controlling the frequency in microgrids. For a further and comprehensive review on EVs considered for supporting the power grids, the readers are referred to [19,20]. The aforementioned studies confirm that there is a great research activity over the globe for developing new methods that model and use EV demand for providing ancillary services to power systems with high penetration of renewable energy sources. Since these methods will be considered in practical and real-world power systems for providing sensitive services that directly impact the stability and security in overall the grid, their security, including cybersecurity becomes of great importance and crucial. Therefore, the cybersecurity in power systems and more specifically, in aggregators and cyber layers related to virtual power plants, EVs aggregators, and different types of DR providers needs to be highly taken into account in future studies.

2.3 An overview of demand side cybersecurity

The implementation of DR has become a reality. In 2018, the USA utilities used more than 4.5% of the peak load capacity for DR services. It is estimated that the percentage will increase to reach 20% in 2030, resulting in operational cost savings of more than 15 billion US dollars [21]. Considering these great benefits and the concern regards the stability and security of power systems enriched with DR programs, the cybersecurity in DR becomes a challenge due to its direct impact on the overall stability and security of energy systems over the world. In this regard, the DR cybersecurity topic started recently attracting research funding bodies, governments, policymakers, and researchers for suggesting DR programs resilient against cyber threats.

Open automated DR (OpenADR) is widely used and implemented for the communication between end-users and DR aggregators [21,22]. The open automated DR standard specifies two ends of a communication channel. These communication channels are the virtual top node (VTN) and the virtual end node (VEN). Generally, OpenADR is responsible for issuing the digital certificates to the nodes for authenticated communication and encrypting information exchanged between the nodes [21]. Despite the standardization and the use of industry-grade encryption, cyber threats in ADR prevail because DR customers lack industry-grade cyber defense and hygiene on their devices [23]. It has been shown in [23] that cyberattackers can exploit vulnerabilities in the communication technologies, such as WiFi and ZigBee, to compromise smart meters and energy management systems used in DR aggregators. Similarly, cellular networks and power line communication used in modern industry DR aggregators are vulnerable to man-in-the-middle cyberattacks [24].

Due to its importance, demand-side cyberattacks on power grids launched by manipulating DR programs built based on EVs and other smart appliances has been recently considered in [21,25,26]. It has been verified in [21,25] that the EVs charging stations used for providing ancillary services can be subjected to a remote cyberattack that could affect the grid security leading to massive blackout in future power systems. The study was done based on a real-data from the New York power grid and transportation system in USA. The main result in the study showed that an intelligent cyberattack on only 1,000 EVs in New York could result in over-frequency leading to cascading events, putting the stability and security of the power grid in danger of blackout. Another study presented in [26] has shown that a demand-side cyberattack that aims to affect the operation of high-voltage transmission systems could also result in massive power cuts. A study in Ref. [27] shows that a cyberattack aims to manipulate the behavior of end-users, who provide DR services, could deplete the reserve used for frequency support in the transmission level and corrupt the voltage profile in the distribution level. Similarly, a new study has explored vulnerabilities of AI-based DR learning and designed a data-driven attack strategy informed by DR data collected from the New York University campus buildings [28]. The aforementioned recent studies focused on the manipulations of DR smart appliances during events affect the stability in the grid. In the cyber defense topic, Alhelou and Cuffe have developed a dynamic observer in [29] that could detect and isolate cyberattacks on power systems with EVs used for supporting the frequency. The results showed that dynamic state estimation methods are promising for developing cyberattack on DR detection techniques. Similarly, Alhelou *et al.* in [30] proposed a novel functional observer that could control the frequency considering unknown inputs and cyberattacks in complex systems with high penetrations of renewable energy sources and DR providers. The aforementioned studies show that there is a research gap related to DR cybersecurity that needs to be filled with new methods for detecting cyber threats and building DR and EVs aggregators that are resilient against cyberattacks.

2.4 Modeling power system with DR

For investigating the capability of DR programs, including EVs aggregators in frequency support in modern power systems, a suitable frequency response (SFR) model should be adopted. In this chapter, we introduce a new model developed based on the recently published benchmark by Alhelou *et al.* in [14]. This new benchmark considers the well-modeling of demand-side in SFR model. Power systems are usually divided into several subsystems called power regions or control areas. These areas are usually connected together using major transmission lines known as tie-lines, which can be HVDC or controlled HVAC transmission links. Likewise, the areas could be a large geographical area in the country or even a representative of the whole country when connected with other power systems in other countries. Figure 2.1 shows the block diagram of the ith area SFR model considering EV aggregation for supporting the frequency. This model assumes that the system is dynamically weak due to the high penetration of renewable energy sources. For more detailed models with high order dynamics of synchronous generating units and their PSS, VAR, and governor-turbine controllers, readers are referred to recently published work by Alhelou *et al.* [14]. Based on the blockdiagram, the SFR model can be described by a set of algebraic–differential equations as follows:

$$\Delta\dot{f}_i = \frac{1}{2H_i}\Delta P_{mi} + \frac{1}{2H_i}\Delta P_{ev_i} - \frac{1}{2H_i}\Delta P_{di} - \frac{D_i}{2H_i}\Delta f_i - \frac{1}{2H_i}\Delta P_{tie,i} \quad (2.1)$$

$$\Delta\dot{P}_{gi} = \frac{K_{ti}K_{ri}}{T_{ti}T_{ri}}\Delta X_{gi} + \frac{T_{ti} - K_{ri}}{T_{ti}T_{ri}}\Delta P_{ri} - \frac{1}{T_{ri}}\Delta P_{gi} \quad (2.2)$$

$$\Delta\dot{P}_{ri} = \frac{K_{ti}}{T_{ti}}\Delta X_{gi} - \frac{1}{T_{ti}}\Delta P_{ri} \quad (2.3)$$

$$\Delta\dot{X}_{gi} = -\frac{K_{gi}}{R_i T_{gi}}\Delta f_i - \frac{1}{T_{gi}}\Delta X_{gi} + \frac{K_{gi}\alpha_{gi}}{T_{gi}}\Delta P_{ci} \quad (2.4)$$

$$\Delta\dot{P}_{ev_i} = -\frac{1}{T_{ev_i}}\Delta P_{ev_i} + \frac{K_{ev_i}\alpha_{ev_i}}{T_{ev_i}}\Delta P_{ci} \quad (2.5)$$

$$\Delta ACE_i = b_i\Delta f_i + \Delta P_{tie_i} \quad (2.6)$$

$$\Delta P_{tie,i} = \sum_{j=1}^{N}(P_{tie,act_{ij}} - P_{tie,sched_{ij}}) \quad (2.7)$$

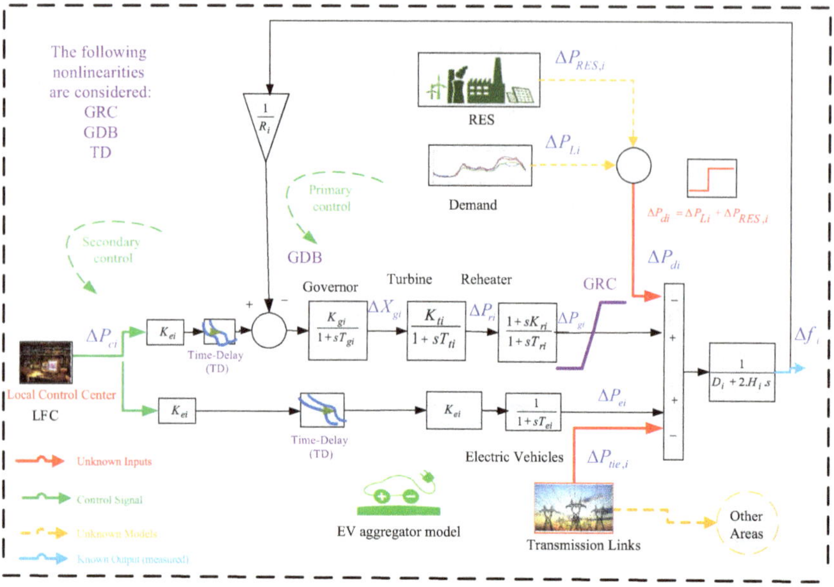

Figure 2.1 System frequency response model for interconnected power systems considering EVs as a DR and high penetration of RESs

where Δf, ΔP_g, ΔP_r, and ΔX_g, are the frequency deviation, active power deviation, output mechanical power deviation from turbine, and valve position deviation, respectively; K_g, K_r, and K_t represent the gains of the governor, mechanical turbine, respectively, while T models the time constant associated with the corresponding subscript.

In the above equations, ΔP_{ev} is the DR power from EV aggregator, K_{ev} is the EV aggregator-based DR gain, and T_{ev} is and EV aggregator-based DR time constant. Figure 2.2 shows the aggregator of EVs that provides the required primary and secondary reserve used for controlling the frequency based on market negotiations with the ISO responsible for operating the system. In this aggregation technique, each individual EV agent sends its information to the concentrator agent in its area, and the concentrator agent sends the information of EVs interested in the provision of the primary reserve to the aggregator agent. The aggregator calculates the reserve available from the EVs in the power system. To this end, the aggregator agent needs to obtain the following information about EVs: the initial SOC, the required SOC for the next trip, their rated charging power, and departure time. Then, based on the required primary reserve announced to the aggregator agent by ISO, the aggregator agent determines the share of each EV in primary and/or secondary frequency response. For more detailed information, this complete procedure is developed by Alhelou *et al.* in [7,8].

The most important factors related to the parameters that model the dynamic behavior of the system, i.e., H, D, and R, are the total rotating inertia, damping

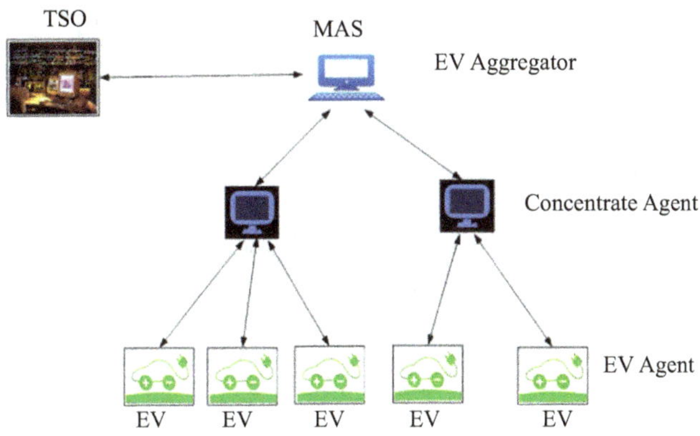

Figure 2.2 EV aggregator-based DR for supporting frequency in power systems

coefficient, and the droop coefficient of the governing system, respectively. ΔACE_i is the area control error (ACE), and b_i is the bias of frequency in area i. It is worth mentioning that the subscripts i and j are indices of the ith and jth areas, respectively. In Figure 2.1, ΔP_{di} models the variations in the generated power from RES and the fluctuations in the consumed power due to stochasticity and uncertainty [14].

In (2.7), the transferred power deviation, which is of great importance in stability and frequency control, is calculated as the difference between the actual measured power, $P_{tie,act_{ij}}$, in real-time compared with the scheduled transfer power, $P_{tie,sched_{ij}}$, based on the decisions from power dispatching operator. The above well-described system can be represented in a linear format considering the fact that load frequency control provided by DR deals with small-frequency variations around the operating point, therefore, the system can be represented by

$$\begin{aligned} \dot{x}(t) &= Ax(t) + Bu(t) + Ed(t) \\ y(t) &= Cx(t) \end{aligned} \tag{2.8}$$

where $x \in \Re^{n\times 1}$ is the state variable vector, $u \in \Re^{r\times 1}$ is the input vector, $y \in \Re^{m\times 1}$ is the output vector, and $d(t) \in \Re^{q\times 1}$ is the disturbance vector. Matrices A, B, C, and E are the state matrix, input matrix, output matrix, and disturbance matrix, respectively. According to Figure 2.1, the state vector, the known input vector, and the unknown input vector for the ith area are [14]:

$$\begin{aligned} x_i &= [\Delta f_i \; \Delta X_{gi} \; \Delta P_{ri} \; \Delta P_{gi} \; \Delta P_{ei} \int ACE]^T \\ u_i &= [\Delta P_{ci}],\; d_i = [\Delta P_{di} \; \Delta P_{tie,i}]^T \end{aligned} \tag{2.9}$$

Based on the aforementioned dynamic model of the power system, including the EV dynamic aggregator, an unknown input observer can be designed for detecting cyberattacks on the cyber layer as described in the method developed recently by Alhelou and Cuffe in [29]. The method could be developed to detect the cyberattack on the different parts of the aggregator resulting in a higher resiliency against cyber threats.

2.5 Discussions on the results of cyberattacks on EV aggregator

It has been shown in the previous section that the EV aggregator could be built based on multi-agent systems. The aggregator has different types of agents, namely individual EV agents, EV concentrate agents, and EV aggregator agents. As regards the power and information flows, there are different layers, as shown in Figure 2.3. For information communication between different agents and between EV aggregator and TSO, there are three layers, where the cyberattack on each of these layers would result in different consequences. For instance, the most severe impact on the power system stability and security would be a result of an attack on the higher information or communication layers since they would result in high-frequency fluctuations.

In what follows, we present the impact of cyberattacks on EV aggregator used for supporting the frequency in power system based on the dynamic model shown in Figure 2.3. The system parameters are as follows: the system frequency $F = 5$ Hz, the frequency bias for each area is 0.425 p.u. MW/Hz, the inertia constant is 5 sec, the thermal turbine time constant is 0.3 sec, the governor time constant is 0.08 sec, the tie-line between the two-areas synchronous coefficient is 0.545 sec, the damping coefficient is 0.0083 p.u. MW/Hz, the governor droop is 2.4 Hz/p.u. MW and the same value used for tuning EVs reserve for supporting frequency response in power systems, the gains are 1.0 p.u., the time constant for a solar PV system is 1.8 sec, the time constant for modeling the wind turbine power plant is 1.5 sec, the

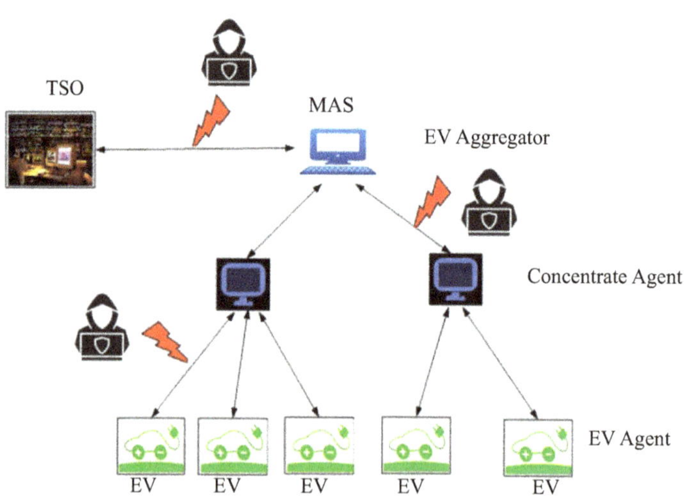

Figure 2.3 Possible cyberattacks on EV aggregator

reserve share from EVs is 0.3 p.u., and the reserve share from conventional power plants is 0.7 p.u.

In the first scenario, it is considered that the power system is equipped with LFC controllers for controlling the frequency in each area of the system (two-area power system). In this scenario, it is considered that both EV aggregator as a DR and conventional power plants contribute to secondary frequency control for regulating the frequency. It is considered that the power system was under normal operation situation, and cyberattack on the EV aggregator occurs suddenly. The cyberattack is of data-injection cyberattack, and it changed the scheduled EV demand through the next 40 sec slightly as shown in Figure 2.4. This minor cyberattack on the scheduled EV demand changed the demand around 0.01 p.u. and the accumulated change reached 0.04 after 25 sec, as can be seen from Figure 2.4. Although this change is minor, its consequences on power system were severe as shown in Figure 2.5. This figure shows that the frequency fluctuates, which affects the over stability in the system. By comparing the frequency of area 1 where the cyberattack is occurred with the frequency in the second area, as shown in Figure 2.6, one can see that this attack resulted in local frequencies, meaning that the frequency in each area is different from the other one, resulting in high threat on the stability of the system and if the power system stabilizer would not act correctly, this could result in cascading events and blackouts.

In the second scenario, a denial of service (DoS) cyberattack is considered. As shown in Figure 2.3, in this scenario, three cases are considered. In the first case, the system is considered under an event as a sudden change in the demand by suddenly connecting a large load, or trapping a generator equal to 0.3 p.u. of

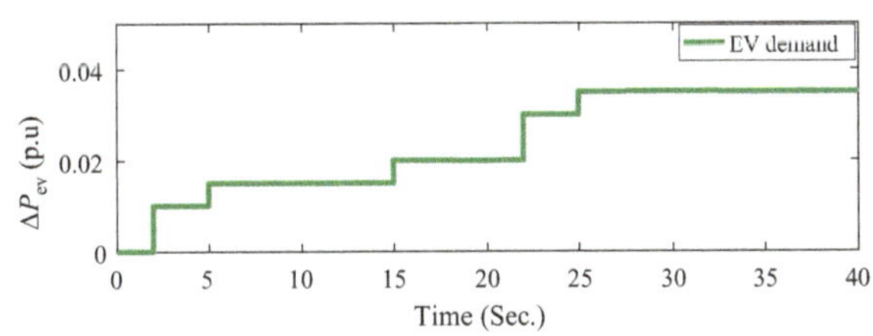

Figure 2.4 Cyberattack on the demand of aggregated EVs

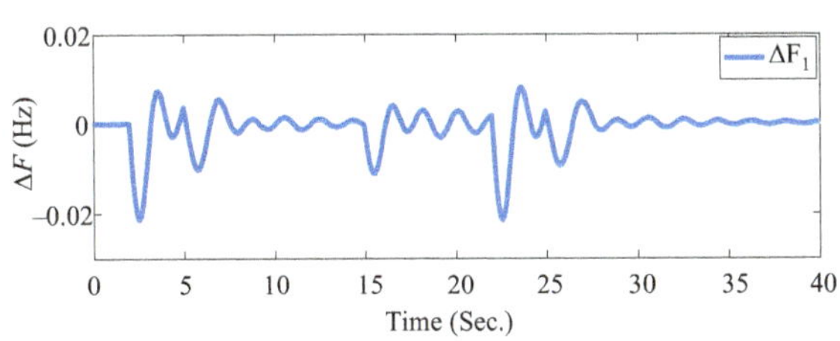

Figure 2.5 Frequency response in area 1 due to cyberattack on EVs demand

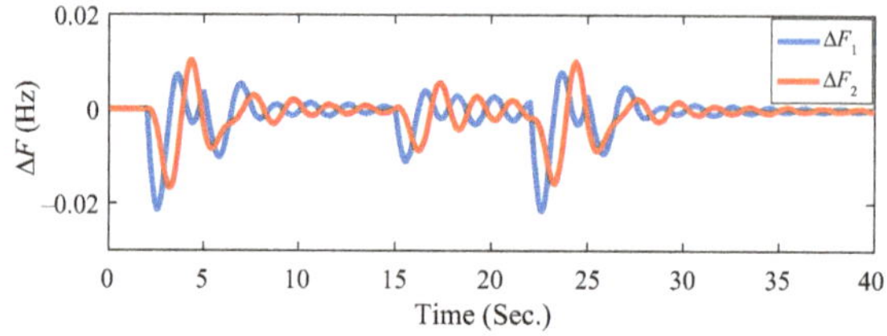

Figure 2.6 Frequency response in areas 1 and 2 due to cyberattack on EVs demand

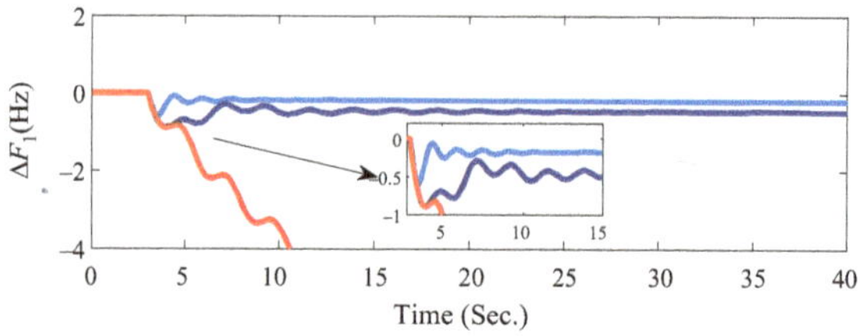

Figure 2.7 Frequency response due to DoS cyberattack on different parts of the EV aggregator

generated power. Figure 2.7 shows that in such case, the frequency will deviate from its nominal value but the system will keep operated in a safe mode since the conventional power plants and EV aggregator will support the primary frequency response from their primary active power reserve (the frequency response is shown in the light blue curve in Figure 2.7). In the second case, it is assumed that the system will be under cyberattack on the EV concentrate agent of area 2 leading to more frequency deviation affecting the stability as shown in the frequency response depicted by the blue curve in Figure 2.7. In the third DoS cyberattack, it is considered that the attacker would attack the EV aggregator agent and prevent EV aggregator from supporting the grid during frequency fluctuations. This case would result in dramatic frequency decline, as shown in the red curve in Figure 2.7, leading to cascading events and blackout due to the DoS cyberattack. The aforementioned scenarios confirm the importance of cybersecurity in ancillary services providers, especially those who provide service for supporting the frequency in power systems with high-power share from RESs, such as DR programs, virtual power plants, battery energy storage systems used for supporting the frequency, and EV aggregators.

2.6 Conclusion

This chapter highlights the importance of cybersecurity in power systems, especially in the providers of frequency support ancillary services. The importance of DR programs for enabling the transition of transitional power systems toward the smart grid concept was first discussed. Likewise, the studies on DR and electric vehicle-based

DR programs were reviewed to highlight research gaps in this important topic. Furthermore, a dynamic system frequency response model considering DR and electric vehicle aggregation was introduced, which is suitable for studying the impact of cyberattacks on power system stability and security from its frequency viewpoint. Moreover, the multi-agent system-based EV aggregator for providing primary and secondary frequency support ancillary services was discussed, followed by presenting two cyberattack scenarios on the aggregator. The scenario studied two different types of cyberattacks, namely false data-injection cyberattack and DoS cyberattack. The results confirmed the importance of considering cyberattack in designing DR programs and EV aggregators to avoid cascading events and blackouts in future power systems.

References

[1] Associates CR. Primer on Demand-Side Management; 2005. [Online; accessed 10-09-2021]. https://silo.tips/download/primer-on-demand-side-management.

[2] Administration UEI. Industrial Sector Energy Consumption; 2016. [Online; accessed 10-09-2021]. https://www.eia.gov/outlooks/ieo/pdf/industrial.pdf.

[3] Shoreh MH, Siano P, Shafie-khah M, *et al.* A survey of industrial applications of demand response. Electric Power Systems Research. 2016;141:31–49.

[4] Shafie-khah M, Siano P, Aghaei J, *et al.* Comprehensive review of the recent advances in industrial and commercial DR. IEEE Transactions on Industrial Informatics. 2019;15(7):3757–3771.

[5] Siddiquee SMS, Howard B, Bruton K, *et al.* Progress in demand response and it's industrial applications. Frontiers in Energy Research. 2021;9:330. Available from: https://www.frontiersin.org/article/10.3389/fenrg.2021.673176.

[6] Izadkhast S, Garcia-Gonzalez P, Frías P. An aggregate model of plug-in electric vehicles for primary frequency control. IEEE Transactions on Power Systems. 2014;30(3):1475–1482.

[7] Alhelou HH, Golshan M. Hierarchical plug-in EV control based on primary frequency response in interconnected smart grid. In: 2016 24th Iranian Conference on Electrical Engineering (ICEE). IEEE; 2016. pp. 561–566.

[8] Alhelou HSH, Golshan M, Fini MH. Multi agent electric vehicle control based primary frequency support for future smart micro-grid. In: 2015 Smart Grid Conference (SGC). IEEE; 2015. pp. 22–27.

[9] Jia H, Li X, Mu Y, *et al.* Coordinated control for EV aggregators and power plants in frequency regulation considering time-varying delays. Applied Energy. 2018;210:1363–1376.

[10] Vafamand N, Arefi MM, Asemani MH, *et al.* Decentralized robust disturbance-observer based LFC of interconnected systems. IEEE Transactions on Industrial Electronics. 2022;69(5):4814-4823.

[11] Alhelou HH, Hamedani-Golshan ME, Heydarian-Forushani E, *et al.* Decentralized fractional order control scheme for LFC of deregulated nonlinear power

systems in presence of EVs and RER. In: 2018 International Conference on Smart Energy Systems and Technologies (SEST). IEEE; 2018. pp. 1–6.

[12] Alhelou HH, Golshan MEH, Hatziargyriou ND. Deterministic dynamic state estimation-based optimal LFC for interconnected power systems using unknown input observer. IEEE Transactions on Smart Grid. 2019;11(2): 1582–1592.

[13] Alhelou HH, Golshan MH, Askari-Marnani J. Robust sensor fault detection and isolation scheme for interconnected smart power systems in presence of RER and EVs using unknown input observer. International Journal of Electrical Power & Energy Systems. 2018;99:682–694.

[14] Alhelou HH, Golshan MEH, Siano P. Frequency response models and control in smart power systems with high penetration of renewable energy sources. Computers & Electrical Engineering. 2021;96:107477.

[15] Khan M, Sun H, Xiang Y, *et al.* Electric vehicles participation in load frequency control based on mixed H2/H. International Journal of Electrical Power & Energy Systems. 2021;125:106420.

[16] Liu H, Qi J, Wang J, *et al.* EV dispatch control for supplementary frequency regulation considering the expectation of EV owners. IEEE Transactions on Smart Grid. 2016;9(4):3763–3772.

[17] Pham TN, Trinh H, *et al.* Load frequency control of power systems with electric vehicles and diverse transmission links using distributed functional observers. IEEE Transactions on Smart Grid. 2015;7(1):238–252.

[18] Rodrigues YR, de Souza AZ, Ribeiro PF. An inclusive methodology for plug-in electrical vehicle operation with G2V and V2G in smart microgrid environments. International Journal of Electrical Power & Energy Systems. 2018;102:312–323.

[19] Amjad M, Ahmad A, Rehmani MH, *et al.* A review of EVs charging: From the perspective of energy optimization, optimization approaches, and charging techniques. Transportation Research Part D: Transport and Environment. 2018;62:386–417.

[20] Peng C, Zou J, Lian L. Dispatching strategies of electric vehicles participating in frequency regulation on power grid: a review. Renewable and Sustainable Energy Reviews. 2017;68:147–152.

[21] Acharya S, Dvorkin Y, Karri R. Causative cyberattacks on online learning-based automated demand response systems. IEEE Transactions on Smart Grid. 2021;12(4):3548-3559.

[22] Holmberg DG, Ghatikar G, Koch EL, *et al.* OpenADR advances. ASHRAE Journal. 2012;54(LBNL-6055E).

[23] Qi J, Kim Y, Chen C, *et al.* Demand response and smart buildings: a survey of control, communication, and cyber-physical security. ACM Transactions on Cyber-Physical Systems. 2017;1(4):1–25.

[24] Seijo Simó M, López López G, Moreno Novella JI. Cybersecurity vulnerability analysis of the plc prime standard. Security and Communication Networks. 2017; 2017.

[25] Acharya S, Dvorkin Y, Karri R. Public plug-in electric vehicles+ grid data: Is a new cyberattack vector viable? IEEE Transactions on Smart Grid. 2020;11(6):5099–5113.
[26] Soltan S, Mittal P, Poor HV. BlackIoT: IoT botnet of high wattage devices can disrupt the power grid. In: 27th {USENIX} Security Symposium ({USENIX} Security 18); 2018. pp. 15–32.
[27] Raman G, Peng JCH, Rahwan T. Manipulating residents' behavior to attack the urban power distribution system. IEEE Transactions on Industrial Informatics. 2019;15(10):5575–5587.
[28] Mieth R, Dvorkin Y. Online learning for network constrained demand response pricing in distribution systems. IEEE Transactions on Smart Grid. 2019;11(3):2563–2575.
[29] Alhelou HAH, Cuffe P. A dynamic state estimator based tolerance control method against cyberattack and erroneous measured data for power systems. IEEE Transactions on Industrial Informatics. 2021. doi:10.1109/TII.2021.3093836.
[30] Alhelou HH, Golshan MEH, Hatziargyriou ND. A decentralized functional observer based optimal LFC considering unknown inputs, uncertainties, and cyber-attacks. IEEE Transactions on Power Systems. 2019;34(6):4408–4417.

Chapter 3

Recurrent neural networks for electrical load forecasting to use in demand response

Nils Jakob Johannesen[1], Mohan Lal Kolhe[1] and Morten Goodwin[2]

Electric load forecasting is a fundamental technique to understand end-user behavior and therefore a crucial factor in the design of demand response (DR) programs. Load forecasting will also identify the appropriate design of DR programs. In this chapter, a range of different machine learning applications are studied to represent the influential factors for electrical load demand forecast in a DR context, with a variety of different data scenarios, temporal and technical scenario. This chapter explores and compares the load prediction analysis through basic recurrent neural networks (RNNs); Vanilla RNN, gated recurrent units (GRU), and long short-term memory (LSTM), using principal component analysis (PCA). It is found that PCA can be used to reduce the number of principal components for Vanilla RNN, GRU, and LSTM networks. Reducing the number of principal components using PCA is one of the techniques that is used in dimensionality reduction. Reduction in dimensionality will relieve the computational burden. In this work, the dimensionality reduction improves the predictive output. It is observed that for electric load demand forecasting, the preferred technique is GRU, trained with a principal component. The performance is evaluated through mean absolute percentage error (MAPE), which is relatively lower than other techniques.

3.1 Introduction

It is important to have more accurate electrical demand forecasting in urban area network for finding the opportunities in industrial demand response (DR) programs. Due to market energy price dynamics, urban area electric energy consumption patterns may change and hence the DR programs. Therefore, it is necessary to have accurate demand prediction considering user patterns as well as impact of meteorological

[1]Department of Engineering & Science, University of Agder, Kristiansand, Norway
[2]Department of ICT, University of Agder, Kristiansand, Norway

parameters [1]. The state-of-the-art research in electrical load demand forecasting focuses on three main aspects in order to make sound predictions, these aspects are related to weather parameters, holidays, and time of day. The relations among these aspects have been found equally important throughout the techniques used in electrical load forecasting, including time series analysis and simpler instance based machine learning models to the more complex black box neural networks [2–4]. Considerations of these aspects are important in short-term [2,5,6], mid-term [7], and long-term forecasting [8]. The impact of external weather parameters has proven also to be important for forecasting on limited data, such as for households and buildings [9], as well as seasonal holiday demands [10].

In the electrical demand forecasting, it is necessary to analyze the load patterns with reference to time of use. A typical diurnal load profile for urban area of New South Wales is given in Figure 3.1. The automatic metering infrastructure (smart energy meters) is helping in collecting the aggregated energy data with time for analyzing the load patterns for finding opportunities of DR program. The enhancement of information structure enabled by automatic metering infrastructure has transformed the grid to a smarter grid that ties consumers directly into managing the grid through Demand Side Management (DSM). An overview of DSM and some of the available incentives in DR programs are illustrated in Figure 3.2. Within DSM, there are two main strategies for alternating the loads, shifting electric load in peak demand (Figure 3.3) [11,12]. The DR program takes care of the first strategy, namely load shifting, where customer load is transferred from daily peak demand to other times of the day when the consumption is less. By shifting the diurnal load demand curve, the DR facilitates more electricity to be produced by less expensive base load generators. A successful DSM flattens the load curve [13].

This chapter is organized in the following sections; Section 3.2 are going to cover the role of electrical load forecasting in DR programs. In Section 3.3, a review of load forecasting is given, and in Section 3.4, the use of RNN's in electrical load forecasting is given. Section 3.5 explains PCA, and in Section 3.6, a case study using RNN's and

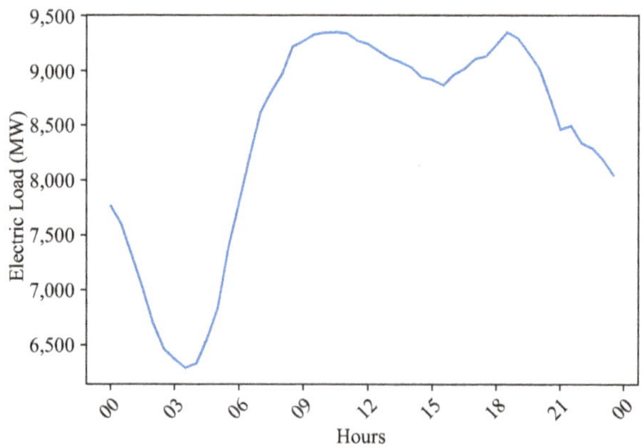

Figure 3.1 Typical diurnal load profile for New South Wales

PCA on the New South Wales power system is given. The results are presented in Section 3.7, and finally the conclusions in Section 3.8.

3.2 DR programs

DSM is divided in DR programs and energy efficiency and it is illustrated in Figure 3.2. Energy efficiency involves a storage capacity, whilst DR programs are actions to actively manage the electrical loads. The dispatchable energy sources involve the sources of electricity that can be controlled by the power grid operators and hence, dispatched at their control. The non-dispatchable energy sources cannot be managed by the operators and time-based incentives are made to engage customers to adjust

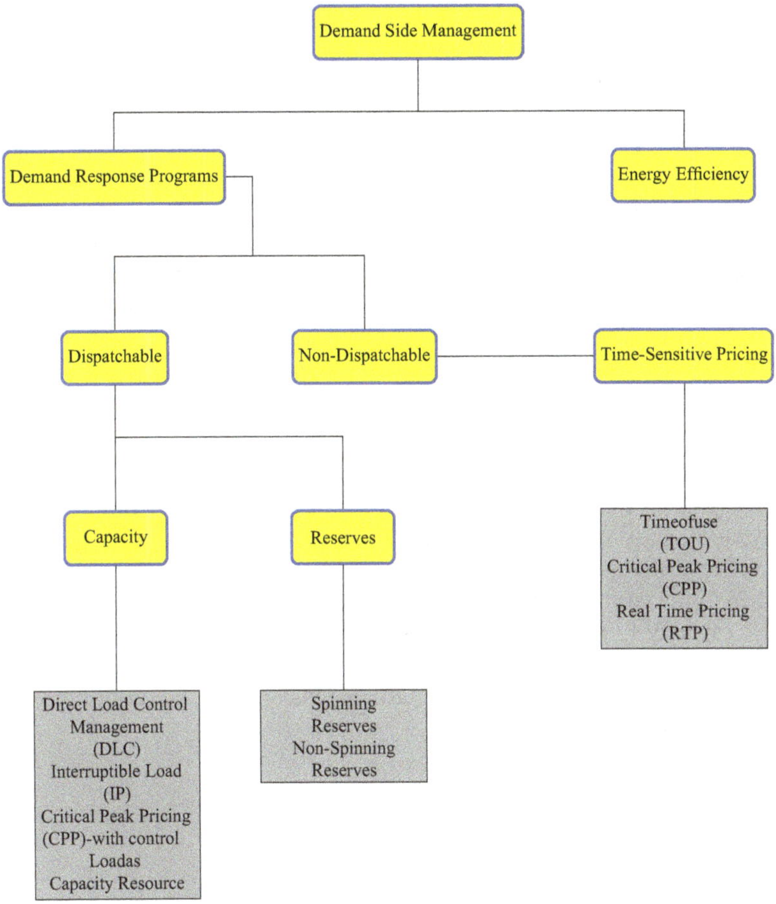

Figure 3.2 Overview of demand side management and representative incentives in DR programs

their energy demand accordingly. Price-based DR programs generally include time of use (TOU), real-time pricing (RTP), critical peak pricing (CPP), and peak time rebates (PTR). TOU is price rates of electricity that either varies diurnally or seasonally. RTP is dynamic pricing that reflects real-time scenarios with prices shifting more frequently than TOU, and up to hourly price variations. CPP is also dynamic, where short-term prices are preadjusted to reach a specified load demand. In RTP, prices are adjusted on short notice and are not fixed at the time the tariff is effectuated. The incentive based actions generally involve direct load control (DLC), interruptible or curtailable rates (I/C), demand bidding (DB)/buy back (BB), emergency DR program (EDRP), capacity market program (CMP), and ancillary service market (ASM). The DLC refers to the control actions taken directly by the power grid operators, normally by remote accessing the customers appliances (e.g. water heaters, electric vehicle charging) and manage and reschedule these loads. While DLC is for the residential sector, I/C is for industrial and large commercial actors that are asked to curtail their consumption for shorter periods when the grid is congested, in return they receive discounted electricity bills. In DB/BB, customers themselves ask or bids at a price they are willing to reduce their load consumption as opposed to DLC where they are asked by the power companies to reduce their loads. Similar to DLC, I/C is designed for large-scale consumers that can plan their operation. For large-scale consumers participating in EDRP-programs, the ISO announces amounts up to 10 times the off-peak electricity price, to cut their consumption. When contingencies arise, costumers that are in the CMP programs commit to curtail their consumption. Ancillary services are support services in the power system and are essential in maintaining power quality and reliability. There are typically three types of ancillary service products that DR can participate in. From the faster to the slower acting, these are regulation, spinning reserve, and nonspinning reserve [14–19]. DR strategies based on changing load patterns are illustrates in Figure 3.3.

3.2.1 Load forecasting in DR

Electric load forecasting is a fundamental technique to understand end-user behavior and therefore a crucial factor in the design of DR programs. Load forecasting will also help in identifying the appropriate design of DR programs, and these programs are described in Section 3.2. Robust load prediction can help determine the DR events and DR capacity [20]. In the DR program, decision making the impact in terms of user comfort have been analyzed, in consumption aware analytical DR scheme, considering factors such as appliance adjustment factor, appliance priority index, and appliance curtailment priority [21].

3.3 Review on load forecasting

Time series analysis has traditionally been performed in meteorology, energy, and economics [22]. Elements from time series analysis are used to find the parametric values in autoregression (AR) and moving average (MA) and later Auto Regression Moving Average (ARMA) [23]. The AutoRegressive Integrated Mean Average (ARIMA)

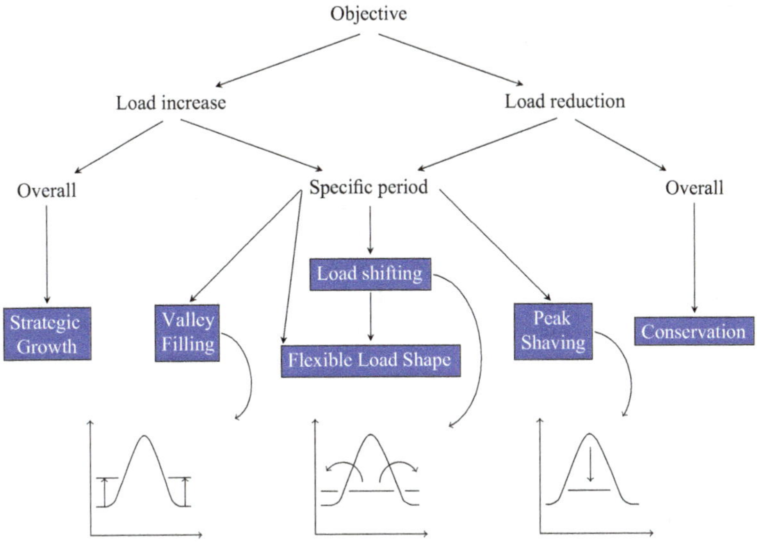

Figure 3.3 Main strategies for alternating electrical load demand

model is useful for modeling time series behavior by removing the influence of trend in the data. In multivariate cases, the exogenous variable aids the predictive outcome in AutoRegressive Integrated Moving Average with Exogenous variables (ARIMAX). This is further developed into Seasonal AutoRegressive Moving Average with Exogenous variables (SARIMAX), that also accounts for seasonal behavior [24]. These methods are useful for the modeling of time series and aid the electrical load analysis. Cycles, trends, and periodicity can be found through tests provided by time series analysis [25]. Medium-term electric load forecasting has been performed in Bruce County in Canada using data from 2010 to 2018 comparing Random Forest Regression, Support Vector Regression and Recurrent Neural Networks (RNN) achieving a MAPE of 4–10% [26]. In short-term electric load demand forecasting, RNN by Levenberg-Marquardt and Bayesian regularization on 30 min predictions had achieved a mean absolute percentage error (MAPE) of an average in one week 1.4792 [27]. One hour ahead prediction has been performed on hourly power consumption in Toronto Canada using Long Short-Term Memory (LSTM), achieving a MAPE of 2.639, which was an improvement of the Vanilla RNN of 3.712 MAPE [28]. The Resnetplus model for the ISO-NE dataset proposed a day-ahead load forecasting model based on deep residual networks. A basic structure of several fully connected layers to produce preliminary forecasts of 24 h. A forecast is then made on the residuals of the preliminary forecast provided with a formulation of Monte Carlo dropout for probabilistic forecasting, achieving an average MAPE of 1.447 [29]. Gated recurrent unit (GRU) was used to predict the electricity market in Singapore. Multi-features input models of different time structural architecture named Multi-GRU has been used to give 30 min predictions [30].

Hybrid forecast combining neural networks with autoregression has proven to aid in tracing the curvature of the peak in the volatile electricity markets [31]. A range of different machine learning applications have been tested to research the influential factors for electrical load demand forecast in a DR context, with a variety of different data scenarios, temporal, and technical scenarios. The authors of [32] tested Echo State Network (ESN), Extreme Learning Machine, Linear Regression and Support Vector Machine, with incrementally adding data input and mixing the factors to test for the most influential factors on the load demand. The authors state [32] that the output from their findings may be used in modeling building characteristics such as ventilation load or as an informed input to the data selection process for modeling. It is found that the most influential factors on load prediction, based on data from an office building in Denmark, are weather and time of day. When including weather forecasts and occupancy did not improve the forecast accuracy. The authors of [20] also used ESN, that uses reservoir computing, a form of recurrence to capture influence of past states in the decision making, and compared them to a general Feed Forward Neural Network, in electrical load forecasting in DR programs.

Based on investigating the electrical load profile of a university campus building and analyzing the weather parameters, data is divided into clusters using k-means-clustering, into tree types of clusters based on types of day. This gives meaningful input to DR applications since some clusters depicts a higher load demand than others, and hence is more useful for shifting loads. K-means clustering is a type of data mining technique that together with k-Nearest Neighbors and ARIMA were used in a case study carried out in Tulkarm District, Palestine [33]. Based on case study objectives, the electricity consumption is shifted through DR programs to periods of lower demand on a weekly basis.

3.4 RNNs in electric load forecasting

The traditional deep neural networks (DNN's) learn patterns on the assumption that inputs and outputs are independent of each other. The first DNN's used stacked generalization to develop deep learning based on the concept of a perceptron [34]. A perceptron mimics the behavior of neurons in the brain for decision making. When the sum of weights and inputs reach a certain threshold value, neurons fire off, similar to learning paths in the brain. In neural networks, an activation function decides upon the state of activation. The output from the activation function is compared to the real value from the targeted response vector in a loss function, as shown at Frame 1 in Figure 3.4. The output from the loss function is used to trace the global minima through stochastic gradient descent. The information is backpropagated through the network and use to adjust the weighted input of the network, and this process is found when training and validation losses converge and a stop criterion terminates the process. In Frame 2 of Figure 3.4, a feed forward neural network (FFNN) as a black box representation, with input, black box, and a learn output is illustrated.

A RNN depends on the prior elements within the sequence, to perform its decision making. The RNNs used in this work are all based on Keras [35]. RNNs were first developed in natural language processing and the Vanilla RNN is a fully-connected

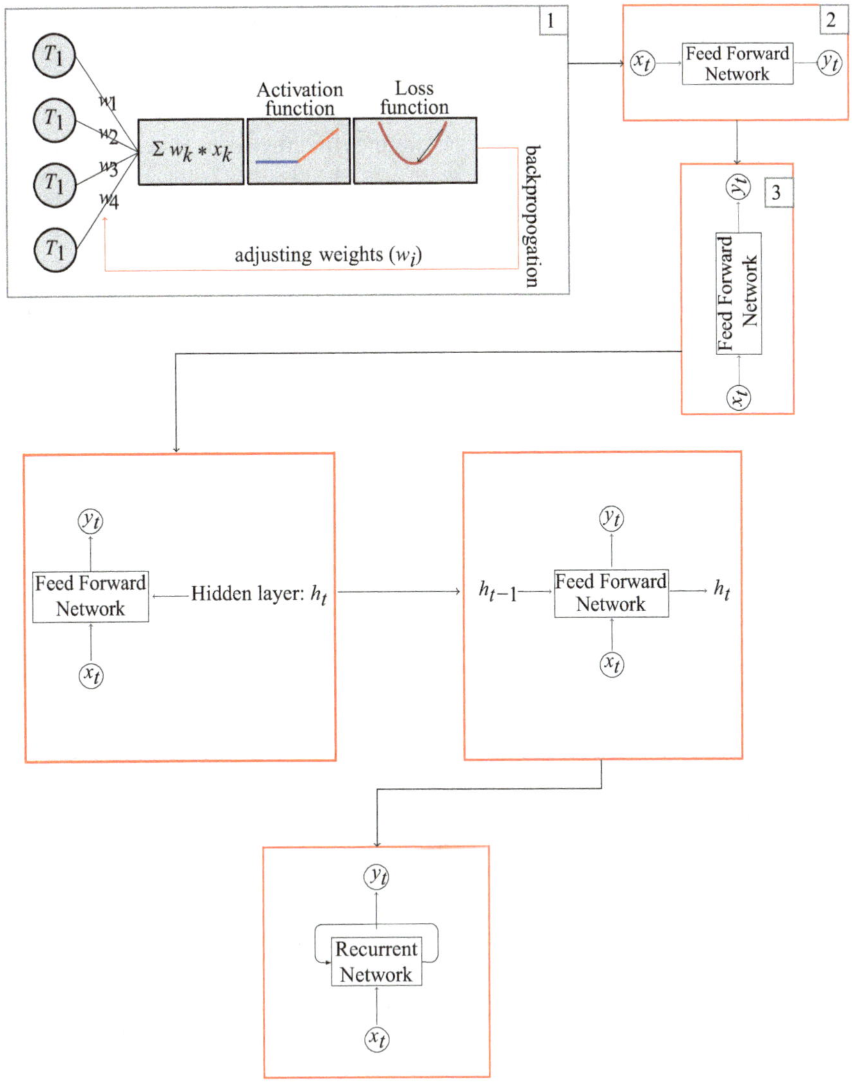

Figure 3.4 RNN explained themathically

RNN where the output from previous time step is to be fed to the next time step by an additional set of units. The units have also proven to be successful in other time series application, and for all problems constituted by sequences, such as electrical load demand.

Frame 3 in Figure 3.4 shows a FFNN transposed to its vertical axis, to show the key concept of units in RNN's.

Recurrence that provides the key concepts behind RNNs, the key idea, is that the RNN's remain the internal state, h_t, that is updated for each timestep, and keep the sense of recurrence in the network. The update is defined mathematically in equation as shown in (3.1):

$$ht = f_w(x_t, h_{t-1}) \tag{3.1}$$

This internal state, h_t, is a hidden layer used to define the state. When computed in the network is used as shown in [36]:

$$h_{i,t} = \sigma_t(Uh_{i,t-1} + Wx_{i,t}) \tag{3.2}$$

In (3.2), σ is the activation function, U and H are learned weighted parameters for hidden states and input vectors. The process then composes a set of learned weighted parameters in matrix V, which for a regression problem uses a linear activation function σ_y to give the result in the output layer:

$$y_{i,t} = \sigma_y(Vh_{i,t}) \tag{3.3}$$

3.4.1 Scaling data, normalizing

Data is scaled. The general method of calculation is to determine the distribution mean and standard deviation for each feature. Next we subtract the mean from each feature. Then we divide the values (mean is already subtracted) of each feature by its standard deviation:

$$x'_{ij} = \frac{x_i - \hat{x}_j}{\sigma_j} \tag{3.4}$$

x'_{ij} is the value of the input variable of row i and column j, $\hat{x}_j$ is the mean of the values in column j, and finally σ_j is the standard deviation of the values in column j [37]. Findings from auto-correlation and cross-correlation the most important time and external factors on the targeted vector are found, and will help in the appropriate design of the networks time-lag vector. The multivariate case is a 3D-vector, containing the amount of data (samples), lags, and number of inputs (features). Equally on the output, it aims for the target vector. In the training phase, this is the next step ahead relative to the input vector.

In the further feature engineering, a lower indicator variable is designed to differentiate over working-days/non-working days with a binary switch [38]. The RNNs purposefully search in a higher category space to find meaningful relations between the vectors, and therefore the time input is coded using circular coding. The circular coding identifies the time of day according to the unit circle, giving both a sine and cosine coordination as its parameters. They are used as training inputs for the target vector, the electric load demand.

3.5 PCA for electrical load forecasting

Appropriate feature engineering reduces the dimensionality by reducing number of attributes, while preserving the variation in the data. The reduced dimensions in data relieve the algorithm some of its computational burden, hence the training time is reduced, as well as reducing noise in the dataset. Another benefit from dimensionality reduction is mitigating the problem of overfitting, since many features of data means that the models become more complex. This is treated in the bias variance trade off, where the most suitable model is found.

Dimension reduction techniques are both linear and non-linear. The state-of-the-art dimension reduction techniques include Deep Convolutional Autoencoder (DCAE), t-Distributed Stochastic Neighbor Embedding (t-SNE) and Uniform Manifold Approximation and Projection (UMAP), and Principal Component Analysis (PCA) [39].

PCA is a multivariate technique that can be applied to many fields for feature reduction. To find the intrinsic nature of linguistic representation, PCA has been performed on the hidden unit activation patterns to reveal that the network solves the task by developing complex distributed representations which encode the relevant time relations and hierarchical constituent structure [40]. It is the underlying patterns and structures in the dataset that can be discovered from the high-dimensional data, through PCA. PCA is extracting the important information for later to represent it in a new set of orthogonal vector input constituting the principal components. This is done by averaging the data and shift the origo to the mean of the data. A best fitted line going through origo is found then by a linear regression of minimizing each inputs square distance to the line or maximizing the sum of squared distances from the projected points to the origo.

These principal components are linear transformation of the data so that the first coordinate explains the most of the variation, the second coordinate the second most, and so on. The components are found through the eigen-decomposition and singular value decomposition [41,42].

To perform PCA, the input matrix is transposed and crossed with its non-transposed version, stored in matrix L. By diagonalzng L, find a matrix M and diagonal matrix W:

$$L = M^T WM \tag{3.5}$$

The feature space is reduced by restricting inputs based on the number of columns that sums up M to make a rotated matrix. The eigenvalues from W are related to the variance of the principal components.

The score value for an observation is the distance from the observation to the origo, along the best fitted line (loading point).

3.6 Load data pre-processing with time organization and training, validation and testing: case study of Urban Area of New South Wales

In this work, New South Wales, Sydney region load profile data set [43] is used, which includes meteorological parameters (e.g. DryBulb and WetBulb Temperature, Humidity, weekday and time of use) [44]. Power system network of New South Wales

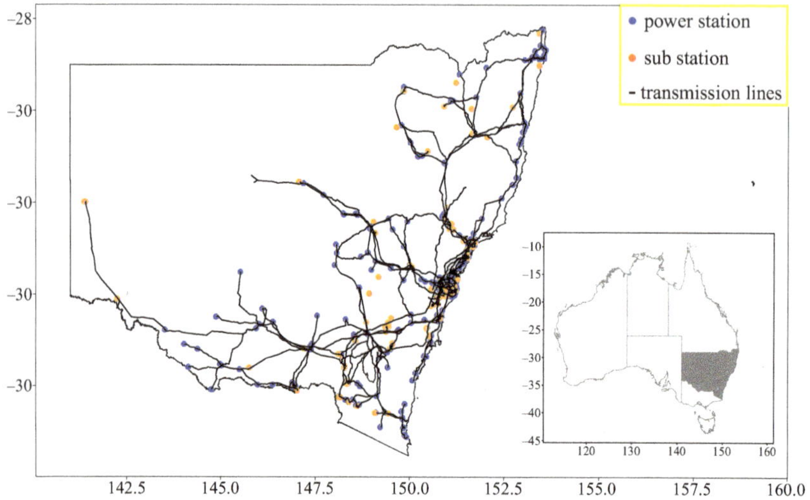

Figure 3.5 New South Wales, Australia, with the main power plants, substations and transmission lines

is illustrated in Figure 3.5. Map made with geojson source file [45] in Geopandas [46], and source files from [47].

Time-dependent structures are composed as vectors and fed as inputs to the RNNs. To avoid biases and overfitting, the data is to be divided amongst training, validation and testing. In particular, the algorithm must capture trends and seasonal variations. If the time series can claim to be stationary, no means needs to be taken. To prove stationarity, a search for no trend, constant variance and constant autocorrelation is conducted. Testing for stationarity is done by introducing the null hypothesis H_0: time series is non-stationary due to trend. By the Augmented Dickey-Fuller (ADF) test, if certain criteria are met the null hypothesis is rejected and the time series is assumed to be stationary. The ADF basically searches for trends in the dataset by evaluating mean and variance over time. Based on this assumption that the time series is stationary, a division into training, validation and test set are done The Sydney Region data with load measurements for every 30 min from 2006 to 2010 (Figure 3.6). The training set ranges from the beginning of the recorded data on 01.01.2006 until 31.12.2008. The entire 2009 is used for validation and finally 2010 is for testing.

The proposed model in this work finds suitable training, validation and test-sets by searching for stationarity through Augmented Dickey Fuller Test. The original training set is then reduced feature space and variation representation by performing its PCA, reducing the principal components from an offset features of 9 to be represented by 8 principal components according for 99% of the variance. The training set has then been scaled, and trained on three different RNNs, Vanilla RNN, GRU, and LSTM, see Figure 3.7. These different models have been tested for different seasons to analyze how they assimilate for seasonal variations. Finally the models using PCA are compared to a version that does not reduce its feature space through PCA.

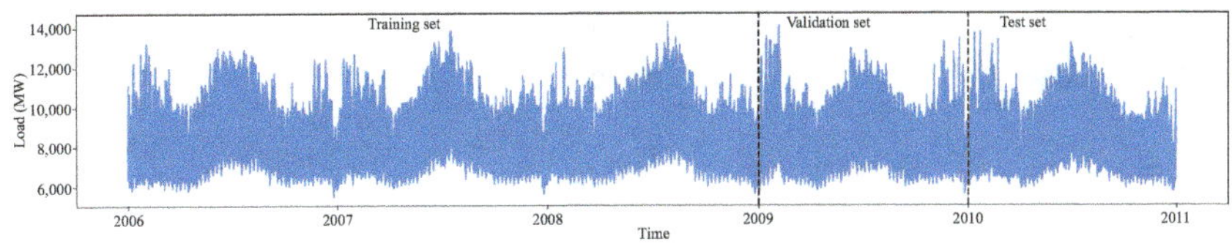

Figure 3.6 The Sydney Region data with load measurements for every 30 minutes from 2006 to 2010. Dashed black lines indicate the separation into train-, validation-, and test-set.

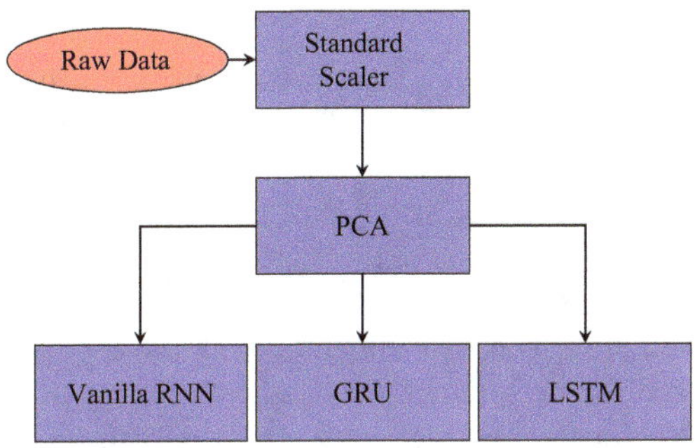

Figure 3.7 The model applied scales the raw Sydney Data, and through PCA predicted by different RNNs

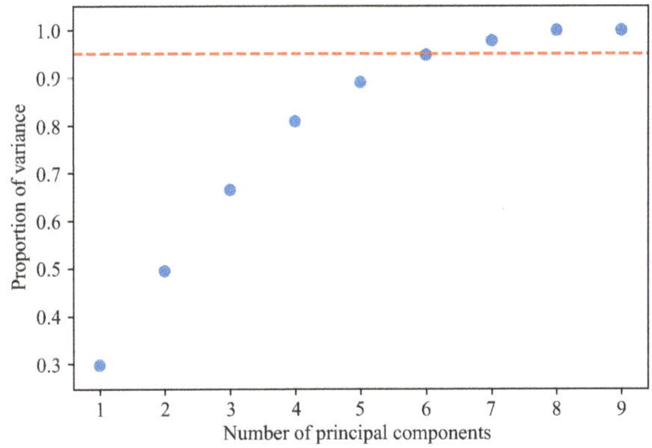

Figure 3.8 A scree plot displays the cumulative variance per introduced principal components, with red dashed line indicating 95% variance

There are several methods to illustrate PCA. A loading plot shows how strongly each loading point influences a principal component, shown by the loading vectors. A scree plot shows how much variation each principal component capture from the data, as in Figure 3.8. The scree plot depicts the proportion of variance that is captured by each number of principal components after feature engineering for the Sydney

Data. The red dashed line signifies that when we include the 6 principal components the PCA-process capture 95% of the variance.

3.7 Results and discussion

It is observed from training the RNNs with PCA that during 50 epochs of training and validation, the training loss and validation loss decrease to a point of stability with a minimal gap between the two final loss values; Figure 3.9 illustrates with the GRU with PCA, for the Vanilla RNN and LSTM, the loss curves show the same convergence.

The RNNs have been tested for a week in January, April, July, and October (see Table 3.1), and MAPE has been averaged. The results show that all of the RNNs are capturing the inherent structure of the electric load demand quite well, resulting in an acceptable MAPE around 1–2% through all seasons, see Table 3.2. In the case of GRU networks, the results for all the seasons are improved through PCA concluding with 99% of the variation captured by the 8 principal components, see Table 3.7. Also for the Vanilla RNN, there is a benefit from reduced number of principal components in a lesser MAPE, and for the summer test on a week in July (Figure 3.10), it scores best of all RNNs. Yet for the LSTM, it does not benefit from an improved MAPE from the PCA. The best results are measured in January when also the electrical load demand is at the highest (Figure 3.10), and the impact of external weather parameters is influencing greatly on the load demand. The curvature of the load profile is dominated by a high peak at noon, and GRU captures this very good.

The results from the week of April (Figure 3.10) have a lower load demand than January. In January, the load demand is highly correlated to the weather parameters

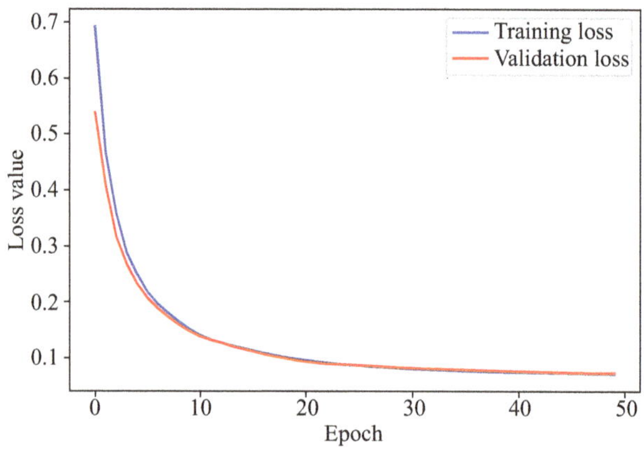

Figure 3.9 GRU networks training and validation loss decreases to a point of stability

Table 3.1 Seasons and test week

Season	Name	Test week
Season 1	Winter	11th–17th of January 2010
Season 2	Spring	10th–16th of May 2010
Season 3	Summer	12th–18th of July 2010
Season 4	Autumn	11th–17th of October 2010

Table 3.2 Performance (MAPE)

MAPE	Vanila RNN	GRU	LSTM
January	0.95	0.87	0.90
April	1.45	1.21	1.25
July	1.84	1.64	1.30
October	1.38	1.26	1.24

readings in winter season. In April, as in January, GRU with PCA achieves the best forecast MAPE result for the week in April, yet with a slightly higher MAPE than for January. This can be explained by the lower load demand in April and that correlations to weather parameters are usually lower in spring and autumn. In the winter season, the correlations to weather parameters are higher than other seasons, as well as in general the winter season has a higher load demand. These are factors explaining the lower MAPE in winter season as opposed to other seasons.

In the test week of October (Figure 3.10), which has the same range in load demand (6,000–10,000 MW), it is also GRU with PCA that scores best with a MAPE of 0.94, see Table 3.3.

When comparing the results, the MAPE is in the same range for Vanilla RNN (1.45 for April and 1.38 for October), GRU (1.21 for April and 1.26 for October), and LSTM (1.25 for April and 1.24 for October), in Tables 3.2 and 3.3. The similarity in results from spring (observed from the test results for the week in April) and autumn (observed from the test results for the week in October) can be explained by similar load range and meteorological conditions. In the case of Vanilla RNN and GRU, the explanations of the compared results indicate the same when investigating the results on the RNNs tested with PCA. The exception is the LSTM tested with PCA that shows a higher MAPE. It is observed that LSTM is a more complex algorithm, than the Vanilla RNN and GRU.

When it is trained with relatively lesser data, although it is analyzed using its principal components, it is not able to improve the predictions. It is observed that for the for the week in July with the lowest load demand, the simplest RNN (Vanilla RNN) with reduced principal components achieves the preferred MAPE, amongst all of the predictors.

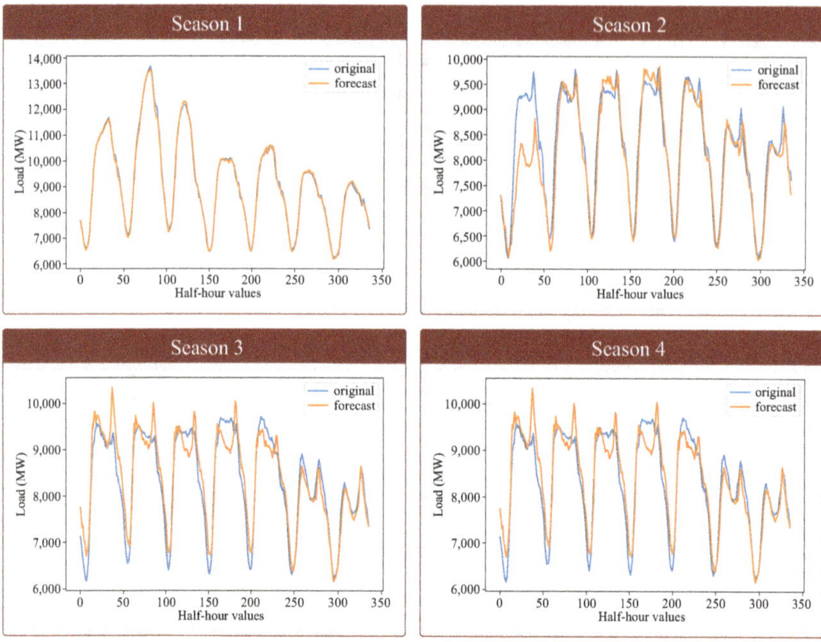

Figure 3.10 Results from RNN on electric load demand in New South Wales

Table 3.3 Performance using PCA(MAPE)

MAPE	Vanila RNN	GRU	LSTM
January	0.87	0.75	0.89
April	1.11	1.16	1.60
July	1.39	1.53	1.75
October	1.06	0.94	1.27

3.8 Conclusion

This chapter explores and compares the load prediction analysis through basic RNNs; Vanilla RNN, GRU, and LSTM, using PCA. The winter season load behavior is more influenced by weather parameters, which explains why in the winter season the RNNs scores relatively higher than in other seasons. It is found that PCA can be used to reduce the number of principal components for Vanilla RNN, GRU, and LSTM networks. Reducing the number of principal components using PCA is one of the techniques that is used in dimensionality reduction. Reduction in dimensionality will relieve the computational burden. In this work, the dimensionality reduction improves the predictive output. For the electric load demand forecasting, the preferred RNN is GRU trained with a principal component of 8, and it is shown through MAPE. After comparing with the version without PCA, the results show that MAPE is reduced when using PCA. For the 30 min forecasting, GRU with PCA performs best MAPE of 0.74%. This work will benefit the reliable forecasting to anticipate the events involved in dispatching, control and management of the operating grid.

References

[1] Johannesen NJ, Kolhe M, Goodwin M. Comparison of regression tools for regional electric load forecasting. In: 2018 3rd International Conference on Smart and Sustainable Technologies (SpliTech). IEEE; 2018. pp. 1–6.

[2] Johannesen NJ, Kolhe M, Goodwin M. Relative evaluation of regression tools for urban area electrical energy demand forecasting. Journal of Cleaner Production. 2019;218:555–564. Available from: https://www.sciencedirect.com/science/article/pii/S0959652619301192.

[3] Johannesen NJ, Kolhe ML, Goodwin M. Comparing recurrent neural networks using principal component analysis for electrical load predictions. In: 2021 6th International Conference on Smart and Sustainable Technologies (SpliTech); 2021. pp. 1–6.

[4] Songpu A, Kolhe M, Jiao L. External parameters contribution in domestic load forecasting using neural network. IET Conference Proceedings. 2015 January. pp. 6–6. Available from: https://digital-library.theiet.org/content/conferences/10.1049/cp.2015.0344.

[5] Aguilar Madrid E, Antonio N. Short-term electricity load forecasting with machine learning. Information. 2021;12(2). Available from: https://www.mdpi.com/2078-2489/12/2/50.

[6] Laouafi A, Mordjaoui M, Haddad S, *et al.* Online electricity demand forecasting based on an effective forecast combination methodology. Electric Power Systems Research. 2017;148:35–47. Available from: https://www.sciencedirect.com/science/article/pii/S0378779617301165.

[7] Mirasgedis S, Sarafidis Y, Georgopoulou E, *et al.* Models for mid-term electricity demand forecasting incorporating weather influences. Energy. 2006;31(2):208–227. Available from: https://www.sciencedirect.com/science/article/pii/S0360544205000393.

[8] He Y, Jiao J, Chen Q, *et al.* Urban long term electricity demand forecast method based on system dynamics of the new economic normal: the case of Tianjin. Energy. 2017;133:9–22. Available from: https://www.sciencedirect.com/science/article/pii/S0360544217308630.

[9] Lusis P, Khalilpour KR, Andrew L, *et al.* Short-term residential load forecasting: impact of calendar effects and forecast granularity. Applied Energy. 2017;205:654–669. Available from: https://www.sciencedirect.com/science/article/pii/S0306261917309881.

[10] Johannesen NJ, Kolhe M, Goodwin M. Load demand analysis of Nordic rural area with holiday resorts for network capacity planning. In: 2019 4th International Conference on Smart and Sustainable Technologies (SpliTech); 2019. pp. 1–7.

[11] Javed F, Arshad N, Wallin F, *et al.* Forecasting for demand response in smart grids: an analysis on use of anthropologic and structural data and short term multiple loads forecasting. Applied Energy. 2012;96:150–160. Smart Grids. Available from: https://www.sciencedirect.com/science/article/pii/S0306261912001213.

[12] Ding Y, Neumann MA, Kehri Ö, *et al.* From load forecasting to demand response – a web of things use case. In: WoT'14; 2014.

[13] Davito B, Tai H, Uhlaner R. The smart grid and the promise of demand-side management. McKinsey on Smart Grid. 2010;3:8–44.

[14] Kiliccote S, Piette MA, Ghatikar G. 15– Improved energy demand management in buildings for smart grids: the US experience. In: Bessède JL, editor. Eco-Friendly Innovation in Electricity Transmission and Distribution Networks. Oxford: Woodhead Publishing; 2015. pp. 15–338. Available from: https://www.sciencedirect.com/science/article/pii/B978178242010100015X.

[15] Aalami HA, Moghaddam MP, Yousefi GR. Demand response modeling considering Interruptible/Curtailable loads and capacity market programs. Applied Energy. 2010;87(1):243–250. Available from: https://www.sciencedirect.com/science/article/pii/S030626190900244X.

[16] Aalami H, Yousefi GR, Parsa Moghadam M. Demand response model considering EDRP and TOU programs. In: 2008 IEEE/PES Transmission and Distribution Conference and Exposition; 2008. pp. 1–6.

[17] Li D, Chiu WY, Sun H. Demand side management in microgrid control systems. In: Mahmoud MS, editor. Microgrid. Oxford: Butterworth-Heinemann; 2017. pp. 203–230 (Chapter 7). Available from: https://www.sciencedirect.com/science/article/pii/B9780081017531000073.

[18] Li Y, Gao W, Ruan Y, *et al.* Demand response of customers in Kitakyushu smart community project to critical peak pricing of electricity. Energy and Buildings. 2018;168:251–260. Available from: https://www.sciencedirect.com/science/article/pii/S0378778817334977.

[19] Khan I. Energy-saving behaviour as a demand-side management strategy in the developing world: the case of Bangladesh. International Journal of Energy and Environmental Engineering. 2019;10:493–510.

[20] Mansoor M, Grimaccia F, Leva S, et al. Comparison of echo state network and feed-forward neural networks in electrical load forecasting for demand response programs. Mathematics and Computers in Simulation. 2021;184:282–293. ELECTRIMACS 2019 ENGINEERING – Modelling and computational simulation for analysis and optimisation in electrical power engineering. Available from: https://www.sciencedirect.com/science/article/pii/S0378475420302391.

[21] Jindal A, Singh M, Kumar N. Consumption-aware data analytical demand response scheme for peak load reduction in smart grid. IEEE Transactions on Industrial Electronics. 2018;65(11):8993–9004.

[22] Box GEP, Jenkins GM. Time Series Analysis: Forecasting and Control. Holden-Day; 1976.

[23] Vemuri S, Huang WL, Nelson DJ. On-line algorithms for forecasting hourly loads of an electric utility. IEEE Transactions on Power Apparatus and Systems. 1981;PAS-100(8):3775–3784.

[24] Rahman S, Bhatnagar R. An expert system based algorithm for short term load forecast. IEEE Transactions on Power Systems. 1988;3(2):392–399.

[25] Johannesen NJ, Kolhe ML. Application of regression tools for load prediction in distributed network for flexible analysis. In: Flexibility in Electric Power Distribution Networks. London: CRC Press; 2021.

[26] Shirzadi N, Nizami A, Khazen M, *et al.* Medium-term regional electricity load forecasting through machine learning and deep learning. Designs. 2021;5(2). Available from: https://www.mdpi.com/2411-9660/5/2/27.

[27] Yahya MA, Hadi SP, Putranto LM. Short-term electric load forecasting using recurrent neural network (study case of load forecasting in Central Java and special Region of Yogyakarta). In: 2018 4th International Conference on Science and Technology (ICST); 2018. pp. 1–6.

[28] Zhang L, Yang L, Gu C, *et al.* LSTM-based short-term electrical load forecasting and anomaly correction. In: E3S Web of Conferences. vol. 182. EDP Sciences; 2020. p. 01004.

[29] Chen K, Chen K, Wang Q, *et al.* Short-term load forecasting with deep residual networks. IEEE Transactions on Smart Grid. 2019;10(4):3943–3952.

[30] Li W, Logenthiran T, Woo WL. Multi-GRU prediction system for electricity generation's planning and operation. IET Generation, Transmission & Distribution. 2019;13(9):1630–1637. Available from: https://ietresearch.onlinelibrary.wiley.com/doi/abs/10.1049/iet-gtd.2018.6081.

[31] Johannesen NJ, Kolhe M, Goodwin M. Deregulated electric energy price forecasting in NordPool market using regression techniques. In: 2019 IEEE Sustainable Power and Energy Conference (iSPEC); 2019. pp. 1932–1938.

[32] Wollsen MG, Kjærgaard MB, Jørgensen BN. Influential factors for accurate load prediction in a demand response context. In: 2016 IEEE Conference on Technologies for Sustainability (SusTech); 2016. pp. 9–13.

[33] AbuBaker M. Household electricity load forecasting toward demand response program using data mining techniques in a traditional power grid. International Journal of Energy Economics and Policy. 2021;11(4):132–148.

[34] Hinton GE, Osindero S, Teh YW. A fast learning algorithm for deep belief nets. Neural computation. 2006;18(7):1527–1554.

[35] Gulli A and Pal S. *Deep Learning with Keras*. Birmingham: Packt Publishing; 2017.

[36] Wu Z, Rincon D, Gu Q, *et al.* Statistical machine learning in model predictive control of nonlinear processes. Mathematics. 2021;9(16). Available from: https://www.mdpi.com/2227-7390/9/16/1912.

[37] Serrano-Guerrero X, Escrivá-Escrivá G, Roldán-Blay C. Statistical methodology to assess changes in the electrical consumption profile of buildings. Energy and Buildings. 2018;164:99–108. Available from: https://www.sciencedirect.com/science/article/pii/S0378778817333248.

[38] Johannesen NJ, Kolhe ML, Goodwin M. Smart load prediction analysis for distributed power network of Holiday Cabins in Norwegian rural area. Journal of Cleaner Production. 2020;266:121423. Available from: https://www.sciencedirect.com/science/article/pii/S0959652620314700.

[39] Ali M, Borgo R, Jones MW. Concurrent time-series selections using deep learning and dimension reduction. Knowledge-Based Systems. 2021;233:107507. Available from: https://www.sciencedirect.com/science/article/pii/S0950705121007693.

[40] Elman JL. Distributed representations, simple recurrent networks, and grammatical structure. Machine learning. 1991;7(2):195–225.

[41] Wold S, Esbensen K, Geladi P. Principal component analysis. Chemometrics and Intelligent Laboratory Systems. 1987;2(1):37–52. Proceedings of the Multivariate Statistical Workshop for Geologists and Geochemists. Available from: https://www.sciencedirect.com/science/article/pii/0169743987800849.

[42] Abdi H, Williams LJ. Principal component analysis. WIREs Computational Statistics. 2010;2(4):433–459. Available from: https://onlinelibrary.wiley.com/doi/abs/10.1002/wics.101.

[43] AEMO. National Electricity Market Data – NEM. AEMO; 2021. Available from: https://aemo.com.au/energy-systems/electricity/national-electricity-market-nem/data-nem/aggregated-data.

[44] Dehalwar V, Kalam A, Kolhe ML, *et al.* Electricity load forecasting for urban area using weather forecast information. In: 2016 IEEE International Conference on Power and Renewable Energy (ICPRE); 2016. pp. 355–359.

[45] Hogan R. Australian States. GitHub; 2014. Available from: https://github.com/rowanhogan/australian-states.

[46] Jordahl K. GeoPandas: Python tools for geographic data. URL: https://github com/geopandas/geopandas. 2014.

[47] Australia G. Major Power stations, Data and Publications. AEMO; 2017. Available from: http://pid.geoscience.gov.au/dataset/ga/82326.

Chapter 4

Optimal demand response strategy of an industrial customer

Arvind Kumar Jain

Demand response (DR), which is an important feature of the smart grid, can play a vital role by making the demand side more responsive to the varying gap between demand and supply. DR is utilized by power utilities to maintain system reliability, security and stability while customers utilize it to reduce the electricity cost by increasing or decreasing the load during valley or peak demand periods. Industries consume huge amounts of electricity; therefore, DR strategies are required to be implemented by industrial customers to enhance the saving. Further, industrial customers can provide DR by employing many different technologies or strategies to achieve shifts in demand in the following ways: (i) reducing or interrupting consumption temporarily with no change in consumption in other periods, (ii) shifting consumption to other time periods, and (iii) temporarily utilizing onsite generation in place of energy from the grid.

Recently, the installation of smart meters at the customer premises, for measuring the actual amount and time of energy consumption, and automatic switching on/off of appliances as per the day-ahead market schedule, has been made possible due to the technological advancement. This enables large industrial customers to change their demand in response to the market price as well as shift the demand from high price periods to comparatively low-price periods.

During recent years, some research work has been carried out to investigate the impact of the DR on electricity markets, with and without transmission congestion. In the existing double-sided electricity markets, an industrial customer can participate directly in the day-ahead market by submitting hourly price-quantity bids. The demand side load curtailment bids can be modeled in the market clearing process in terms of the demand benefit function of buyers. Although these bids influence the market price, these are limited in certain periods and are unable to recover the loss of load that occurred during the high price periods. Further, if the line flow constraints are not included in the bidding formulation problem, locational marginal prices (LMPs) will be the same at all the buses. However, if any line flow is constrained, LMPs will vary from bus to bus or zone to zone.

The LMP methodology is being used as a tool for pricing and congestion management in day-ahead and real-time spot markets of PJM interconnection, California, New England, and New York. In the majority of the existing double-sided electricity markets, congestion management and system security are analyzed after the day-ahead market is settled. This approach may not be suitable for an industrial buyer, as it cannot afford to curtail the energy consumption required for the completion of the production process.

Therefore, the prime aim of this chapter is to propose a novel optimization formulation for developing the optimal DR strategy of an industrial buyer. This will be incorporated into the bidding strategy by formulating it as a stochastic linear programming problem, comprising of two sub-problems, viz. market-clearing sub-problem and maximization of the purchase cost saving sub-problem. In addition, the impact of the DR strategy of the industrial customer on market clearing price and on the other market participants will be investigated. The effectiveness of the proposed methodology will be established with the help of a case study.

4.1 Demand side management categories

DSM was introduced by electric power research institute (EPRI) in the 1980s. Demand side management (DSM) can play a vital role to maintain the reliable, secure, and efficient operation of the modern grid. Under restructured power systems, DSM is a customer-driven activity. Depending on the timing and the impact of the applied method on the customer process, DSM categorization is shown in Figure 4.1.

DR is being considered as an integral part of the market operations. At present, although demand responsiveness in electricity markets is low as compared to the other commodities markets, even a small amount of DR can make difference in enhancing the market efficiency and system security. During June 2000 California crisis, rolling

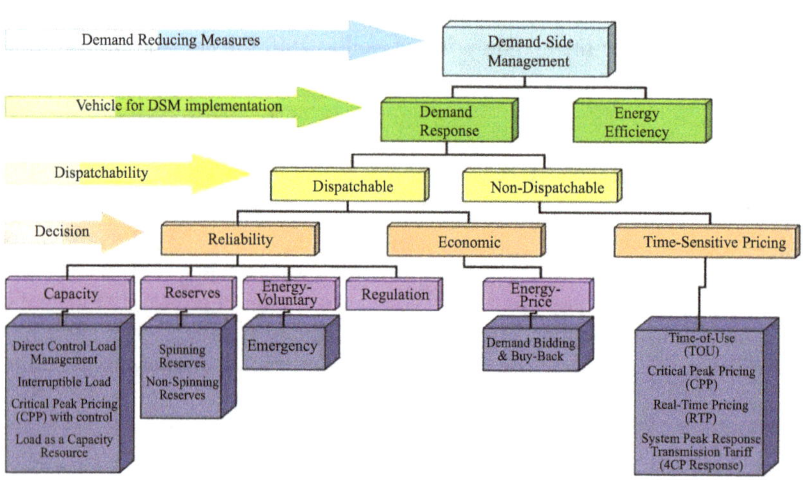

Figure 4.1 DSM categories (NERC-2011)

blackouts had taken place due to a shortage of 300 MW in a system of 50,000 MW [1]. This example underlines the need of demand-side participation in the electricity market.

Further, the introduction of DR in constrained electricity networks can significantly lower the peak energy costs and can potentially act as a check against the exercise of market power by generators [2]. DR also has the potential to increase the long-run efficiency of the energy market. DR programs are expanding in response to rapid load growth, as well as the cost and time required in bringing the new generation into service [3].

4.2 What is DR?

Electricity cannot be stored in bulk. Therefore, the supply and demand of electricity must remain in balance for secure and stable system operations. When electricity demand goes up during peak hours or supply decreases due to the generator and/or network outages, utility and grid operators have the followings options: (i) buy electricity from the real-time market which is expensive, (ii) fire up the next peaking power plant if not already running, (iii) dispatch a DR network and (iv) risk a blackout. Instead of adding more generations to the system, DR may be implemented to maintain the balance.

According to Federal Energy Regulatory Commission, DR is defined as: "Changes in electric usage by end-use customers from their normal consumption patterns in response to changes in the price of electricity over time, or to incentive payments designed to induce lower electricity use at times of high wholesale market prices or when system reliability is jeopardized." The end-user customers can also be the customers who are participating in the wholesale market operations.

A customer can incorporate the demand responsiveness by curtailing the energy consumption, time-shifting of the energy consumption and shifting to own source of energy during peak hours. Since consumers become more sensitive and responsive to price signals, DR results in an increase in the elasticity of demand.

4.3 Why DR?

DR is utilized by power utilities to maintain system reliability, security, and stability by increasing or decreasing the load during valley or peak demand. Further, DR led to a reduction in power system infrastructure requirement to meet the peaking load. It may control the market price by shifting the load from high price hours to low price hours. During forced outage of the generator, it helps to avoid brownouts, blackouts, or utilization of expensive generators. In recent years, the load-modifying capacity of DR is used to enhance grid flexibility through the provision of ancillary services. These services may facilitate the integration of renewable energy sources by balancing between supply and demand in real-time.

4.4 DR classification

DR resources can be classified into non dis-patchable resources and dis-patchable resources.

i Non dis-patchable resources: Non dispatchable resources participate in price-based DR programs such as real-time pricing, critical peak pricing, and time-of-use-tariffs. As smart meters are installed, price-based DR provides an opportunity to customers, mainly residential and commercial, to reduce their energy consumption during high price periods. Further, residential and small commercial customers may provide dis-patchable services to system operators at contact price or market price through aggregators.

ii Dis-patchable resources: Dis-patchable resources may influence the market price by participating in competitive bidding. In regions, where the electricity market does not exist, distribution utilities implement DR to avoid high peak prices and maintain system reliability by balancing consumption and supply.

Further, DR may be classified into competitive, non-competitive, and incentive-based DR.

4.4.1 Competitive DR

In competitive DR, aggregators/distribution utilities/industrial buyers participate in competitive day-ahead/real-time electricity market and submit their price-quantity bid to the system operator. System operator clears the market utilizing suppliers and buyers bids. The equilibrium point gives market price and quantity. In the market clearing process, DR plays a vital role to set the market price with power balance.

4.4.2 Non-competitive DR

In non-competitive DR, aggregators/distribution utilities/industrial buyers participate in day-ahead/real time electricity market and submit their modifiable load and price bid to system operator but do not participate in the setting of market-clearing price. However, during the high price periods system operator utilizes the DR resources who submitted bid price less than market price.

4.4.3 Incentivebased DR

If the DR provider is not interested in participating in an electricity market or the electricity market does not exist, then the DR provider informs the system operator about the curtailment of load in one day advance. The system operator utilizes the DR for obtaining system-wide techno-economic advantages and pays some monetary incentives to the DR provider.

4.5 Benefits of DR

DR provides various techno-economic benefits to participants by flattening the demand curve. The major benefits of DR are the followings:

- i Construction of new power plants to meet the peak load may be curtailed, which will reduce the adverse the environment impact.
- ii DR can be activated within 5 min, which may help in improving the reliability of the system during an outage of the generator or line.
- iii Distribution system operator (DSO) can use DR for managing congestion, voltage, and quality of supply at the distribution level.
- iv Incentive-based DR may reduce the financial risk of electricity retailer, who sells the electricity at a flat rate, during high price periods.
- v Distributed generation under smart grid paradigm motivates the inclusion of DR to achieve tecno-financial benefits.
- vi Market integrated price-based DR can mitigate the market power of suppliers by shifting the demand from peak hours to off-peak hours.
- vii System operator may use DR to avoid brownouts and black-outs by load modifications.

4.6 Challenges in DR implementation

There are several challenges in implementing DR in a modern electricity grid. The major challenges are mentioned as below:

- i Information communication technologies (ICT) play a pivotal role in implementing the DR. Therefore, sufficient metering and communication equipment are required to be integrated into the power system to check the system performance and DR. Further, training of distribution companies (DisCos) is highly required to implement the DR.
- ii Participation of customers in DR is a major challenge which may be dealt with by educating the customers about the benefits of DR.
- iii Installation of smart recording equipment at customer's location involves financial risk. A cost–benefit analysis to recover the cost within the stipulated time may motivate the customers to participate in DR.
- iv DR being a comparatively new technology has uncertainty regarding system-wide techno-economic performance, which leads to low acceptance.
- v Policy and standard are required to be developed by regulators to set the reference for measuring DR contribution to ensure desired performance and compensation.
- vi To incorporate the DR in the electricity market, barriers pertaining to the minimum size of DR and type of participants may be modified to enhance the effective implementation of DR.

4.7 DR provisions

Industrial, commercial, and residential customers can provide DR by employing many different technologies or strategies to achieve shifts in demand in the following ways,

- Reducing or interrupting consumption temporarily with no change in consumption in other periods.
- Shifting consumption to other time periods.
- Temporarily utilizing onsite generation in place of energy from the grid.

In addition, ancillary services like frequency regulation and load-following can be provided by DR. When surplus electric energy is available, DR with a storage facility may be used for pumping water, charging batteries, compressing air, etc. while maintaining the power balance between load and generation automatically.

4.8 Applications of DR

The main applications of DR are given below:

i *Reduction in power system infrastructure and environment issues*
Peak energy demand is less than 10% of total energy demand. In order to supply this load demand, huge power system infrastructure like generation, transmission, and distribution is required which has huge financial implications as well as environmental issues. DR can mitigate these problems by flattening the load profile.

ii *Reduction in market price*
In a competitive energy market suppliers' objective is to maximize their profit by submitting a price higher than the marginal price. This phenomenon is called market power. DR can mitigate the market power through participation in the electricity market and shifting the demand from peak hours to off-peak hours.

iii *Renewable energy integration*
Integration of renewable energy generation in the power grid is known to pose a challenge in the reliable operation of the grid due to its associated intermittent and uncertain nature. DR can be used to deal with these challenges introduced by renewable energy generation by changing the demand in accordance with the energy generated by the renewable sources.

iv *Ramping*
DR can be used to reduce ramp-up and ramp-down requirements of the generators. This can be achieved by increasing the demand to fill the valley of the load curve and by decreasing it for the peak shaving. This will help in saving the ramping cost of the generators apart from managing the system operation.

v *Voltage and frequency control*
The stability of the power system can be ensured by maintaining the voltage–frequency variations within limit while changes in demand occur. Traditionally, it is achieved by load-following approach where the power plant

increases/decreases their generation according to the load variation. In the modern power system, the generation following approach can be used by integrating DR to control voltage and frequency.

vi *Spinning reserve*

Spinning reserve is utilized to meet the load requirement during generator outages and/or peak load hours. Economically, DR participation led to less committed generating units in energy scheduling and accordingly less operation cost. Technically, within peak load hours the share of generator participation in providing reserve capacity becomes lower than the share of DR participation, while in other hours the opposite trend is observed.

4.9 Motivation about DR

DR would have a significant impact on the operation of the competitive electricity market. It has been observed that the absence of the DR is the prime reason for causing price spikes, shortages, and exercise of the market power. Kirschen *et al.* [4] discussed that the overall benefit, that is derived from trading, is optimal when suppliers and consumers in a competitive market are allowed to operate freely and the price settles at the intersection of the supply and demand curves. Although, demand responsiveness in the electricity markets is low as compared to the other commodities markets, even a small increase in the demand elasticity can improve the market performance [5]. In the literature, most of the work has focused on studying the impact of the demand price elasticity on the market. Some algorithms have been proposed to consider it in the auction mechanism [6]. However, few efforts have been made to develop the optimal bidding strategy of the buyers or load-serving entities using either load adjustment without recovery or load curtailment to model the price elasticity [7]. This approach may not be suitable for an industrial buyer as it cannot afford to curtail the energy consumption required for the completion of the production process. During recent years, a few research works have been carried out to investigate the impact of the price-based DR on electricity markets, with and without transmission congestion [8] and [9]. However, no efforts have been done to develop the optimal bidding strategy of a buyer considering the price-based DR.

4.10 DR of an industrial buyer

In the existing double-sided electricity markets, an industrial customer/Disco can participate directly in the day-ahead market by submitting hourly price–quantity bids [10]. The demand side load curtailment bids can be modeled in the market clearing process in terms of the demand benefit function of buyers. Although these bids influence the market price, these are limited in certain periods and are unable to recover the loss of load that occurred during the high price periods [11]. This approach may not be suitable for an industrial buyer, which cannot afford to curtail the energy consumption required for the completion of the production process. Also, it is not suitable for a

Disco/load-serving entity as this may lead to monetary loss and /or bad reputation among the retail customers due to the reduction of their demands from the scheduled contract.

Recently, the installation of smart meters at the customer premises, for measuring the actual amount and time of energy consumption, and automatic switching on/off of appliances as per the day-ahead market schedule [12], has been made possible due to technological advancement. This enables large industrial customers, Discos and group of small customers/domestic consumers to change their demand in response to the market price as well as shift the demand from high price periods to comparatively low price periods.

The prime aim of the work carried out in this chapter is to propose a new optimization formulation for developing the optimal bidding strategy of a buyer considering price responsive demand shifting (PRDS) under uniform price, hourly day-ahead electricity markets. This has been incorporated into the bidding strategy by formulating it as a stochastic linear programming problem, comprising of two sub-problems, viz. market clearing sub-problem and maximization of the purchase cost saving sub-problem.

In addition, the impact of the PRDS based bid strategy of the buyers on market clearing price and on the other market participants have been investigated. The effectiveness of the proposed methodology has been presented through a case study. Results obtained with the PRDS based bidding strategy have been compared with those obtained with a conventional price quantity (CPQ) bidding of a buyer.

4.11 Problem formulation

Development of an optimal bidding strategy requires the determination of the optimal bid price as well as the optimal bid quantity of the buyer. In a uniform price market, though the buyers, who are willing to pay a higher price, are given priority, all the selected buyers are required to pay uniform market clearing price (MCP), irrespective of their bid price. Therefore, in the uniform price market, the strategy of the buyer would be to bid higher than the MCP in order to avoid the risk of not getting selected in the market. Thus, the optimal bidding strategy of an industrial buyer gets reduced to determine the optimal bid quantity, which maximizes their purchase cost-saving over the scheduling horizon. For computing the purchase cost-saving, a buyer needs to predict the MCP either by forecasting it or by simulating the market clearing process. In this work, MCP has been estimated by simulating the market clearing process for the double-sided bidding, which requires the bid quantity of the participants. Thus, the MCP and the bid quantity of the participants are interrelated. Therefore, the optimal bidding strategy problem of a buyer has been formulated as a stochastic linear optimization problem. The nature of the proposed PRDS-based optimization model is stochastic due to the uncertainty involved in the prediction of electricity price and demand.

The proposed formulation comprises two sub-problems. The first sub-problem deals with the market clearing and the second sub-problem is formulated as the purchase cost saving maximization problem. These two sub-problems are explained below.

4.11.1 Market clearing sub-problem

In double-sided bidding, the system operator receives bids from the suppliers as well as from the buyers. The market is cleared by maximizing the social welfare subject to the power balance, while satisfying minimum and maximum generation limits, minimum and maximum load requirements of buyers. Mathematically, it can be expressed as

$$Max \sum_{t=1}^{T} \left(\sum_{j=1}^{N_l} \rho_{jd}^t * P_{jd}^t - \sum_{i=1}^{N_g} \rho_{is}^t * P_{is}^t \right) \tag{4.1}$$

subject to,

$$\sum_{i=1}^{N_g} P_{is}^t - \sum_{j=1}^{N_l} P_{jd}^t = 0, \quad \forall\, t \tag{4.2}$$

$$P_{is\,\min}^t \leq P_{is}^t \leq P_{is\,\max}^t, \quad \forall\, i, \forall\, t \tag{4.3}$$

$$P_{jd\,\min}^t \leq P_{jd}^t \leq P_{jd\,\max}^t, \forall j, \forall t \tag{4.4}$$

where ρ_{is}^t and ρ_{jd}^t are the bid prices of supplier-i and buyer-j at time t, respectively, P_{is}^t and P_{jd}^t are the bid quantity of supplier-i and buyer-j at time t, respectively, $P_{is\,\min}^t$ and $P_{is\,\max}^t$ are the minimum and the maximum generating capacity of supplier-i at time t, $P_{jd\,\min}^t$ and $P_{jd\,\max}^t$ are the minimum and the maximum demand requirement of buyer-j at time t. N_g and N_l represent the number of suppliers and buyers, respectively and T denotes the scheduling horizon.

The solution to the above problem determines the market-clearing price as well as the optimal generation and consumption schedule for each trading hour.

4.11.2 Proposed purchase cost-saving optimization sub-problem

In the existing double-sided electricity markets, the system operator (SO) accepts price-quantity bids from the participants. The generators submit offer price and quantity whereas, the buyers submit the load requirement and the price they are willing to pay. While submitting the bids, generators' motive is to maximize their profit, due to which they may bid higher than the marginal cost and may lead to MCP away from the competitive level. From the buyers' perspective, instead of bidding as per the actual load requirement, a buyer, who can shift its consumption pattern, may bid strategically for demand shifting. This shifting of demand would change the price and load pattern over the scheduling horizon, leading to a reduction in the MCP. Thus, the objective of the buyer would be to maximize the difference between the purchasing cost incurred without demand shifting and with demand shifting while satisfying the demand requirement over the scheduling horizon. Mathematically, it can be expressed as,

$$Max \sum_{t=1}^{T} \left(MCP_0^t * P_{d0}^{j,t} - MCP^t * \left(P_{d\min}^{j,t} + P_{dshift}^{j,t} \right) \right) \tag{4.5}$$

subject to,

$$\sum_{t=1}^{T} \left(P_{d0}^{j,t} - P_{d\min}^{j,t} - P_{dshift}^{j,t} \right) = 0 \tag{4.6}$$

$$P_{dshif\min t}^{j,t} \leq P_{dshift}^{j,t} \leq P_{dshift\max}^{j,t}, \quad \forall t \tag{4.7}$$

where,

$$\left. \begin{aligned} P_{dshift\max}^{j,t} &= \alpha * \sum_{t=1}^{T} P_{d0}^{j,t} \\ P_{dshift\min}^{j,t} &= 0 \end{aligned} \right\} \tag{4.8}$$

MCP_0^t and MCP^t are the market clearing prices at time t without demand shifting and with demand shifting, respectively, $P_{d0}^{j,t}$ is the actual demand requirement of buyer-j at time t, $P_{d\min}^{j,t}$ is the minimum demand of buyer-j at time t, $P_{dshift}^{j,t}$ is shifted demand of buyer-j at time t, $P_{dshift\min}^{j,t}$ and $P_{dshift\max}^{j,t}$ are the minimum and the maximum demand to be shifted by buyer-j at time t, α is the value of the price responsive demand.

The output of the above sub-problem is the optimal value of the demand of buyer-j, which can be shifted during each trading hour. Based on it, a buyer can obtain the optimal consumption pattern for the scheduling horizon.

The demand, which is to be shifted in any particular hour, has a certain minimum and maximum bounds. In a period with higher MCP, the buyer would like to meet only the minimum demand required and, hence, the demand to be shifted in this hour is zero. Thus, the minimum bound on the demand, which can be shifted, is kept zero. However, in periods with lower MCP, the buyer would like to shift the whole of the shift-able demand. Thus, the maximum bound on the shift-able demand is the total shift-able demand over the scheduling horizon.

4.12 Proposed solution algorithm

The solution of the purchase cost-saving maximization sub-problem, as described in Section 4.11.2, requires the initial guess of the MCP. This has been obtained by simulating the market clearing process considering the actual load requirement of the buyer. Based on this MCP, the optimal bidding strategy of the buyer is obtained by solving the purchase cost-saving maximization sub-problem. The main purpose of the proposed simulation is to develop PRDS-based bidding strategy of buyer(s). Buyer (s) may shift the demand from peak demand periods to off-peak demand periods for maximizing the purchase cost saving by modifying the MCP. To consider this, the proposed methodology is simulated for a sufficient number of iterations, which

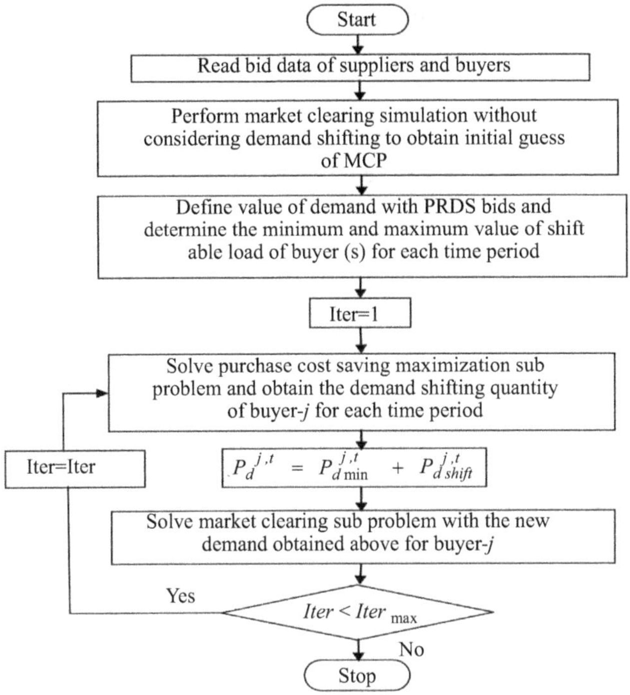

Figure 4.2 Flowchart of the proposed demand shifting based bidding strategy of a buyer

is taken as 25 in the present work after experimentation. Therefore, the maximum number of iterations has been considered as the stopping criterion in the proposed work as suggested in [13]. A flowchart describing the proposed solution algorithm is shown in Figure 4.2.

4.13 Case study

The effectiveness of the proposed methodology for developing the optimal bidding strategy of an industrial buyer, considering PRDS, has been tested on the 5-bus, system, whose details are given in [14]. The proposed stochastic linear optimization problem has been solved using MATLAB optimization toolbox. To ensure that the bid of an industrial buyer, whose strategy has to be developed, will be accepted and the minimum quantity required at each hour is satisfied, the bid price of the industrial buyer is taken to be higher than the marginal cost offer of the most expensive generator. Thus, the optimal bidding strategy of the industrial buyer is to determine the optimal bid quantity over the scheduling horizon, which is taken as 24-h, to maximize its purchase cost saving. In this work, Gencos are assumed to submit a multi-block bid, whereas the industrial buyers are assumed to give an hourly price–quantity bid for each trading hour. The value of the demand, with PRDS bids, has been assumed to be 5% for both systems during each hour. The results obtained on the two systems are described below.

The 5-bus system comprises of three Gencos and two buyers. The optimal bid strategy of buyer-1 has been developed. Table 4.1 lists the offer prices of the Gencos whereas, the forecasted system demand is shown in Table 4.2. The maximum and minimum demands are 440 MW and 310 MW, respectively, and the total generation capacity is 375 MW. The total demand of both the buyers, over the scheduling horizon of 24-h, is 4,433.8 MW each. The PRDS-based bidding strategy of the industrial buyer has been developed considering the following cases:

- Case I: Buyers 1 and 2 both offer CPQ bids in each hour.
- Case II: Industrial buyer-1 offers PRDS bid and buyer-2 offers CPQ bid in each hour.
- Case III: This case is the same as the case II except that the buyer-1 cannot shift the demand during hours 6–10.
- Case IV: Buyers 1 and 2 both offer PRDS bids in each hour.

The load profiles of both the buyers are assumed to be the same in order to show the impact of the demand shifting bidding strategy of the buyer-1 on the buyer-2. The bid prices of the buyer-1 and the buyer-2 for the entire scheduling horizon are taken

Table 4.1 Offer prices of generators in 5-bus system

Gen.	Bid quantity in MW				Marginal cost in $/MWh		
	Block 1	Block 2	Block 3	Total	Block 1	Block 2	Block 3
1	75	25	50	150	7.827	8.342	8.856
2	50	50	25	125	9.166	9.731	10.18
3	40	30	30	100	8.292	8.567	8.916

Table 4.2 Forecasted system demand in 5-bus system

Hour	Forecasted system demand in MW	Hour	Forecasted system demand in MW	Hour	Forecasted system demand in MW
1	340	9	385	17	372.5
2	320	10	415	18	420
3	390	11	430	19	430
4	385	12	440	20	390
5	340	13	427.5	21	312.5
6	320	14	395	22	310
7	330	15	370	23	330
8	340	16	355	24	320

Table 4.3 Optimal bid strategy of buyer-1 considering demand shifting in 5-bus system

Hour	Forecasted demand (MW)	Min. demand (MW)	Demand shift (MW)			Optimal bid quantity (MW)		
			Case II	Case III	Case IV	Case II	Case III	Case IV
1	170.00	161.50	−8.50	−8.50	6.27	161.50	161.50	176.27
2	160.00	152.00	36.29	36.48	6.77	196.29	196.48	166.77
3	195.00	185.25	−9.75	−9.75	−9.75	185.25	185.25	185.25
4	192.50	182.87	−9.62	−9.62	5.15	182.87	182.87	197.65
5	170.00	161.50	−8.50	−8.50	6.27	161.50	161.50	176.27
6	160.00	152.00	36.29	0.00	6.77	196.29	160.00	166.77
7	165.00	156.75	−8.25	0.00	6.52	156.75	165.00	171.52
8	170.00	161.50	−8.50	0.00	6.27	161.50	170.00	176.27
9	192.50	182.87	−9.62	0.00	5.15	182.87	192.50	197.65
10	207.50	197.12	−10.37	0.00	−10.37	197.12	207.50	197.12
11	215.00	204.25	−10.75	−10.75	−10.75	204.25	204.25	204.25
12	220.00	209.00	−11.00	−11.00	−11.00	209.00	209.00	209.00
13	213.75	203.06	−10.68	−10.68	−10.68	203.06	203.06	203.06
14	197.50	187.62	−9.87	−9.87	−9.87	187.62	187.62	187.62
15	185.00	175.75	−9.25	−9.25	5.52	175.75	175.75	190.52
16	177.50	168.62	−8.87	−8.87	5.90	168.62	168.62	183.40
17	186.25	176.93	−9.31	−9.31	5.46	176.93	176.93	191.71
18	210.00	199.50	−10.50	−10.5	−10.50	199.50	199.50	199.50
19	215.00	204.25	−10.75	−10.75	−10.75	204.25	204.25	204.25
20	195.00	185.25	−9.75	−9.75	−9.75	185.25	185.25	185.25
21	156.25	148.43	36.45	36.31	6.96	192.70	192.56	163.21
22	155.00	147.25	36.78	36.41	7.02	191.78	191.41	162.02
23	165.00	156.75	−8.25	−8.25	6.52	156.75	156.75	171.52
24	160.00	152.00	36.29	36.16	6.77	196.29	196.16	166.77

to be 14.17 \$/MWh and 13.15 \$/MWh, respectively. The total demand to be shifted over the whole scheduling horizon is 221.69 MW.

The optimal bidding strategy of the buyer-1 obtained using the proposed methodology, for the cases II, III and IV is shown in Table 4.3. From Figures 4.3–4.5, it can be observed that when the MCP is highest, only the minimum demand of the buyer-1 has been met and the rest of the quantity is shifted during comparatively low MCP periods. In case II, during the hours 2, 6, 21–22, 24, the demand of the buyer-1 is maximum because during these hours, MCP is low. During the remaining hours, buyer-1 has the minimum demand. In case III, the demand pattern is the same as in the case II, except during hours 6–10. During these hours, it is assumed that buyer-1 cannot shift their demand due to some technical reasons. For cases II and III, during the hours 5, 7, 8, 15–17, 23, though the MCP is comparatively low, it can be seen from Figures 4.3 and 4.4, that the demand shift is negative in spite of sufficient generation

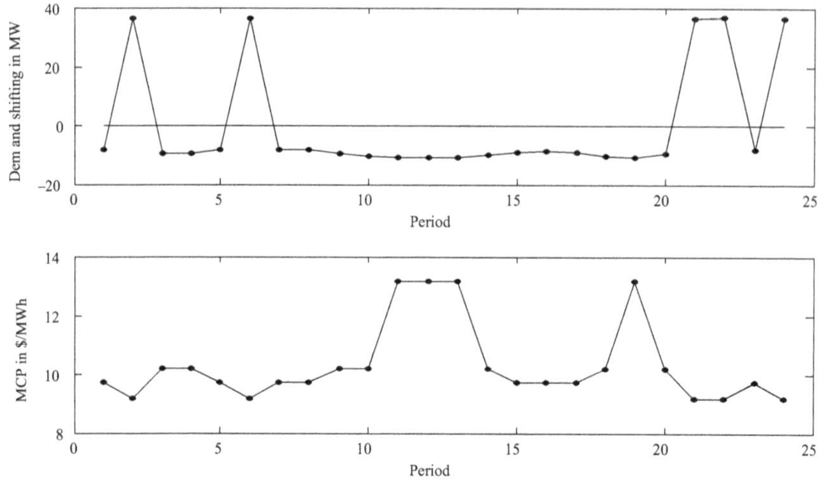

Figure 4.3 Response of demand shifting of buyer-1 to MCP for case II

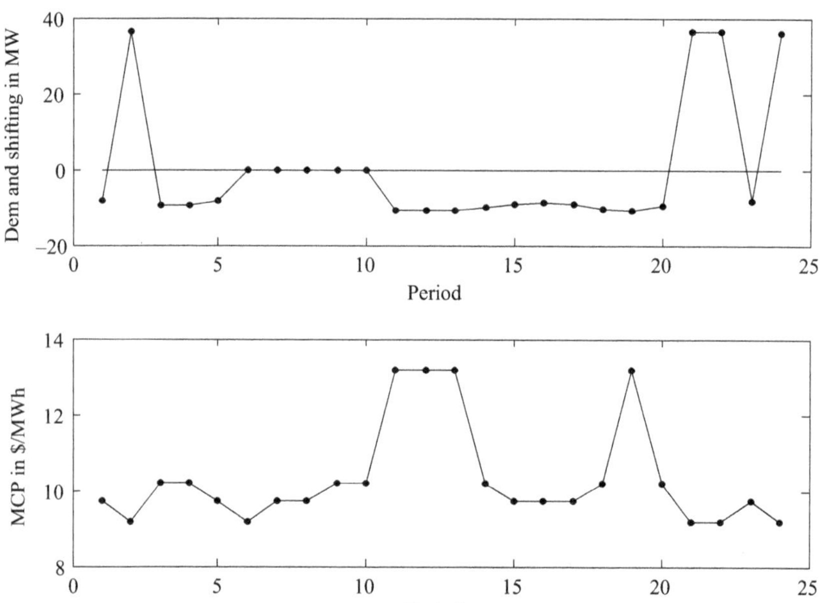

Figure 4.4 Response of demand shifting of buyer-1 to MCP for case III

capacity. It has happened to avoid a further increase in the MCP. PRDS of buyer-1 and buyer-2 for case IV are shown in Figure 4.5. In this case, fluctuations in MCP are less as compared to cases II and III, because both the buyers are offering PRDS bids in each hour. Day-ahead MCPs, for the four cases, are given in Table 4.4. From

Table 4.4 Day-ahead MCP under four cases in 5-bus system

Hour	Case I	Case II	Case III	Case IV	Hour	Case I	Case II	Case III	Case IV
1	9.73	9.73	9.73	9.73	13	13.15	10.18	10.18	10.18
2	9.16	9.73	9.73	9.73	14	10.18	10.18	10.18	9.73
3	10.18	10.18	10.18	9.73	15	9.73	9.73	9.73	9.73
4	10.18	9.73	9.73	10.18	16	9.73	9.73	9.73	9.73
5	9.73	9.73	9.73	9.73	17	9.73	9.73	9.73	9.73
6	9.16	9.73	9.16	9.73	18	10.18	10.18	10.18	10.18
7	9.73	9.16	9.73	9.73	19	13.15	10.18	10.18	10.18
8	9.73	9.73	9.73	9.73	20	10.18	10.18	10.18	9.73
9	10.18	9.73	10.18	10.18	21	9.16	9.73	9.73	9.73
10	10.18	10.18	10.18	10.18	22	9.16	9.73	9.73	9.73
11	13.15	10.18	10.18	10.18	23	9.73	9.16	9.16	9.73
12	13.15	13.15	13.15	10.18	24	9.16	9.73	9.73	9.73

Table 4.5 Impact of demand shifting bid strategy of buyer-1 in 5-bus system

	Case I	Case II	Case III	Case IV
Purchase cost of buyer-1in $	46,279	44,390	44,480	43,874
Purchase cost of buyer-2 in $	45,655	44,337	44,427	43,860
Satisfied demand of buyer-1 in MW	4,433.8	4,433.8	4,433.8	4,433.8
Satisfied demand of buyer-2 in MW	4,386.2	4,424.8	4,424.8	4,433.8
Total system load fulfilled in MW	8,820	8,858.6	8,858.6	8,867.6
Saving of buyer-1 in $ using PRDS bid	–	1,889	1,799	2,405

this table, it can be seen that the PRDS based bidding normalizes MCPs by reducing the peak values and increasing the off-peak values.

Comparison of the CPQ bids with the PRDS bids and the impact of the PRDS bidding strategy of the buyer-1 on various factors are shown in Table 4.5. From this table, it can be observed that due to the PRDS bidding strategy, the purchase cost of buyer 1 has been reduced in cases II, III and IV, as compared to the case I. However, the minimum purchase cost is obtained in the case IV. Further, it has been observed that due to fixed demand during hours 6–10, saving of buyer1 has reduced in case III as compared to the cases II and IV, since it cannot take advantage of the low MCPs during these hours by shifting the loads.

The total demand requirement of the buyer-1 is fulfilled in all the cases as its bid price is higher (Figure 4.5). When both the buyers are bidding PRDS bids, in case IV, the total unsatisfied demand of the buyer-2 has reduced to zero MW. This is because, now, both the buyers have shifted their consumption pattern in response to the price to fulfill their demand, as shown in Figure 4.5. In case IV, the saving of buyer-1 is $2,405, which is higher than the cases II and III. The reason is that, in this case, both the buyers are offering PRDS bidding, which has resulted in a reduction

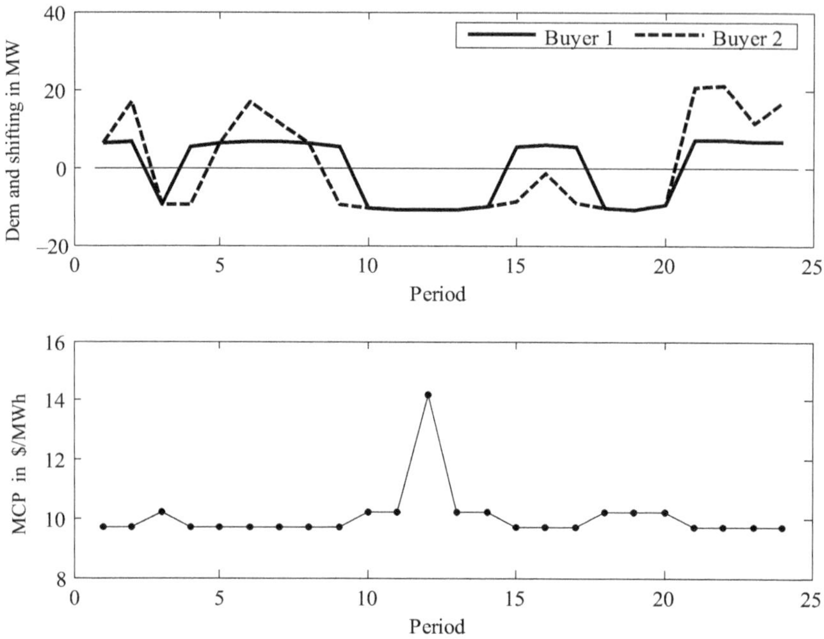

Figure 4.5 Response of demand shifting of buyer-1 and buyer 2 to MCP for case IV

in the peak load demand and MCP. The total system load fulfilled is depicting the increasing trend from case I to case IV. It shows that the PRDS bidding is beneficial to all the market participants, as the demand fulfilled has increased and the MCP is comparatively low.

4.14 Conclusion

In this chapter, a stochastic linear optimization problem has been proposed to develop a price responsive demand shifting (PRDS) based bidding strategy of an industrial buyer. The results obtained with the PRDS based bidding strategy have been compared with those obtained with the conventional price quantity (CPQ) bidding strategy of the buyer. Effect of the demand shifting has been observed on the MCP, purchase cost and satisfied demand of the buyers, system demand fulfilled and savings of a buyer, whose bid strategy has been developed. The results on the 5-bus systems reveal that the savings of a buyer depend on their ability to reduce their demand during the peak price periods. Further, it is observed that the PRDS strategy improves the economic efficiency of the day-ahead market by reducing the electricity price. The proposed bidding strategy is generic in nature and can be utilized by the buyers in any electricity market, which allows the demand side bidding.

References

[1] Hunt S.: *Making competition work in electricity*, New York, NY: Wiley; 2002.

[2] *Quantifying demand response benefits in PJM, a report for PJM interconnection and mid Atlantic distributed resource initiative (MADRI)*, 2007. [Online]. Available from http://www.energetics.com/madri/pdfs/BrattleGroupReport.pdf.

[3] *Federal energy regulatory commission, assessment of demand response & advance metering staff report*; 2008.

[4] Strbac G. and Kirschen D.: "Assessing the competitiveness of demand-side bidding", *IEEE Transactions on Power Systems*, Feb 1999, vol. 14, pp. 120–25.

[5] Bompard E., Ma Y., Napoli R., and Abrate G.: "The demand elasticity impacts on the strategic bidding behavior of the electricity producers", *IEEE Transactions on Power Systems*, Feb 2007, vol. 22, pp. 188–97.

[6] Kian A.R., Cruz J.B., and Thomas R.J.: "Bidding strategies in oligopolistic dynamic electricity double sided auctions", *IEEE Transactions on Power Systems*, Feb 2005, vol. 20, pp. 50–58.

[7] Oh H.S and Thomas R.J.: "Demand side bidding agents: Modeling and Simulation", *IEEE Transactions on Power Systems*, Aug 2008, vol. 23, pp. 1050–56.

[8] Su C.L. and Kirschen D.S.: "Quantifying the effect of demand response on electricity markets", *IEEE Transactions on Power Systems*, Aug 2009, vol. 24, pp. 1199–207.

[9] Singh K., Padhy N.P., and Sharma J.D.: "Influence of price responsive demand shifting bidding on congestion and LMP in pool-based day-ahead electricity markets", *IEEE Transactions on Power Systems*, May 2011, vol. 26, pp. 886–96.

[10] Alvarez C., Gabaldon A., and Molina A.: "Assessment and simulation of the responsive demand potential in end-user facilities: Application to a university customer", *IEEE Transactions on Power Systems*, May 2004, vol. 19, pp. 1223–31.

[11] Borghetti A., Gross G., and Nucci C.A.: "Auctions with explicit demand side bidding in competitive electricity markets", in *The Next Generation of Electric Power Unit Commitment Models, Norwell*, M.A. Kluwer, 2001, pp. 53–74.

[12] Black J.W. and Ilic M.: "Survey of technologies and cost estimates for residential electricity services", *IEEE PES Summer Meeting, Vancouver*, BC, Canada, vol. 1, July 2001, pp. 255–60.

[13] de la Torre S., Conejo A.J., and Contreras J.: "Simulating oligopolistic pool based electricity markets: a multi period approach", *IEEE Transactions on Power Systems*, Nov 2003, vol. 18, pp. 1547–55.

[14] Jain A.K. and Srivastava S.C.: "Optimal bidding strategy under transmission congestion using genetic algorithm", *Proceeding of International Conference on Genetic and Evolutionary Methods (GEM2007)*, 25–28 June 2007, Las Vegas, USA, pp. 88–94.

Chapter 5

Price-based demand response for thermostatically controlled loads

K.S. Swarup[1] *and Devika Jay*[1]

Smart grid enables active participation of consumers daily operation of the grid through Demand Response (DR). DR refers to the actions initiated from contracted customers by changing their demand in response to price signals, incentives, or directions from grid operators. In this chapter, industrial DR suitable for frequency regulation is discussed. For this, a mathematical model of price-based DR from thermostatically controlled loads (TCL) for controlling the temperature of the chillers in large academic complex environment is presented. A probabilistic model of the density function of aggregated TCL loads is discussed. The variation of the thermostat set point demand temperature an increase in the price is presented. In order to match the power demand and power supply, a new method for dynamic demand control (DDC) with automatic generation control (AGC) in smart grid environment is proposed. A load frequency control using DDC was modeled in this study. The load frequency control model was simulated for a step load change of 0.01. The frequency deviation was compared with the frequency deviation obtained when generation control, using PI controller, alone was implemented for frequency control. Thus, DDC alone is required to maintain the system frequency, during small load variations. DDC will play a major role in reducing these losses caused to the GENCOs under a Smart Grid environment.

5.1 Demand response

In order to increase efficiency and reduce costs, a shift from supply-side-only viewpoint to an integrated demand- and supply-side viewpoint was considered, in which customers are considered as a new utility planning option. This leads to the concept of Demand Side Management (DSM). In [1], DSM has been defined as "the planning, implementation, and monitoring of those utility activities designed to influence customer use of electricity in ways that will produce desired changes in the utility's load

[1]Department of Electrical Engineering, IIT Madras, Chennai, India

shape." Load shape means the load profile or the demand of the customer on time-of-day, day-of-week basis. The important DSM techniques relevant to even traditional electric power systems are as follows [1]:

1. Peak clipping
   This involves reduction of peak load by using direct load control. Another example of peak clipping is the use of curtailable rates for industrial and commercial customers.
2. Valley filling
   This involves building off-peak loads. This is suitable when long-run incremental cost is less than the average price of electricity. One of the many ways to achieve this is by adding new thermal energy storage in place of loads served by fossil fuels.
3. Load shifting
   This is shifting load from on-peak to off-peak periods. Use of storage water heating, coolness storage, and customer load shifts, etc. falls in this category of DSM.

Smart grid environment facilitates for an increased role of DSM in planning and operation of the grid. This is because with the advent of Information Communication Technology, the gap between the various utilities has been bridged effectively. Thus the dimension of DSM widens and many more load management techniques can be implemented in a smart grid. These techniques have been classified based on the time scale as follows [2]:

1. Energy efficiency
2. Time of use
3. Demand response (DR)
4. Spinning reserve – primary and secondary control of grid frequency

The time frame for these techniques is as shown in Figure 5.1.

Energy efficiency refers to energy conservation techniques that are implemented at the customer level. Time of use DSM is change in load profile during peak and off peak hours, and this can be initiated by change in market price. Load can act as a "virtual" spinning reserve during grid frequency restoration, by responding to the grid conditions like frequency etc. Among this, DR is gaining much importance, which means a response from demand side to explicit requests to shut off or change consumption pattern from grid operators.

It has been observed that electricity price mechanism plays a major role in initiating DSM [3,4]. Very low elasticity of the demand causes large price spikes and

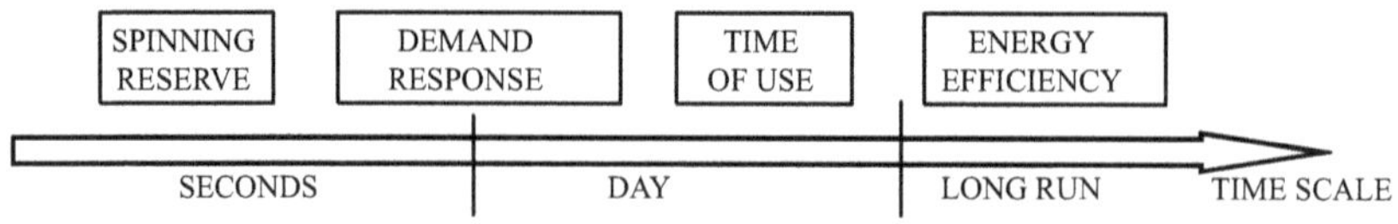

Figure 5.1 DSM techniques

thus allowing generating companies to deliberately reduce their generation. A price-based DSM will be useful to curtail price spikes and the price mechanism should include demand side also [5]. Thus DR turns out to be the most important among DSM techniques.

For a sustainable operation of the grid, renewable energy resources will play a larger significant role in electric power systems. But renewable power generation is only in parts plan and adjustable. This means that in the future, the share of "easily" adjustable power generation will decrease posing challenges to future energy management system. A solution to this problem is a paradigm shift from "generation follows load" to "load adapts to generaton" [6].

DR means response of customer consumption of electricity to supply conditions, for example, having electricity customers reduce their consumption at critical times or in response to market prices. DR mechanisms respond to explicit requests to shut off. Dynamic demand devices passively shut off when grid operator issues instruction when the grid is stressed. DR can be mainly divided into two schemes:

1. Energy management: this means the energy balance needs to be achieved in each charging period, generally 15–60 min.
2. Near real-time power management: this means energy needs to be balanced at all times.

Energy management DR, also known as Market DR, can be classified as

1. Flat rate pricing schemes
2. Slab rate pricing schemes
3. Incentive-based DR
   (a) Direct control
   (b) Load curtailment
   (c) Market-based demand bidding
   (d) Emergency DR
4. Price-based DR
   (a) Time of use
   (b) Critical peak pricing
   (c) Extreme day pricing
   (d) Real-time pricing

Real-time DR or physical DR can be achieved using emergency signals issued by the system operator. This can include real-time price-based DR schemes also.

For effective DR, a transparent pricing mechanism is required to ensure active participation of customers. Communication channel is the backbone of DR as information on current load and generation, and real-time measurements are requirements for DR. Also forecast of these quantities will help in active DR in the grid. An efficient DR scheme can be achieved in residential and building energy management, when there are manageable loads like Heating Ventilation and Air Conditioner loads, Distributed Energy Resources and storage devices.

5.2 Smart grid control

The electric power industry has been working to improve the functionality, efficiency, and availability of electricity so as to meet the ever increasing demand. The development of human society and industry demanded a revolution of electric power systems. The advancements in technology, the electrical power industry have transformed the way to generate, deliver, and consume power today. Modern civilization practices result in the release of greenhouse gases to the atmosphere causing global climate change. The global efforts to reduce carbon emission will be incomplete if power sector continues to be in its traditional power generation and operation schemes.

The word "Smart grid" or "Intelligent grid" has been coined and this has created a buzz around the world. The term Smart Grid is used to describe a "digitized" and intelligent version of the present-day power grid. A grid becomes smart when power delivery via a two way digital technology supports secure and effective control over consumers resulting in efficient, economic, and environmental friendly power systems. Smart grid features are still under debate. A complete picture of the various entities in a smart grid, and the features associated with each entity that makes the grid smart is depicted in Figure 5.2.

The emerging technologies under smart grid are described below:

1. DSM
   Advanced metering infrastructure (AMI) plays an important role in Smart grid which provides a 2-way consumption control. This helps in curtailment of load and demand management. Home Area Network also provides DSM facilities. The backbone of DSM is communication infrastructure.
2. Photovoltaic and solar heating

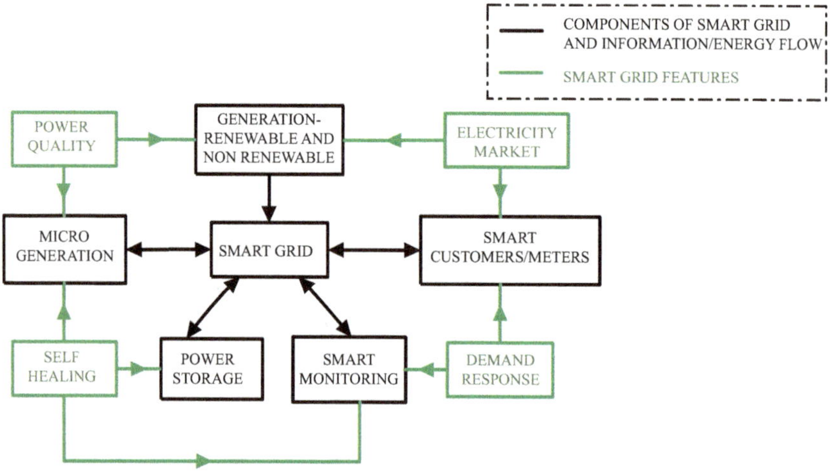

Figure 5.2 Smart grid components and features

Photovoltaic (PV) can serve local loads. But the challenge is inherent uncertainty and the need for energy storage devices.

3. Energy storage
   Solar, thermal, and wind energy plants require storage devices. Lithium–ion, nickel–cadmium, and lead acid batteries are common storage devices. Fly wheel and super capacitors are also being involved.
4. Microgrids
   Distributed energy resources (DER) are an efficient way to reduce environmental pollution, transmission losses and grid congestion. DERs may be referred to PV cells, wind resources, fuel cells, combined heat and power (CHP), etc. Some of these such as PV and wind are intermittent in nature and possess some problems when directly connected to the grid. If these DERs (both intermittent and consistent) of a locality or a distribution network are interconnected together along with necessary storage systems and a control center, can form an autonomous grid termed as a "MicroGrid (MG)." Microgrid can solve some of the problems associated with distributed generators (DG) in grid integration [7]
5. Self-healing
   Self-healing capability of a grid makes the grid smart and intelligent. This requires efficient monitoring, analysis and control of the grid. The advent of Synchronous Phasor Measurement units (PMU) has made monitoring efficient. An enormous amount of data is now available from PMUs. Thus smart monitoring is achieved when data obtained from the PMUs are then converted to useful information. Analysis of the system limits like thermal limit and stability limit must be then carried out using the real time information obtained from the real-time measurements. The analysis can be then validated using simulation-based analysis. As suggested in [8], the analysis should be comprehensive and proactive with look-ahead analysis. A coordinated control based on online security assessment and online restoration plans are necessary to make the grid self-healing.

5.3 Modeling of thermostatically controlled loads (TCL)

Though the concept of "Load Following Supply" existed earlier, DR is now emerging as an important feature in smart grid paradigm with the help of Information Communication Technology (ICT). DR broadly classified into price based and incentive based has been extensively studied and analyzed in various aspects. The focus was mainly on peak clipping, load shifting and demand control schemes. For residential load control, TCLs like heating, ventilating, and air conditioner (HVAC) loads are considered as they cause minimum inconvenience to the users.

The feasibility of such HVAC load control was studied in [9]. Aggregation of these loads and the impact of DR were studied in [10]. In [11], an On/OFF scheme for HVAC units was discussed. A control scheme that manipulates the thermostat set point of such loads to balance fluctuations from intermittent renewable generators was developed in [12]. A change in thermostat set point results in a change in state of the unit and this has been modeled and studied in [13]. Thermostat set point was considered to be varying with market price based on a predefined linear relation. But a scheme that can find the optimal thermostat set point corresponding to market price that varies in real time was not studied from a consumer perspective.

Dynamic characteristics of TCL and aggregated response of such loads had been studied and modeled using a set of ordinary and partial differential equations called Coupled Fokker Planck Equations [14]. "Load Following capability" of these loads when they are under a minimum variance control law was studied in [12]. The coupled Fokker Planck equations are detailed in Appendix A. Later in [15], a transfer function that provides the aggregate response of TCL units to uniform disturbances was derived. But the limitation of the model is the assumption that the magnitude of disturbance applied to the thermostat set point is negligible when compared to the deadband zone of the thermostat set point. But with consumers aiming at minimizing the increase in amount paid by reducing the thermostat set point, this assumption will not be valid.

This section presents a mathematical formulation to determine the optimal thermostat set point corresponding to the market price is provided for price based demand response from a consumer perspective. The scheme being from a consumer perspective aims at finding the optimal thermostat set point that will minimize the increase in amount paid due to increase in price issued by the system operator.

When there is an increase in electric price signal, DR is expected from the customers contracted for price-based DR. Customers can adjust their consumption by adjusting the thermostat set point of HVAC loads (TCL) to exercise DR, instead of load shifting or load shedding. In this section, the change in power consumption in a TCL unit due to change in thermostat set point is derived.

HVAC loads are specified by their energy efficiency ratio (EER) which is defined as

$$EER = COP * 3.413 \tag{5.1}$$

where COP is the coefficient of performance given by [16]

$$COP = \frac{Work \quad Done(Q)}{Electric \quad Power \quad Input(P_{input})} \tag{5.2}$$

Work done (Q) by the HVAC unit is given by

$$Q = m * C_p * (T_{out} - T_{in}) \tag{5.3}$$

where

m is the mass of the coolant; C_p is the specific heat capacity of the coolant; T_{out} is the outside temperature; T_{in} is the inside temperature. Assuming thermostat set point to be the same as inside temperature, change in work done (ΔQ) for a change in thermostat set point (ΔT_{st}) is

$$\Delta Q = -m * C_p * \Delta T_{st} \tag{5.4}$$

A change in work done (ΔQ) results in a change in electric power consumed (ΔP_{input}), given by (5.2). Thus

$$\Delta P_{input} = \frac{\Delta Q}{COP} \tag{5.5}$$

$$\Delta P_{input} = \frac{-m * C_p * \Delta T_{st}}{COP} \tag{5.6}$$

Equation (5.6) establishes a linear relation between change in thermostat set point and change in power consumed by the HVAC unit. An increase in electricity price, initiates DR because customers aim at reducing the increase in total amount paid. It shows that DR contracted customers can adjust their thermostat set point so as to reduce the power consumption. Thus an increase in price can be responded with an increase in thermostat set point so as to reduce the power consumption and in turn reduce the increase in amount paid.

Let ρ_1 \$/W be the current rate at which consumers are charged. Thus amount paid (P_1) for an input power, D_1, of HVAC units is

$$P_1 = \rho_1 * D_1 \tag{5.7}$$

Let ρ_2 \$/W be the new rate, such that $\rho_2 > \rho_1$. Then the amount to be paid ($P_2{}^1$) for the same input power, D_1, is

$$P_2^1 = \rho_2 * D_1 \tag{5.8}$$

Customers aim to minimize the increase in amount paid by exercising DR. This results in change in demand from D_1 to $D_2 = D_1 + \Delta D$. Thus amount paid becomes

$$P_2 = \rho_2 * D_2 \tag{5.9}$$

The increase in amount paid (ΔP) is

$$\Delta P = P_2 - P_1 = (\rho_2 - \rho_1) * D_1 + \rho_2 * \Delta D \tag{5.10}$$

Substituting (5.6),

$$\Delta P = (\rho_2 - \rho_1) * D_1 - \frac{\rho_2 * m * C_p * \Delta T_{st}}{COP} \tag{5.11}$$

Figure 5.3 shows the relation between the increase in amount paid and the change in thermostat set point.

Customer objective is to minimize the difference in amount paid. The optimal difference in amount paid should be zero. The initial difference in amount paid is ΔP_1 and shown as operating point, $A(\Delta T_{st1}, \Delta P_1)$ in Figure 5.4. The operating point corresponding to optimal difference in amount paid is $B(\Delta T_{stdesired}, 0)$.

The optimal operating point is achieved by increasing the thermostat set point to the value given by (5.11). Thus price-based demand response contracted customers can exercise DR, by adjusting the thermostat set point to an optimal value which will minimize the increase in amount paid (due to increase in price signal) to zero. Another advantage of the proposed scheme is that consumers need not turn off their TCL/HVAC loads for DR, making DR scheme much more attractive with more participants.

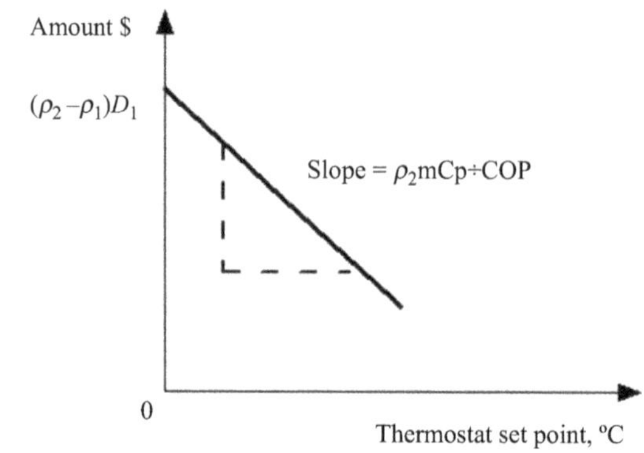

Figure 5.3 Increase in amount paid versus change in thermostat set point

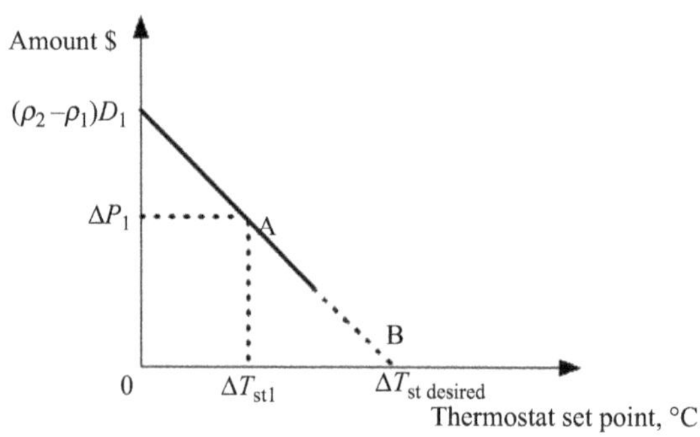

Figure 5.4 The optimal change in thermostat set point for price-based DR

5.4 DR from aggregated TCLs—load model

The price-based DR scheme discussed in the previous section determines the optimal thermostat set point change required to exercise DR. In this section, a transfer function for the response of aggregated TCLs to the change in thermostat set point is derived. The model is based on the Coupled Fokker-Plank equations (CFPE) [14] that describe the aggregated behavior of TCLs. An exact solution for the CFPE was given in [12], so as to model the change in power demand in aggregated population of TCL due to change in thermostat set point under minimum variance control law. Aggregate power response model in a homogeneous population of TCLs was derived in [15] based on the exact solutions of CFPE. In this section, the exact solution of CFPE is utilized to model the change in power consumption in aggregated TCL due to the change in thermostat set point to optimal value as defined in the previous section.

Consider "N" homogeneous TCLs at thermostat set point θ^0 with dead band of Δ. Let P be the power drawn by the TCL. The dynamic characteristic of the TCL is shown in Figure 5.5. θ_-^0 from θ_+^0 is the heating period and θ_+^0 to θ_-^0 is the cooling period.

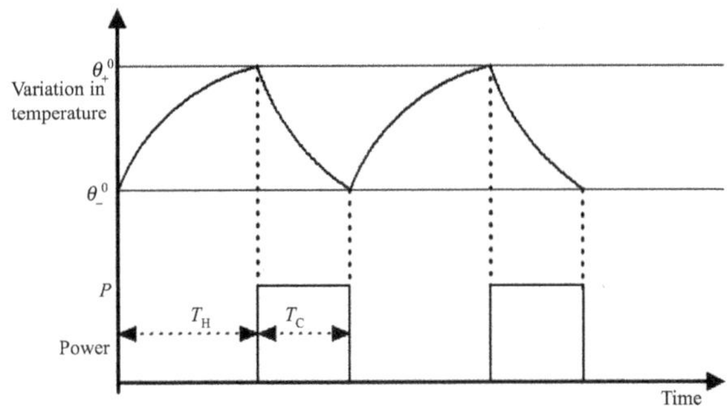

Figure 5.5 Dynamics of thermostatically controlled loads

The expressions for time taken to reach temperature θ_n during cooling period $t_c(\theta_n)$ are [15]

$$t_c(\theta_n) = C * R * ln\left(\frac{PR + \theta_+^0 - \theta_{amb}}{PR + \theta_n - \theta_{amb}}\right) \tag{5.12}$$

and during $t_h(\theta_n)$ is

$$t_h(\theta_n) = C * R * ln\left(\frac{\theta_{amb} - \theta_-^0}{\theta_{amb} - \theta_n}\right) \tag{5.13}$$

where,
C is the thermal capacitance; R is the thermal resistance; θ_{amb} is the ambient temperature.

In steady-state cooling time, constant T_C and heating time constant T_H are defined as [15]

$$T_C = CR * ln\left(\frac{PR + \theta_+^0 - \theta_{amb}}{PR + \theta_-^0 - \theta_{amb}}\right) \tag{5.14}$$

$$T_H = CR * ln\left(\frac{\theta_{amb} - \theta_-^0}{\theta_{amb} - \theta_+^0}\right) \tag{5.15}$$

The ON probability density curve $f_1(\theta)$ and OFF probability density curve $f_0(\theta)$ for aggregated TCLs are shown in Figure 5.6.
Where

$$f_1(\theta) = \frac{C * R}{(T_C + T_H) * (PR + \theta - \theta_{amb})} \tag{5.16}$$

$$f_0(\theta) = \frac{C * R}{(T_C + T_H) * (\theta_{amb} - \theta)} \tag{5.17}$$

These functions define the probability of a TCL load to be in either ON state or OFF state. These are obtained by solving the CFPE equations [12].

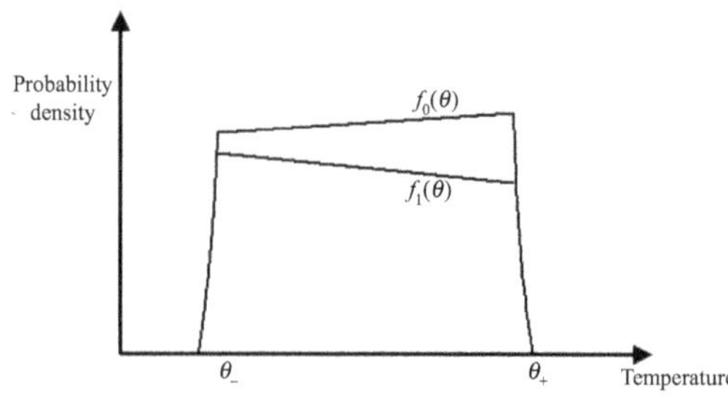

Figure 5.6 Steady-state probability densities

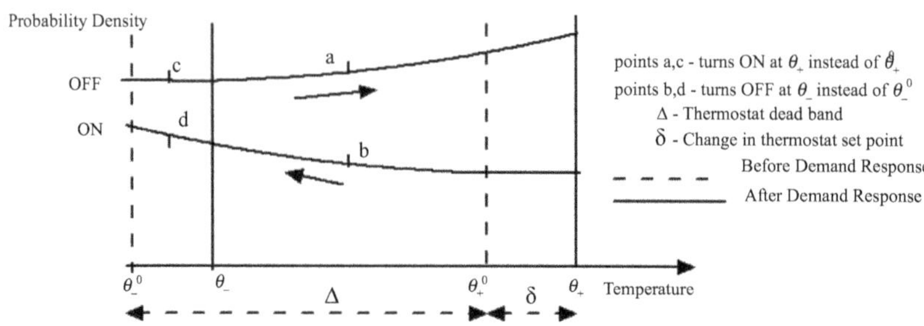

Figure 5.7 Four different operating points of aggregated TCLs in the probability density curve

5.4.1 Transfer function of aggregated response of TCL units

A change in thermostat set point, δ, causes a shift in the temperature band of operation $[\theta_-, \theta_+]$. To study the response due to change in thermostat set point, four different TCL conditions namely a, b, c, and d are considered on the probability density curve, as shown in Figure 5.7.

The power consumption waveforms for the four operating points at the initial thermostat set point and the new thermostat set point are provided in Figure 5.8. Initial waveform is shown using dotted lines and power waveform after demand response is shown in thick lines.

From Figure 5.8, a difference in power consumption can be observed. The ON time is reduced and OFF time is increased when thermostat set point increases for demand response.

The difference in power consumed before and after demand response is the primary concern and thus the power gain at each operating point has to be determined in detail.

The power gain for operating point a is shown in Figure 5.9.

It is observed that the power gain after θ_+ occurs after cooling period which is in hours. Hence for applications like Load Frequency Control, power gain after θ_+

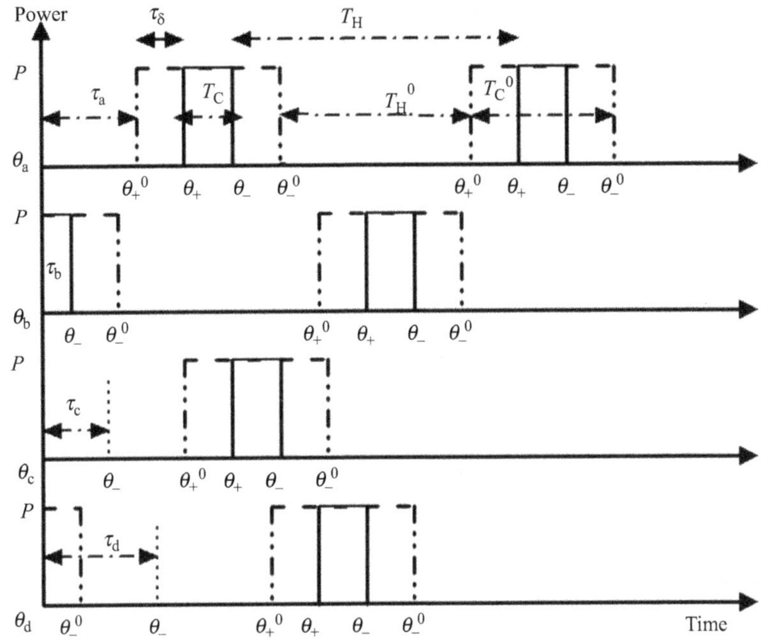

Figure 5.8 Power consumption waveforms at four operating points before and after change in thermostat set point

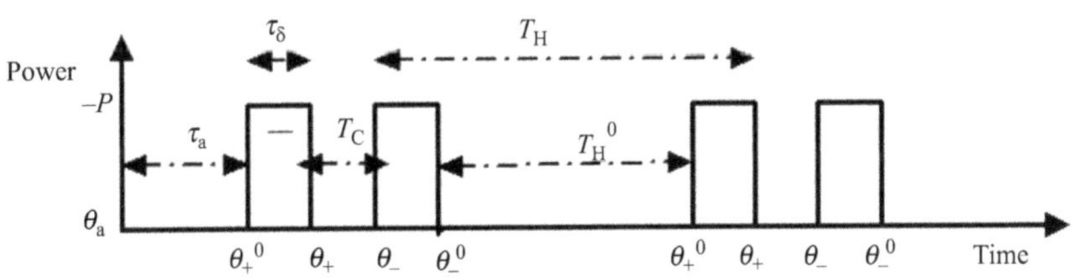

Figure 5.9 Power gain waveform for TCL units at operating point a

need not be considered. Thus the waveform relevant to LFC time frame is the power gain waveform from θ_a to θ_+. The Laplace transform of the relevant waveform is computed and the total power gain for aggregated TCL units operating at point a is calculated by integrating the transform over the distributions f_0. Integration is done over f_0 because the operating point considered lies in the off period.

Thus the total power gain from TCLs at operating point a is

$$P_a(s) = \int_{\theta_-}^{\theta_+^0} f_0(\theta_a) * G_a(s) d\theta_a \tag{5.18}$$

where $G_a(s)$ is the Laplace transform of the difference in power consumption before and after DR of loads at operating point a given by

$$G_a(s) = \frac{P * e^{-s\tau_a} * (e^{-s\tau_\delta} - 1)}{s} \tag{5.19}$$

and $\tau_a = T_H - t_h(\theta_a)$ $\tau_\delta = T_C^0 - t_c^0(\theta_-)$ which can also be defined as $\tau_\delta = T_H - t_h^0(\theta_+^0)$

Operating point b is shown in Figure 5.10.

The difference in power i.e. power gain after θ_-^0 occurs after a time period equivalent to the heating period before DR, which is in hours. Thus the Laplace transform of the relevant waveform from θ_b to θ_-^0 is computed and the total power gain is calculated by integrating over the distributions f_1. The total power gain from TCLs at operating point b is

$$P_b(s) = \int_{\theta_-}^{\theta_+^0} f_1(\theta_b) * G_b(s) d\theta_b \tag{5.20}$$

where

$G_b(s)$ is the Laplace transform of the difference in power consumption of loads at b, given by

$$G_b(s) = \frac{P * e^{-s\tau_b} * (e^{-s\tau_\delta} - 1)}{s} \tag{5.21}$$

and

$\tau_b = t_c^0(\theta_-) - t_c^0(\theta_b)$ which are given by (5.12) with respect to temperature set point before DR.

The power gain waveform for TCL units at operating point c is shown in Figure 5.11. Similar to point a, in operating point c power gain after θ_+ occurs after cooling period. Thus the Laplace transform of the waveform from θ_c to θ_+ is computed and the total power gain is calculated by integrating over the distributions f_0. The total power gain from TCLs at operating point c is

$$P_c(s) = \int_{\theta_-^0}^{\theta_-} f_0(\theta_c) * G_c(s) d\theta_c \tag{5.22}$$

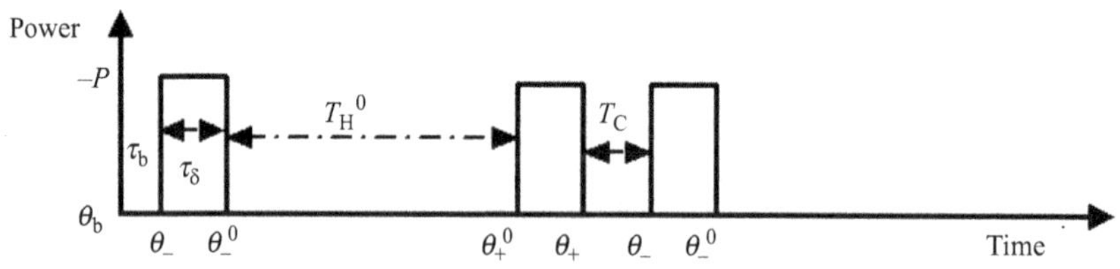

Figure 5.10 Power gain waveform for TCL units at operating point b

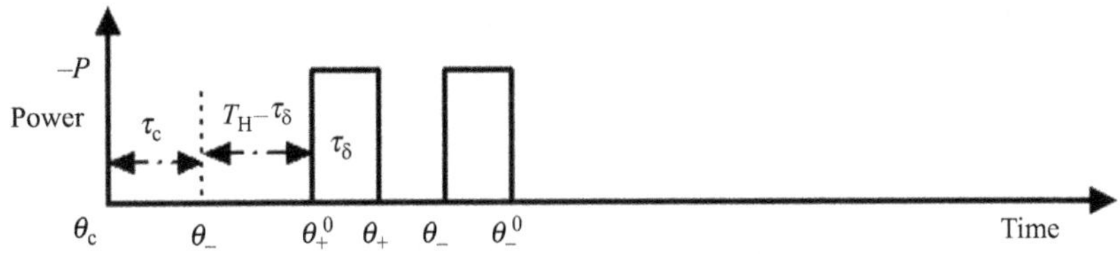

Figure 5.11 Power gain waveform for TCL units at operating point c

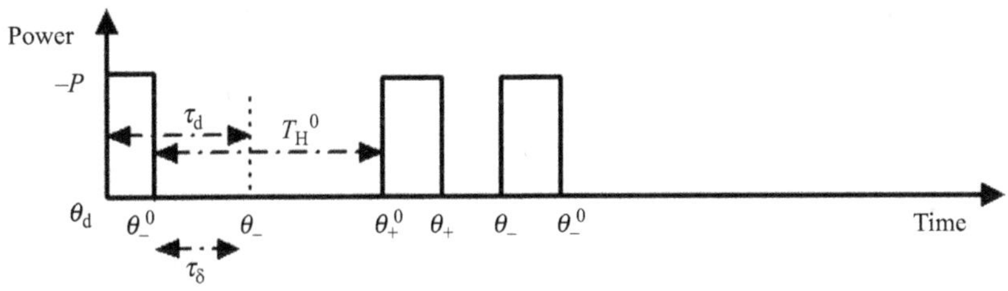

Figure 5.12 Power gain waveform for TCL units at operating point d

where $G_c(s)$ is the Laplace transform of the difference in power consumption of loads at c given by

$$G_c(s) = \frac{P * e^{-s\tau_c} * e^{-sT_H}(1 - e^{s\tau_\delta})}{s} \tag{5.23}$$

and
$\tau_c = t_h^0(\theta_-) - t_h^0(\theta_c)$ with respect to initial temperature set point. In operating point d, the power gain is as shown in Figure 5.12. The Laplace transform of the waveform from θ_d to θ_-^0 is computed and the total power gain is calculated by integrating over the distributions f_1. The total power gain from TCLs at operating point d is

$$P_d(s) = \int_{\theta_-^0}^{\theta_-} f_1(\theta_d) * G_d(s) d\theta_d \tag{5.24}$$

where
$G_d(s)$ is the Laplace transform of the difference in power consumption of loads at d given by

$$G_d(s) = \frac{P * (e^{-s(\tau_d - \tau_\delta)} - 1)}{s} \tag{5.25}$$

and $\tau_d = T_C^0 - t_c^0(\theta_d) + \tau_\delta$

Thus average change in power due to demand response from "N" TCL units is given by

$$P_{avg}(s) = P_a(s) + P_b(s) + P_c(s) + P_d(S) = \frac{-P * (e^{s\tau_\delta^2} - 1)}{s(T_C + T_H)} \tag{5.26}$$

The above equation is the transfer function for the change in power from aggregated thermostatically controlled loads that execute price based demand response by adjusting the thermostat setpoints.

5.5 Automatic generation control (AGC)

Once a generating unit is tripped or a block of load is added to the system, the power mismatch is initially compensated by an extraction of kinetic energy from system inertial storage. This causes system frequency to decline as speed decreases. As system frequency decreases, power taken by loads decreases. Thus equilibrium for large systems is often obtained at the resulting new frequency. If the mismatch is large enough to cause the frequency to deviate beyond the governor deadband of generating units, their output will be increased by governor action. Many governor deadbands are beyond 35 mHz.

Interconnected systems are formed by allowing tie line flows between generation plants. This helps in sharing resources during emergencies and during normal operating conditions in economics of power production. For control analysis, this interconnected system is divided into control areas. Each control area has an Energy Control Centre to monitor the change in system frequency and the change in tie-line flow. Power systems utility in a geographical area has control over certain generator units to provide secondary control that allocates generation. This control scheme is called load frequency control (LFC). As frequency control is achieved by automatically changing generation, it should also be ensured that the change in generation of each plant is economical. Thus LFC followed by economic dispatch (ED) completes the objective of the system and the cycle is called AGC.

The objectives of AGC are to [17]

1. Minimize mismatch between generation and load
2. Maintain system frequency at nominal value
3. Maintain Tie-line flow in interconnected systems (area) at the scheduled value
4. In an area, generation sharing must be at optimal value, i.e. economical

The first three objectives come under the term load frequency control and the last objective comes under economic dispatch. Thus AGC is complete only when load frequency control scheme is followed by economic dispatch. AGC scheme can be considered to have three levels of control:

1. Primary control
2. Secondary control
3. Tertiary control

It should be noted that AGC is not expected to limit the magnitude of the first frequency swing which occurs within the seconds after the loss of a block of generation or load in the system. Generation changes are realized by sending control signals to the units. Thus the design of AGC depends on how the unit responds to the signal [18].

5.5.1 Primary frequency control

Traditional load frequency control is based on the electromechanical dynamics of the system, which is described by the swing equation [19]

$$2H\frac{\Delta\omega}{\Delta t} = (T_m - T_e) \tag{5.27}$$

where,
H is the inertia constant (seconds); ω is the rotor speed; T_m is the mechanical torque; T_e is the electrical torque.

The relationship between power, P, and torque, T, is given by

$$P = \omega_r * T \tag{5.28}$$

Considering a small deviations ΔP, ΔT from initial values P_0 and T_0 respectively, we get

$$P_0 + \Delta P = (\omega_0 + \Delta\omega_r) * (T_0 + \Delta T) \tag{5.29}$$

On further simplification of the above equation,

$$\Delta P = \omega_0 * \Delta T + T_0 * \Delta\omega_r \tag{5.30}$$

ΔP occurs due to difference in mechanical power ΔP_m and electrical power ΔP_e as given in

$$\Delta P = \Delta P_m - \Delta P_e \tag{5.31}$$

At steady state, T_0 is equal to 0. At per unit, $\omega_r = 1$. Then

$$\Delta P_m - \Delta P_e = \Delta T_m - \Delta T_e \tag{5.32}$$

Load can be classified as frequency independent and frequency dependent,

$$\Delta P_e = \Delta P_L + D * \Delta\omega_r \tag{5.33}$$

where,
ΔP_L is the non-frequency-sensitive load change,
D* $\Delta\omega_r$ is the frequency-sensitive load change, and D is the load-damping constant.

Governor and turbine are modeled as the first-order system. The transfer function for Governor with time delay T_g is

$$G(s) = \frac{1}{1 + sT_g} \tag{5.34}$$

Turbine transfer function with time delay T_t is

$$T(s) = \frac{1}{1 + sT_t} \tag{5.35}$$

Speed governors accomplish primary frequency control. The ratio of steady-state frequency deviation Δf to change in valve position so as to change power output ΔP_m is referred to as the speed regulation or droop R,

$$R = \frac{\text{percent change in speed or frequency}}{\text{percent change in power output}} * 100 \tag{5.36}$$

Primary speed control action will result in steady-state frequency deviation. The deviation in frequency is shown in Figure 5.13.

The initial steady-state operating point is A (P_0, f_0). An increase in load results in a deviation in frequency and the droop characteristics of the governor results in an operating point B (P_1, f_1). The entire control area can be modeled as transfer function in Figure 5.14.

The single area scheme for primary frequency control involves governor with the speed droop characteristics, turbine model, and the combined generator load model. This is sufficient to minimize very small variations in system frequency.

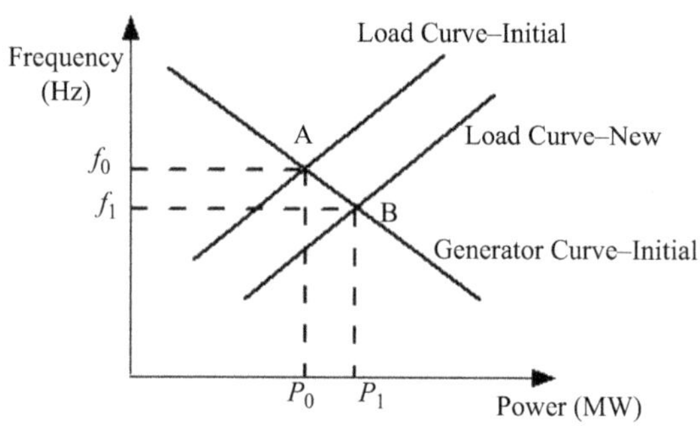

Figure 5.13 Deviation in frequency with increase in load

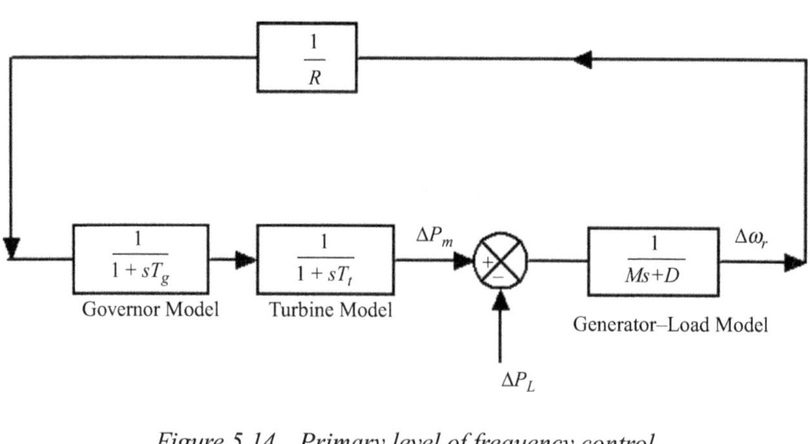

Figure 5.14 Primary level of frequency control

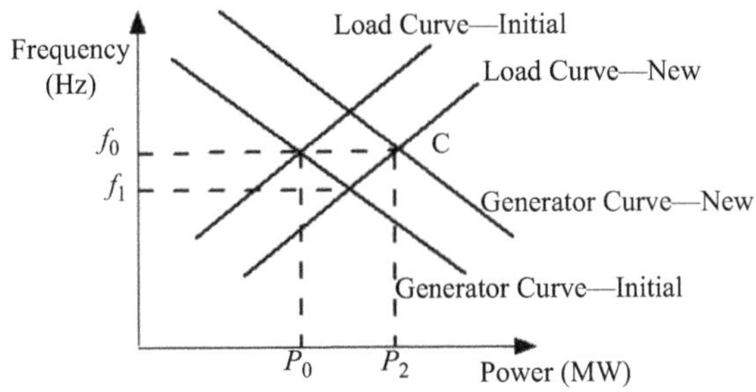

Figure 5.15 Deviation in frequency after secondary control

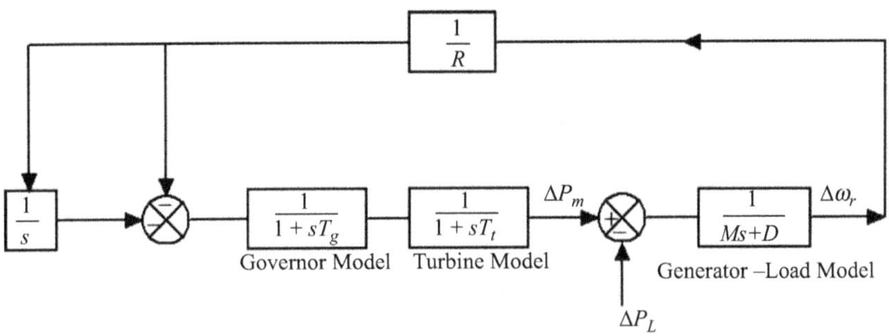

Figure 5.16 Load frequency control scheme for a single area system

5.5.2 Secondary frequency control

The primary control scheme as given in Figure 5.13 shows that steady-state error i.e. $\Delta f \neq 0$. This can be achieved by shifting the generator curve to reach the operating point C as shown in Figure 5.15. Hence, a control action is required to maintain the frequency at nominal value by minimizing the steady state error. This can be achieved using an integral controller. The load frequency control for a single area system is shown in Figure 5.16.

The control signal issued by the controller acts as governor reference set point, which is capable of minimizing the steady-state error to zero. An integral controller integrates the change in frequency over time and produces a signal. This signal is then fed to the speed governor as the reference set point.

5.6 Dynamic demand control (DDC)

The "load follows supply" concept based on frequency adaptive power energy scheduler (FAPER) was introduced in [6]. An active role for intelligent loads in frequency

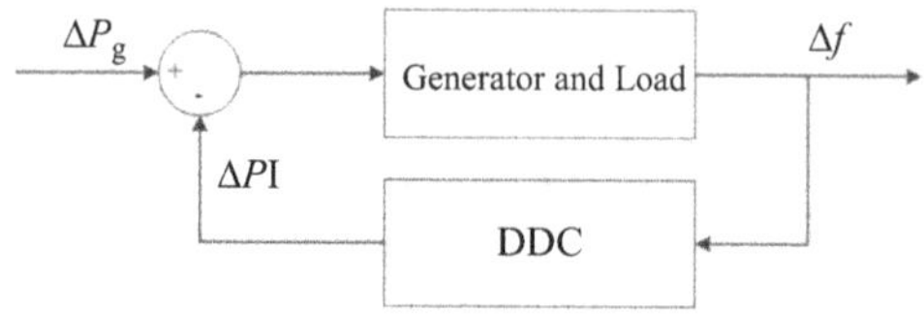

Figure 5.17 DDC model for frequency regulation

control was proposed in [9]. This leads to the concept of DDC. DDC implies an optimal load control strategy suitable for controlling real power demand based on frequency deviation that occurs due to generation-load imbalance. Through DDC the load follows supply concept is found to be effective for regulating frequency [20].

DDC will be useful in cases where wind penetration is high. Loads whose utility to the consumer is a function of the energy consumed over a period of time rather than their instantaneous power consumption are used for DDC. Thus in order to match power demand and power supply, DDC can be used along with AGC. Advantages of DDC are (1) Fewer generators would require speed governor control and thus allowing to work under fixed power loading, thus more efficient power plant operation. (2) Improves system stability. In distributed generation, or smart grid, the variation in system load is a combination of actual load uncertainties and fluctuations in generation from renewable distributed generation. This may lead to situation in which every MW of renewable generation must be backed up by a MW of fossil fuel capacity. A demand-based frequency control eliminates this constraint by reducing the level of generation based frequency control [21]. In this section, the load model derived from the price-based DR model for aggregated thermostatically controlled loads is discussed.

The generator-load model described in Figure 5.16 contains non-frequency sensitive and frequency sensitive component of loads. In addition to this model, the price-based DR model is introduced as shown in Figure 5.17. The price-based DR model represents the DDC model which is based on frequency deviation in the system. The required control in demand is achieved through price-based DR of aggregated model of thermostatically controlled loads in the single area.

Figure 5.17 shows the block diagram of load frequency control in which, in addition to change in generation, change in load is also incorporated. ΔP_l is the change in load and ΔP_g is the change in generation. Δf is the frequency deviation. For small frequency variations, a change in load is used instead of changing generation whereas for large variations a generation change is essential. During contingencies, the load is shed to maintain the system frequency at 50 Hz.

5.7 Simulink model

Let a TCL with coefficient of performance (COP) 2.92 and m, C_p equal to 1, be contracted for price-based demand response. For a change in price signal sensed, the

Table 5.1 Increase in price signal and the corresponding increase in thermostat set point

Increase in price signal	Increase in thermostat set point
3 units to 4 units	0.75
3 units to 5 units	1.2
3 units to 8 units	1.8

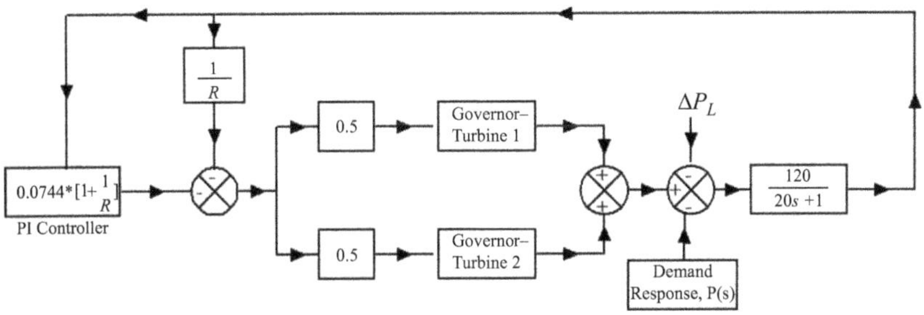

Figure 5.18 AGC with price-based DR

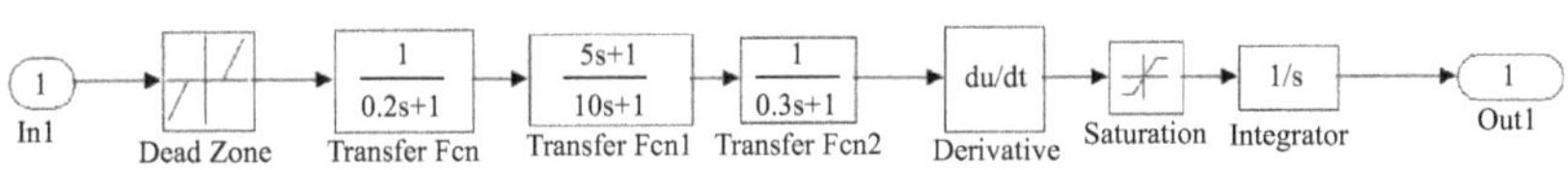

Figure 5.19 Governor—turbine model

corresponding optimal change in thermostat set point required is given in Table 5.1. It can be observed that the scheme which is capable of minimizing the increase in amount paid by the customer cause minimum inconvenience to them, as the increase in thermostat set point required for DR is within the thermal comfort level.

The aggregated TCL DR model is then applied to a single area AGC scheme. A single area system with 2 steam turbine units with reheat was considered as in Figure 5.18.

In this system, the deadband nonlinearity of governor as well as the generation rate constraints was included as shown in Figure 5.19. A single area LFC with real-time pricing scheme based on frequency deviation [22] was considered. Real-time price which is proportional to the initial df/dt value [22] is assumed to be known at the system operator with the help of phasor measurement units. The parameters of the single area system are provided in Appendix B. System operator sends price information signals to the customers that participate in DR. The delay in communication is assumed to be negligible. 1,000 TCLs were assumed to be available for DR.

A single area system with a step load change of 0.01 p.u. was simulated in Matlab/Simulink. The load frequency control scheme with and without price-based DR is compared and the change in frequency and generation are provided in Figure 5.20.

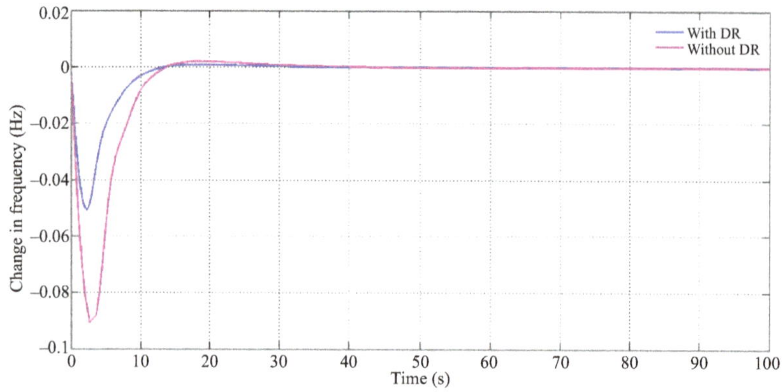

Figure 5.20 Change in frequency for 0.01 p.u. step load change

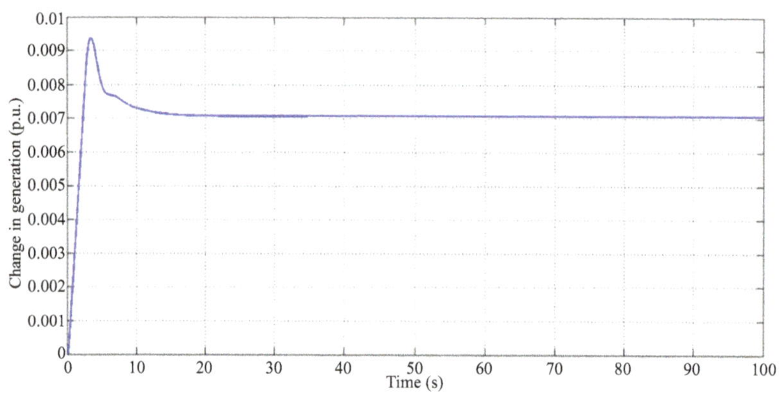

Figure 5.21 Change in generation for 0.01 p.u. step load change

The change in generation for 0.01 p.u. step change in load, when DR was included in the AGC scheme, is given in Figure 5.21.

The system was the simulated for a load change of 0.02 p.u. The change in frequency is given in Figure 5.22.

The corresponding change in generation required when demand response was included in the scheme is given in Figure 5.23.

Simulation for a step load change of −0.03 p.u. was done and the change in frequency is shown in Figure 5.24.

The change in generation required in DR-based AGC is given in Figure 5.25.

A step load change of −0.04 p.u. was also simulated. The frequency deviation in this case is provided in Figure 5.26.

The change in generation in this case is shown in Figure 5.27.

Aggregation of TCL loads that are contracted for price-based DR is modeled to obtain the change in power consumed by these loads when a change in thermostat set

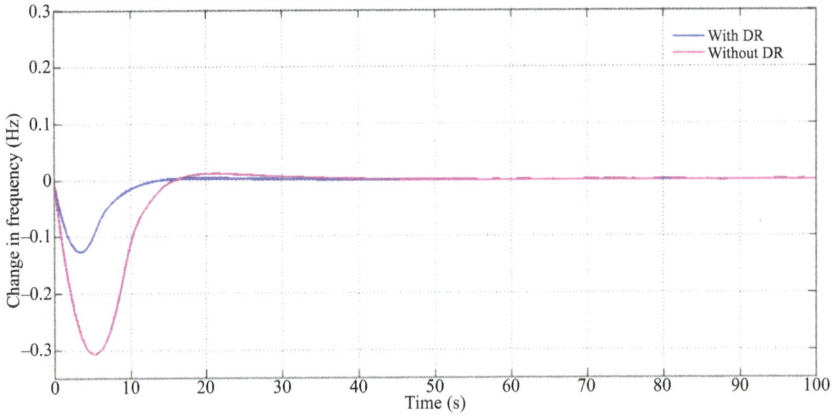

Figure 5.22 Change in frequency for 0.02 p.u. step load change

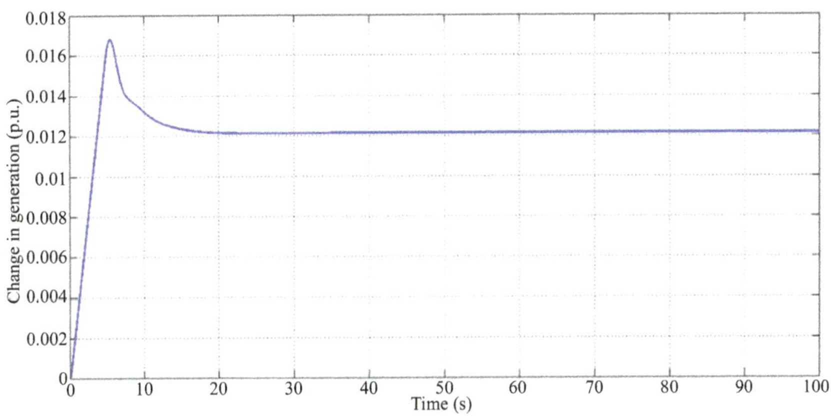

Figure 5.23 Change in generation for 0.02 p.u. step load change

point occurs. The model is then applied to load frequency control scheme for a single area system. The increase in generation required when there is an increase in load can be reduced when demand response is included into LFC scheme. The results are tabulated in Table 5.2.

Aggregation of TCL loads that are contracted for price-based DR is modeled to obtain the change in power consumed by these loads when a change in thermostat set point occurs. The model is then applied to load frequency control scheme for a single area system. The increase in generation required when there is an increase in load can be reduced when demand response is included into LFC scheme.

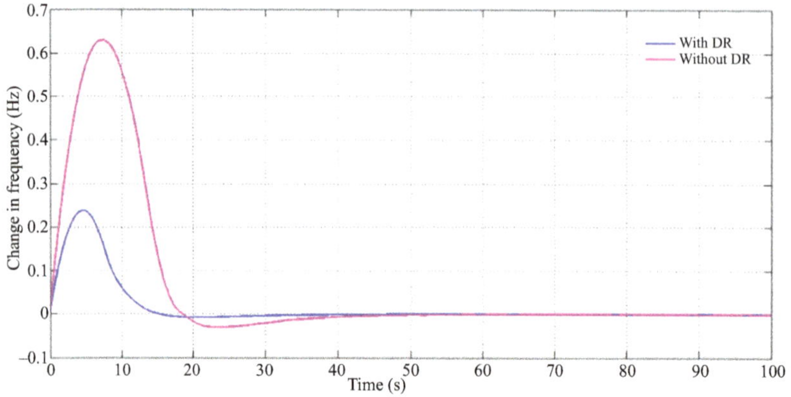

Figure 5.24 Change in frequency for −0.03 p.u. step load change

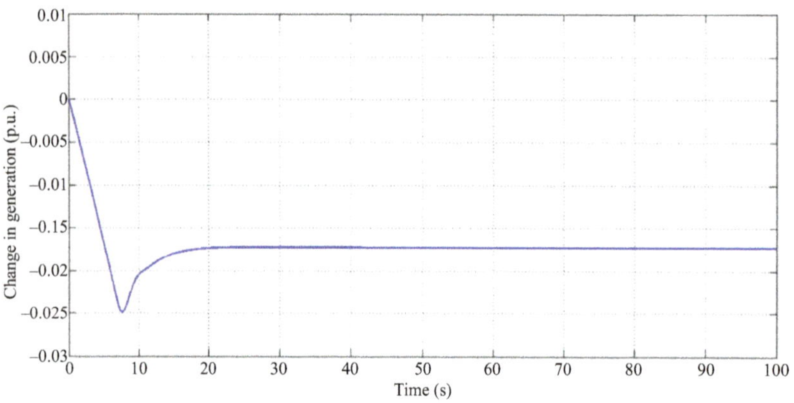

Figure 5.25 Change in generation for −0.03 p.u. step load change

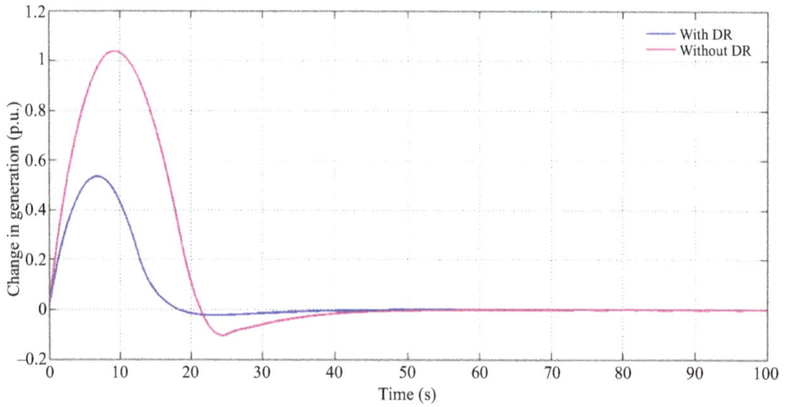

Figure 5.26 Change in frequency for −0.04 p.u. step load change

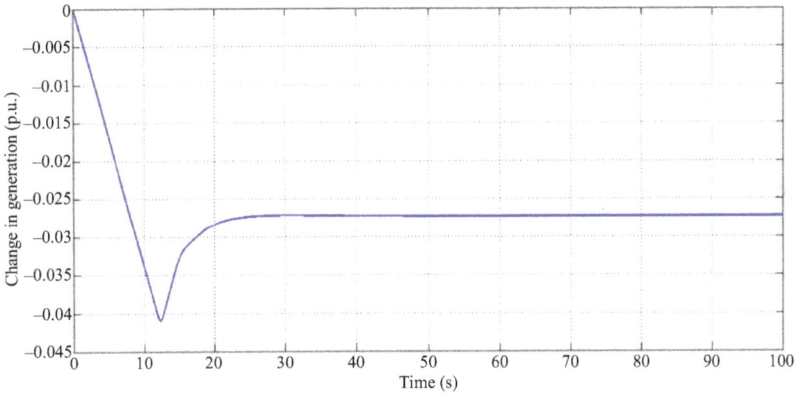

Figure 5.27 Change in generation for −0.04 p.u. step load change

Table 5.2 Increase in price signal and the corresponding increase in thermostat set point

Step change in load	**Change in generation** Without DDC	 With DDC
+0.01	+0.01	+0.00707
+0.02	+0.02	+0.01214
−0.03	−0.03	−0.01731
−0.04	−0.04	−0.02731

Appendix A: Modeling of aggregated TCL loads using coupled Fokker–Planck equations

Considering the probability distribution by temperature of aggregated homogeneous thermostatically coupled loads (TCL), i.e. a population of TCL units, a set of equations called Coupled Fokker–Planck Equations (CFPE) was used to model the dynamics of aggregate TCL units [14].

Coupled Fokker–Planck equations
Let $\theta(t)$ be the temperature of a TCL unit at time (t). Let the probability densities of loads in the on state be $f_1(\theta, t)$ and for off state be $f_0(\theta, t)$. Let C be the thermal capacitance (kWh/deg C), R be the thermal resistance (deg C/kW)
θ_a be the ambient temperature, P is the rate of energy transfer due to operation of TCL. (kW)

Then the CFPE equations are

$$\frac{\partial}{\partial t} f_1 = \frac{\partial}{\partial \theta}\left[\left(\frac{\theta(t)-\theta_a}{CR}+\frac{P}{C}\right) f_1\right] + \frac{\sigma^2}{2}\frac{\partial^2}{\partial \theta^2} f_1 \tag{5.37}$$

$$\frac{\partial}{\partial t}f_0 = \frac{\partial}{\partial \theta}\left[\left(\frac{\theta(t) - \theta_a}{CR}\right)f_0\right] + \frac{\sigma^2}{2}\frac{\partial^2}{\partial \theta^2}f_0 \tag{5.38}$$

Let a represent the region where temperature is less that the deadband (θ_-) and b denote the region within the deadband (θ_- to θ_+). c represents temperature greater than deadband. The boundary conditions at upper and lower limits of thermostat deadband to enforce the conservation of probability are

$$\frac{\partial}{\partial \theta}f_{0a}(\theta_-, t) - \frac{\partial}{\partial \theta}f_{0b}(\theta_-, t) - \frac{\partial}{\partial \theta}f_{1b}(\theta_-, t) = 0 \tag{5.39}$$

$$\frac{\partial}{\partial \theta}f_{1c}(\theta_+, t) - \frac{\partial}{\partial \theta}f_{0b}(\theta_+, t) - \frac{\partial}{\partial \theta}f_{1b}(\theta_+, t) = 0 \tag{5.40}$$

At the limits of deadband, the natural boundary conditions are

$$f_{0a}(\theta_-, t) = f_{0b}(\theta_-, t) \tag{5.41}$$

$$f_{0b}(\theta_+, t) = f_{1b}(\theta_-, t) = 0 \tag{5.42}$$

$$f_{1a}(\theta_+, t) = f_{1b}(\theta_+, t) \tag{5.43}$$

$$f_{1c}(\infty, t) = f_{0a}(-\infty, t) = 0 \tag{5.44}$$

The total probability mass in the system is unity and hence

$$\int_{-\infty}^{\theta_-} f_{0a}d\theta + \int_{\theta_-}^{\theta_+} (f_{0b} + f_{1b})d\theta + \int_{\theta_+}^{\infty} f_{1c}d\theta \tag{5.45}$$

These equations form the complete system of CFPE equations to model the dynamics of a population of homogeneous TCL units

Appendix B: Single area AGC system parameters

The parameters of the single area system used for simulation are as shown in Table 5.3.

Table 5.3 Single area system parameters

Parameter	Value
Inertia constant, H	5 sec
Damping constant, D	$8.33*10^{-3}$ p.u. MW/Hz
Turbine time constant, T_t	0.3
Steam turbine reheat constant, K_r	0.5
Steam turbine reheat time constant, T_r	10 s
Governor time constant, T_g	0.08
Governor Droop, R	2.4Hz/p.u. MW
Change in load, ΔP_L	0.01 p.u.

References

[1] Gellings CW. The concept of demand-side management for electric utilities. Proceedings of the IEEE. 1985;73(10):1468–1470.

[2] Palensky P, Dietrich D. Demand side management: demand response, intelligent energy systems, and smart loads. IEEE Transactions on Industrial Informatics. 2011;7(3):381–388.

[3] Yusta JM, Dominguez JA. Measuring and modeling of industrial demand response to alternative prices of the electricity. In: Proc. Power Syst. Comput. Conf., Session. vol. 15; 2002.

[4] Faruqui A, George SS. The value of dynamic pricing in mass markets. The Electricity Journal. 2002;15(6):45–55.

[5] Moshari A, Yousefi G, Ebrahimi A, *et al.* Demand-side behavior in the smart grid environment. In: Innovative Smart Grid Technologies Conference Europe (ISGT Europe), 2010 IEEE PES. IEEE; 2010. pp. 1–7.

[6] Schweppe FC, Tabors RD, Kirtley JL, *et al.* Homeostatic utility control. IEEE Transactions on Power Apparatus and Systems. 1980;(3):1151–1163.

[7] Lasseter RH, Paigi P. Microgrid: a conceptual solution. In: Power Electronics Specialists Conference, 2004. PESC 04. 2004 IEEE 35th Annual. vol. 6. IEEE; 2004. pp. 4285–4290.

[8] Zhang P, Li F, Bhatt N. Next-generation monitoring, analysis, and control for the future smart control center. IEEE Transactions on Smart Grid. 2010;1(2):186–192.

[9] Trudnowski D, Donnelly M, Lightner E. Power-system frequency and stability control using decentralized intelligent loads. In: Transmission and Distribution Conference and Exhibition, 2005/2006 IEEE PES. Ieee; 2006. pp. 1453–1459.

[10] Molina-Garcia A, Kessler M, Fuentes JA, *et al.* Probabilistic characterization of thermostatically controlled loads to model the impact of demand response programs. IEEE Transactions on Power Systems. 2011;(99):1–1.

[11] Lu S, Samaan N, Diao R, *et al.* Centralized and decentralized control for demand response. In: Innovative Smart Grid Technologies (ISGT), 2011 IEEE PES. IEEE; 2011. pp. 1–8.

[12] Callaway DS. Tapping the energy storage potential in electric loads to deliver load following and regulation, with application to wind energy. Energy Conversion and Management. 2009;50(5):1389–1400.

[13] Lu N, Chassin DP. A state-queueing model of thermostatically controlled appliances. IEEE Transactions on Power Systems. 2004;19(3):1666–1673.

[14] Malhame R, Chong CY. Electric load model synthesis by diffusion approximation of a high-order hybrid-state stochastic system. IEEE Transactions on Automatic Control. 1985;30(9):854–860.

[15] Kundu S, Sinitsyn N, Backhaus S, *et al.* Modeling and control of thermostatically controlled loads; 2011. http://web.eecs.umich.edu/hiskens/publications/fp355.pdf.

[16] Venkatarathnam G. The coefficient of performance of an ideal air conditioner. International Journal of Refrigeration. 2009;32(8):1929–1931.

[17] Carpentier J. 'To be or not to be modern' that is the question for automatic generation control (point of view of a utility engineer). International Journal of Electrical Power & Energy Systems. 1985;7(2):81–91.

[18] Jaleeli N, VanSlyck LS, Ewart DN, *et al.* Understanding automatic generation control. IEEE Transactions on Power Systems. 1992;7(3):1106–1122.

[19] Kundur P, Balu NJ, Lauby MG. Power system stability and control. vol. 4. New York, NY: McGraw-Gill; 1994.

[20] Jay D, Swarup K. Price based demand response of aggregated thermostatically controlled loads for load frequency control. In: Proceedings of the National Power Systems Conference (NPSC), Varanasi, Uttar Pradesh, India; 2012. pp. 12–14.

[21] Black JW, Ilic M. Demand-based frequency control for distributed generation. In: Power Engineering Society Summer Meeting, 2002 IEEE. vol. 1. IEEE; 2002. pp. 427–432.

[22] Berger AW, Schweppe FC. Real time pricing to assist in load frequency control. IEEE Transactions on Power Systems. 1989;4(3):920–926.

Chapter 6

Electric vehicle massive resources mining and demand response application

Yun Zhou[1] *Donghan Feng*[1] *and Chen Fang*[2]

6.1 Introduction

In 2020, the sales volume of electric vehicles (EVs) in China reached 1.367 million. A rapid growth trend was witnessed by the huge increment of electric vehicles in past the several years. By the end of 2020, China has nearly 5 million new energy vehicles. Meanwhile, China's charging infrastructure has reached 1,681,000 units. It is expected that the global penetration rate of new energy vehicles will exceed 30% in 2030. At that time, the number of electric vehicles in China is prospected to be 80–100 million. As the largest EV market in the world, China has unique conditions to develop and study the interactive application of EVs and power grid. The power system can have a chance of promoting comprehensive innovation thanks to the booming of the EV industry. A smart energy transportation network, that can participate in the grid demand response (DR) timely, would possibly consist of massive EVs, the power grid, renewable energy network and transportation network.

Because of the inherent mobile energy storage characteristics of EVs, flexible large-scale EV clusters have great potential in power load regulation, renewable energy consumption, power quality improvement, etc. Thus EVs can be used to participate in auxiliary services such as peak shifting and valley filling, frequency regulation, emergency support so as to interact friendly with the grid. In recent years, many cities in China have tried to include EVs in the pilot and made positive exploration in vehicle network interaction.

6.2 Development status and trend of EVs and charging infrastructure

Under the background of energy saving, carbon emission peak, and carbon neutrality, the technologies of EVs and public charging infrastructure have developed rapidly and

[1]Department of Electrical Engineering, Shanghai Jiao Tong University, Shanghai, China
[2]Electric Power Research Institute, State Grid Shanghai Municipal Power Company, Shanghai, China

gained significant attention over the last few years around the world. The discussion of this section mainly focuses on the development status of EVs, the construction situation of charging infrastructure, and the governments' supporting policies.

6.2.1 Development status of EVs

As an effective vehicle technology, EVs can reduce the greenhouse gas emissions and gasoline consumption obviously [1]. According to the data from EV Sales Blog, 3,124,793 EVs (battery EVs (BEVs) and plug-in hybrid EVs (PHEVs)) were sold worldwide in 2020, of which more than 68% were pure EVs (PEVs) [2].

The detailed EVs sales worldwide in 2020 are summarized in Table 6.1. Tesla, Volkswagen, SAIC Motor, Renault-Nissan-Mitsubishi, and BMW Group occupied the top five spots in the sales chart. The total share of the top five is 51.7%, and other EVs' electric car brands' shell has taken over 48.3%.

Figure 6.1 shows the ownership of new energy vehicles (NEVs) in China from 2016 to 2021. The number of EVs has been steadily increasing from 2011 to 2021. At the end of June 2021, the number of NEVs nationwide reached 6.03 million, which accounts for 2.06% of the total number of vehicles in China. In addition, there are 4.93 million BEVs among them, occupying 81.68% of all NEVs. The proportion of NEVs in newly registered vehicles has increased to 7.80% in 2021, which means one is a new energy vehicle in per 14 newly registered vehicles [3].

Table 6.1 The sales volume of EVs in 2020

Brands	**Tesla**	**Volkswagen**	**SAIC**	**Renault–Nissan–Mitsubishi Alliance**	**BMW**	**Others**	**Tesla**
Sales	499,535	421,591	272,210	226,975	195,979	1,508,503	499,535
Market share	16%	13%	9%	7%	6%	48.3%	16%

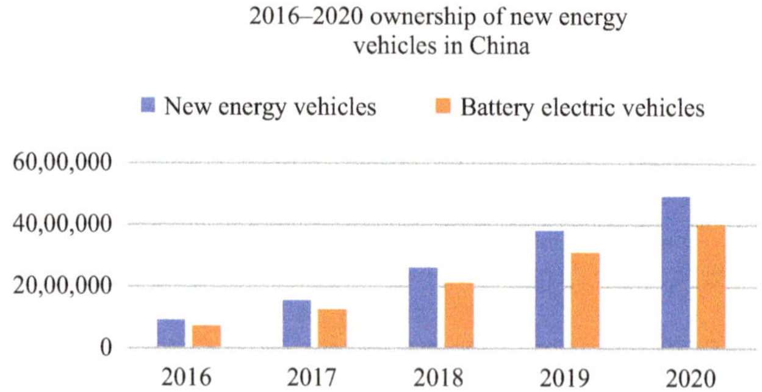

Figure 6.1 2016–20 ownership of new energy vehicles in China

6.2.2 Construction situation of charging infrastructure

The construction of energy supply infrastructures is the foundation and prerequisite for the large-scale promotion and application of EVs. By the end of 2020, there were 4.92 million NEVs on China's roads. At the end of April 2021, there are 0.87 million EV chargers for public use in China, including 41.82% DC charging piles and 58.28% AC charging piles. That means around 6 EVs share one EV charger.

In Figure 6.2, a schematic diagram of a public EV charging station with photovoltaic and energy storage systems (PV-EES-EV charging station) is presented. The EV chargers for public use are mainly located in Guangdong Province, Shanghai City, and Beijing City. At the end of April 2021, Guangdong Province ranks the first with approximately 119,000 public chargers, while Shanghai comes second with 86,000 and Beijing occupies the third with 83,000. The top 10 provinces account for 72.10% of total public EV chargers in mainland China. As for the operator, at the end of February 2020, TELD occupies over 28.7% share of the market. Star Charge occupies approximately 24.5%, and State Grid occupies around 15.9% [4].

Besides the public charging network, expanding private charging infrastructure is also significant support to develop the mobility characteristic of EVs [5]. Taking Shanghai City as an example, for private EV chargers installed in residential areas, EV users pay according to the inexpensive residential electricity price standard. And from 2016, for a newly constructed residential communities in Shanghai, 100% of parking spaces should equip or reserve installation conditions for EV chargers [6].

6.2.3 Governments' supporting policies

Governments around the world took various measures to increase policy support for EVs. The State Council of China published the 'Guidance on accelerating construction

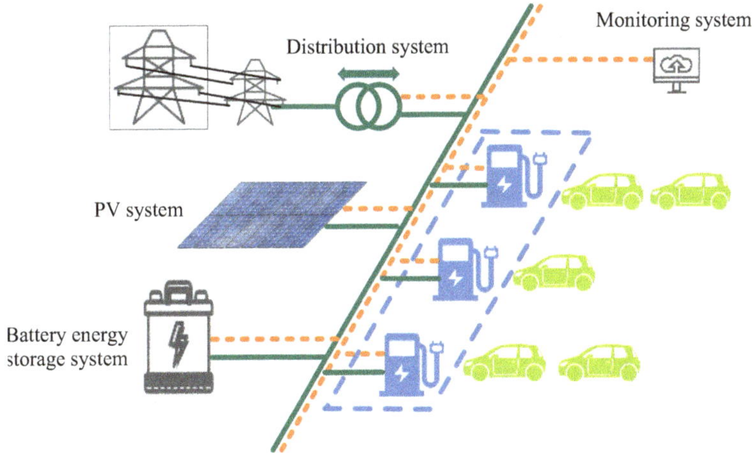

Figure 6.2 Public PV-EES-EV charging stations

Table 6.2 China's supporting policy issued on 09 September 2015

Goal	To 2020, the EV charging infrastructure should meet the charging demand of more than five million vehicles.
Guidance	According to the "piles and station first" requirement, orderly promote the construction, ensuring that the construction scale is moderately ahead.
Planning and design	For new residential, the supporting parking spaces with charging facilities or reserved construction conditions should be 100%; For large public buildings, the ratio between the public parking lot with charging facilities or reserved construction conditions and parking lot without charging facilities is not less than 10%; Every 2,000 EVs at least support the construction of a public charging station.

of the EV charging infrastructure' on 09 September 2015. The goal, guidance, planning and design of the supporting policy are summarized in Table 6.2.

For the latest policy in China, on 2 November 2020, the State Council issued a development plan for the new energy vehicle industry from 2021 to 2035 aiming at accelerating the country into an automotive powerhouse. The plan put forward five important tasks including technological innovation ability, building new-type industry ecosystems, advancing industrial integration and development, perfecting the infrastructure system, and deepening opening-up and cooperation [7].

6.3 EV massive resources digging and DR capability/potential evaluation

With the support of national policies and the development trend of new power systems, EVs have occupied an increasingly important position. Large-scale EVs will bring opportunities and challenges to the grid at the same time. Because of the uncertainty and randomness of the behavior of EV users, it will increase the difficulty of optimization and control of the grid. If the resources of EVs cannot be used reasonably, large-scale EVs are charged during the peak load period, which will aggravate the distance between the peak load and the valley load of the grid and increase the burden on the power system. On the contrary, if the charging load resources of EVs are fully dug, friendly interaction between EVs and the grid can be realized, and the consumption of renewable energy will be promoted.

6.3.1 EV massive resources digging

In order to effectively realize the friendly interaction between large-scale EVs and the power grid, fully exploring the charging load resources of EVs on the basis of considering the uncertainty and randomness of the behavior of EV users is necessary. On the one hand, the charging load resource can be modeled and predicted based on the historical data of EVs, and, on the other hand, the real-time dispatchable energy of EVs can be evaluated by obtaining real-time data of EVs.

6.3.1.1 Prediction of EVs charging resources based on historical data

Based on trip chain theory and EV user behavior, an EV charging load forecasting model is established for quick charge station operators [8]. It is possible to predict the charging load of a large-scale EV group in a wide area, and the result has a high reference value for the construction of charging facilities and the operation of the grid.

First, the characteristics of EVs trips are analyzed based on the travel statistics (such as the 2017 National Family Travel Survey (NHTS) statistics). To abstract the behavior patterns of EV users, a large amount of data should be recorded. For each closed trip chain, all necessary data should be collected, including plug-in time, plug-out time, trip duration, trip length, average velocity, and so on. According to the number of different types of trip chains, the proportion of every type of trip chain can be obtained. Through the kernel density estimation method to process the data, the distribution of these parameters to describe the behavior patterns of EV users can also be obtained [8].

Second, based on the travel characteristics and geographic information of EVs, the Monte Carlo simulation method is an effective way to simulate the travel behavior of each electric vehicle in the city to obtain the spatiotemporal characteristics of the charging behavior of each vehicle. The load is superimposed according to the temporal and spatial characteristics, and the charging load of each location (POI, point of interest) within the city (or part of the administrative division) at each time period is obtained.

Finally, based on the obtained charging load of all locations, the prediction result of the total charging load in a wide area can be obtained. In addition, the kernel density estimation method can be used for smoothing, and thermal analysis of the charging load can be used to grasp the thermal characteristics of the charging load.

Using this method to predict the charging load in the Shanghai area has produced relatively good results in Figure 6.3. The result is better in line with the actual charging load situation in Shanghai.

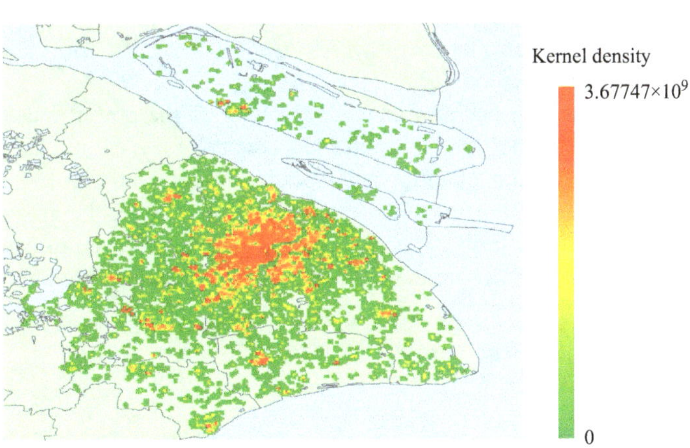

Figure 6.3 Charging load results by the modeling prediction

6.3.1.2 Evaluation of EVs dispatchable capacity based on real-time data

First, evaluate the dispatchable capacity of one EV in real-time information. The real-time data includes the current capacity status of the EV, the end charging time, and the desired capacity status when leaving. These real-time data can be used to obtain the real-time dispatchable capacity of each EV:

$$S_{j,t+\Delta t}^{\text{EVbase}} = S_j^{\text{EVdep}} - \left(t_j^{\text{plug-out}} - (t + \Delta t)\right) \times P^{\max} \times \eta^{\text{ch}} \tag{6.1}$$

$$P_{j,t}^{+} = \min\left\{\frac{S^{\max} - S_{j,t}^{\text{EV}}}{\Delta t \times \eta^{\text{ch}}}, P_{\max}\right\} \tag{6.2}$$

$$P_{j,t}^{-} = \begin{cases} \frac{S_{j,t+\Delta t}^{\text{EVbase}} - S_{j,t}^{\text{EV}}}{\eta^{\text{ch}} \times \Delta t}, & \text{if } S_{j,t+\Delta t}^{\text{EVbase}} > S_{j,t}^{\text{EV}} \\ \max\left\{\frac{S_{j,t+\Delta t}^{\text{EVbase}} - S_{j,t}^{\text{EV}}}{\eta^{\text{dis}} \times \Delta t}, -P^{\max}\right\}, & \text{if } S_{j,t+\Delta t}^{\text{EVbase}} \le S_{j,t}^{\text{EV}} \end{cases} \tag{6.3}$$

where $S_{j,t}^{\text{EV}}$ presents the capacity of the jth EV at the current moment, $S_{j,t+\Delta t}^{\text{EVbase}}$ presents the minimum capacity of the jth EV at the next moment $t_j^{\text{plug-out}}$ presents the end charging time of the jth EV, $P_{j,t}^{+}, P_{j,t}^{-}$ present the power of the jth EV can be adjusted up and down at the current moment respectively, S_j^{EVdep} presents the desired capacity when the jth EV leaves, $P^{\max}, S^{\max}$ present the upper limit of charging power and the upper limit of capacity, respectively, $\eta^{\text{ch}}, \eta^{\text{dis}}$ present the charging and discharging efficiency, respectively.

The meaning of (6.1)–(6.3) is to calculate the power space that the EV can be adjusted up and down during the evaluation period. For each EV, its maximum power does not exceed $P^{\max}$, nor does it exceed the power that can fully charge its battery at the next moment. Considering V2G, EVs also need to ensure that if the EV is off the grid at the next moment, its energy is not lower than S_j^{EVbase} and the maximum discharge power cannot exceed $-P^{\max}$.

In the entire charging cycle of each EV, as long as the power constraints and capacity constraints are met, the dispatchable capacity of the entire charging cycle can be evaluated. In addition, charging and discharging frequently for EVs can cause serious battery life loss [9]. The deeper the depth of discharge, the greater the life loss. Therefore, the lower limit of the EV energy can be specified to calculate the real-time dispatchable capacity of the EV considering the battery life loss:

$$\begin{cases} -P^{\max} \le P_{j,t} \le P^{\max} & t \in \left[t_j^{\text{plug-in}}, t_j^{\text{plug-out}}\right] \\ P_{j,t} = 0, & t \notin \left[t_j^{\text{plug-in}}, t_j^{\text{plug-out}}\right] \end{cases} \tag{6.4}$$

$$S_{j,t+\Delta t}^{\text{EV}} = S_{j,t}^{\text{EV}} + \eta(P_{j,t}) \times P_{j,t} \times \Delta t \tag{6.5}$$

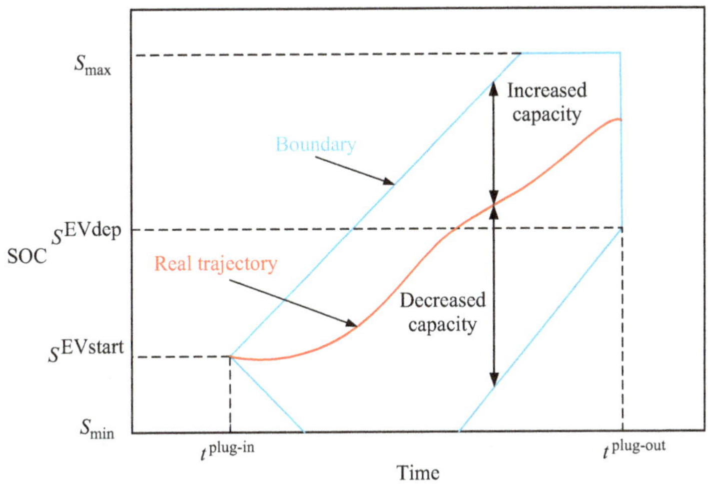

Figure 6.4 Dispatchable capacity of one EV during the entire charging cycle

$$\eta(x) = \begin{cases} \eta^{\mathrm{ch}}, & \text{if } x > 0 \\ 1, & \text{if } x = 0 \\ \eta^{\mathrm{dis}}, & \text{if } x < 0 \end{cases} \tag{6.6}$$

$$S^{\mathrm{EV}}_{j,t_j^{\text{plug-out}}} \geq S_j^{\mathrm{EVdep}} \tag{6.7}$$

$$S^{\min} \leq S^{\mathrm{EV}}_{j,t+\Delta t} \leq S^{\max} \tag{6.8}$$

where $t_j^{\text{plug-in}}$ presents the start charging time of the jth EV. According to the constraints (6.4)–(6.8), the dispatchable capacity of one EV during the entire charging cycle can be analyzed. It is shown in Figure 6.4.

The dispatchable capacity of each EV is insignificant to the grid, so it is necessary to evaluate the dispatchable capacity of large-scale EVs from the perspective of an aggregator. Accumulating the real-time dispatchable capacity of EVs in the corresponding time period can obtain the dispatchable capacity of the aggregator. The dispatching resources can be used to participate in auxiliary services, including frequency regulation and emergency support of the power grid, thereby interacting friendly with the power grid.

6.3.2 EVs in DR capability/potential evaluation

V2G technology can enable mass EVs to realize energy storage functions. As a typical representative of user-side and distributed new energy storage, EVs have important development potential in the application of new power systems.

Large-scale EVs can provide: power from megawatts to gigawatts or more; continuous discharge time with hour level; response speed with a minute and second level; accurate control and the stability at any power point; bidirectional adjustment capability as a load to charge or a power source to discharge.

According to these different characteristics, EVs are the potential resource to participate in the auxiliary services, including peak shifting and valley filling, frequency regulation, emergency support so as to interact friendly with the grid.

6.3.2.1 EVs in peak shifting and valley filling

The characteristics of the high power level and long-term continuous discharge capability can help EVs participate in the auxiliary service of peak shifting and valley filling of the power grid. Generally, the DR method is used to organize EVs to participate in peak shifting and valley filling. The financial incentive is a commonly used method to guide the customers in the demand process [10]. The customers can participate in DR well by controlling their electricity consumption following the need for power grid. Though the capacity of a single EV is small, a large scale of EVs can become one of the most potential DR participators. Among them, private cars are the main body of EVs participating in DR [11]. The travel habits of owners are usually constant, which is plugging in after work and plugging out before leaving home. So the plug-in duration of EVs is so easy to overlap with the peak load time that increases the pressure of the grid. Because of this phenomenon, EVs participating in peak regulation and valley filling is an effective measure to relieve the peak load from the disordered charging of EVs [12].

The DR service is carried out in Shanghai, and the resources of EVs are reasonably regulated through price incentives. In Figure 6.5, it can be seen that the peak charging load of EVs has transferred part of the low charging load at night through the form of DR. Therefore, EVs have the potential to participate in peak shifting and valley filling ancillary services.

6.3.2.2 EVs in frequency regulation

Utilizing the high power level and the ability to quickly and accurately respond to frequency modulation commands can help EVs participate in the frequency regulation auxiliary services of the grid. The auxiliary services generally require the participants to have a second-level responsibility and EVs can meet this requirement well. The dispatchable capacity of one EV is insignificant for the power grid, so a large number of EV resources need to be aggregated to participate in the frequency regulation auxiliary services of the power grid.

With the development of communication technologies such as optical fiber and wireless communication, real-time monitoring of the charging status of large-scale EVs can be realized and these real-time charging data and data digging techniques can be used to improve the accuracy of V2G power capacity evaluation. Through the "decentralized access and centralized control" mode for large-scale EVs, use aggregators for unified management and control, and participate in grid dispatching as a unit of aggregators in Figure 6.6. The aggregator acts as a bridge between the grid and EVs and it mainly includes two parts: data management system and control system [13]. The two parts have their own tasks correspondingly. The data management system is responsible for collecting real-time data and related historical data sent by

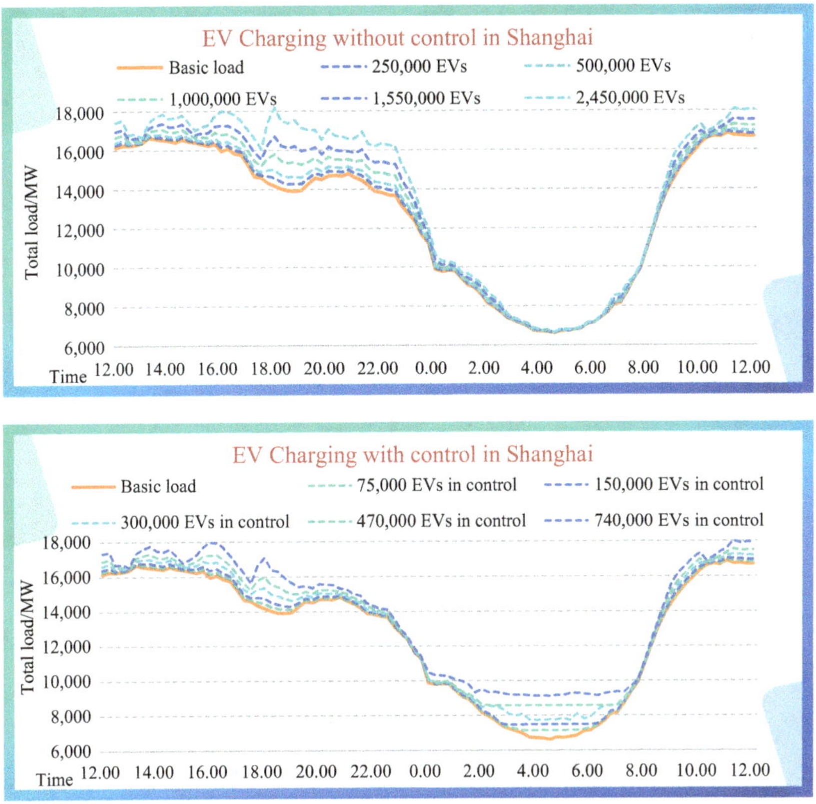

Figure 6.5 The comparison figure of EV charging load with/without control

the battery energy management monitoring system and the power grid dispatching center. The control system is responsible for analyzing the data and instructions, optimizing calculations, and obtaining the charging and discharging power of each EV. The control of the intelligent charging and discharging machine can regulate the behavior of the charging and discharging power of EVs.

6.3.2.3 EVs in emergency support

With the introduction of large-scale EVs, the distribution network as its main carrier gradually shows the characteristics of multi-source initiative, which brings new opportunities to the formulation of power supply restoration strategies for the distribution network.

Utilizing on-stake EVs can participate in emergency support of the power grid. In the islanding mode, when the internal power supply in the power outage area cannot meet the demand for the loads, it can be considered to remove the existing charging loads of EVs with the high SOC level in turn, so that the important power outage

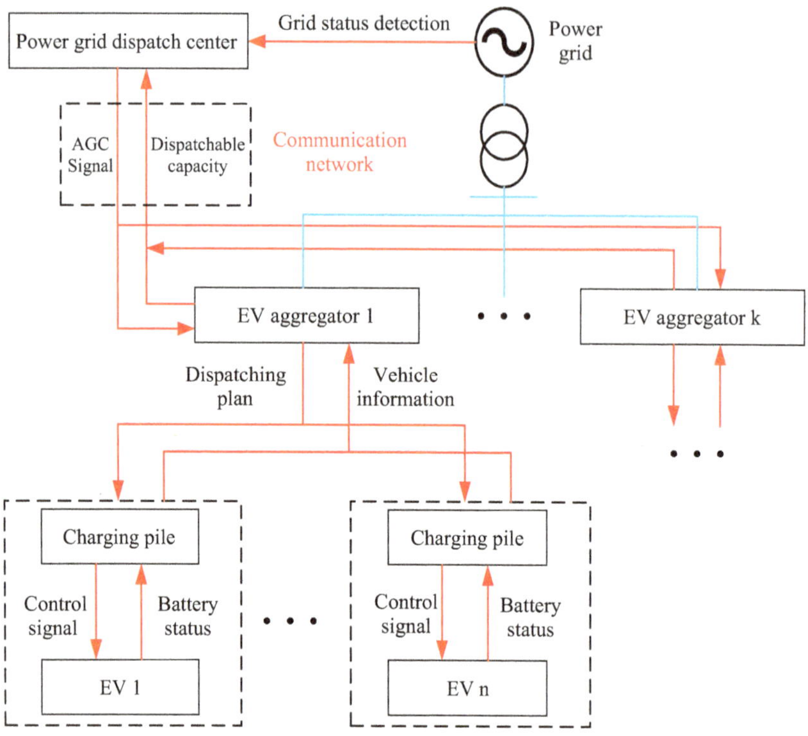

Figure 6.6 Information exchange of EVs in frequency regulation

loads can be supplied quickly to improve the reliability of the power grid. In the grid-connected mode, in addition to cutting off the charging loads of EVs, EVs can exert their V2G characteristics as the power source to supply power to the outage area. This mode digs the potential of EV participating in ancillary services more deeply and can complete the emergency support service better [14].

In addition, it can also play the emergency support function of emergency EVs. Emergency EVs contain large-capacity energy storage batteries. Compared with ordinary EVs, they have more power storage and higher output power. They can be used as power sources to maintain the stability during outages. Because of their mobile feature, they can be allocated to the required nodes for charging and discharging flexibly during failure recovery.

6.4 The mode of EVs participating in DR

In order to improve the power grid's regulation and control of EVs in DR, a large-scale cluster modeling of EVs is proposed in the multi-station and single-station forms.

For the multi-station mode, we propose an efficient organization and management structure, and a charging optimization strategy with the goal of minimizing the overall cost of the grid in the angle of the power grid. For the single-station mode, we propose a charging optimization strategy that aims at the maximum operating profit of the station at the angle of the charging station.

6.4.1 Research on multi-station mode participating in power grid DR

In order to quantify the effect and value of EVs participating in grid DR, this section considers that the output plan of the grid units needs to be formulated in advance. With the goal of minimizing the grid's comprehensive costs, the energy optimization algorithm of the day-ahead stage, as well as efficient solving method are proposed.

6.4.1.1 The DR mechanism for EVs in multi-station mode

In order to improve the efficiency of information transmission and charging control between the power grid and EVs, aggregators are generally used as middlemen to uniformly manage one or more charging stations. The aggregator, on behalf of the grid, issues DR invitations to EV owners before the DR project starts and then obtains control right over EVs charging from those who agree to participate in the DR, which is called *flexible EV load* (Figure 6.7). For those EVs that do not participate in DR (*inflexible EV load*), the aggregator is simply a power supplier [15].

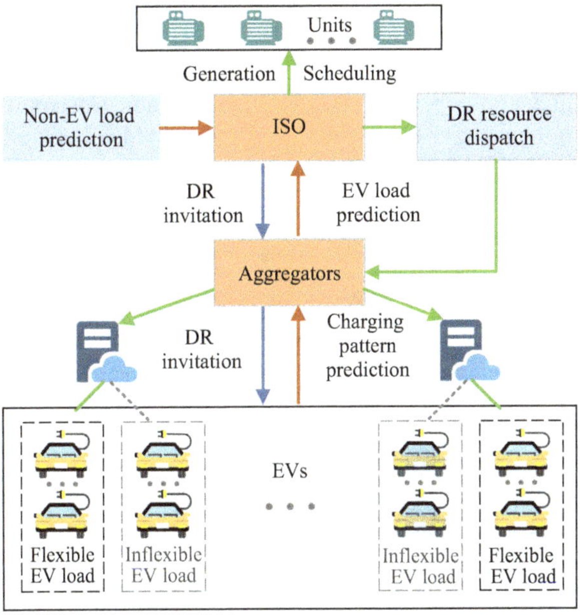

Figure 6.7 Structure of DR mechanism for EVs in multi-station mode

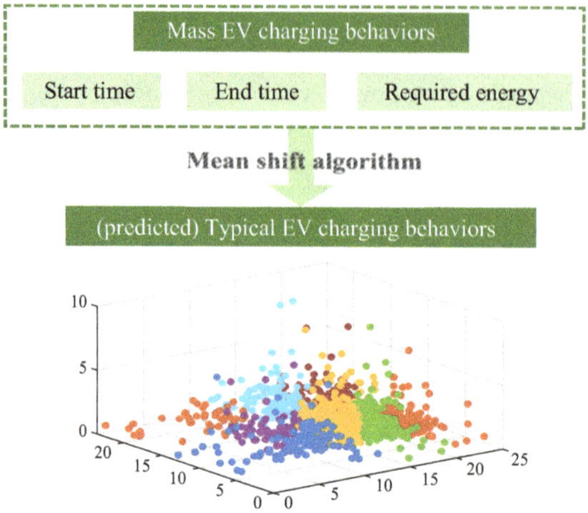

Figure 6.8 Extraction of EV charging behavior cluster

6.4.1.2 Extraction of EVs charging behavior cluster

EV charging behavior can be described by a set of parameters ($t^{\text{plug-in}}, t^{\text{plug-out}}, E^{\text{ch}}$), which are composed of start charging time $t^{\text{plug-in}}$, end charging time $t^{\text{plug-out}}$, and required energy E^{ch} (determined by EV initial SOC, end SOC and battery capacity) (Figure 6.8). This set of parameters can be calculated by aggregators based on historical charging data. EV charging behavior clusters can be extracted from all EV parameters and described by a new set of parameters ($t^{\text{plug-in}}, t^{\text{plug-out}}, E^{\text{ch}}, v$). The main difference lies in the number of EVs added into the cluster (v). N_{EV} EVs are managed by N_{A} aggregators separately and the sth aggregator clusters its EVs into Π_s typical EV charging patterns ($s = 1, 2, \ldots, N_{\text{A}}$). Clustered EV charging patterns are presented as ($t^{\text{plug-in}}_{s,\pi}, t^{\text{plug-out}}_{s,\pi}, E^{\text{ch}}_{s,\pi}, v_{s,\pi}$). Then

$$\sum_{s=1}^{N_{\text{A}}} \sum_{\pi=1}^{\Pi_s} v_{s,\pi} = N_{\text{EV}} \tag{6.9}$$

Extraction of EV charging behavior cluster can significantly decrease the dimensionality of the optimization model, while the necessary information between aggregators and the ISO can be simplified meanwhile.

6.4.1.3 Rolling optimization process

Generation scheduling is generally set 24 h in advance, but EVs are usually charged from night to the next morning, which affects the 2-day unit generation scheduling [16]. In order to jointly optimize the charging process of EVs and unit scheduling for two consecutive days, rolling optimization process is utilized.

Figure 6.9 depicts the timeline of the rolling optimization process for ISO to dispatch *flexible EV load* and generation scheduling. The timeline consists of a moving

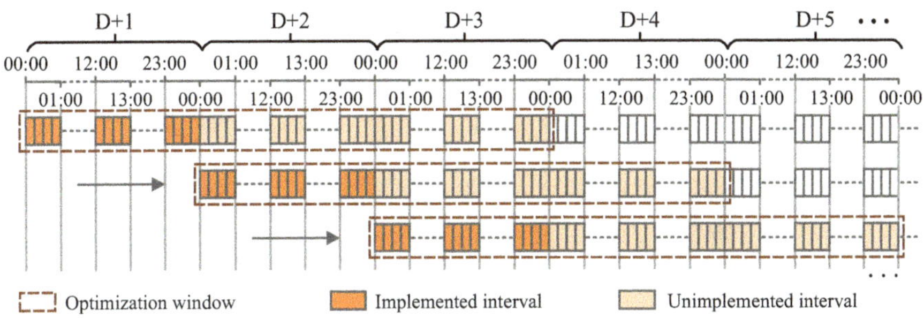

Figure 6.9 Timeline of the rolling optimization process

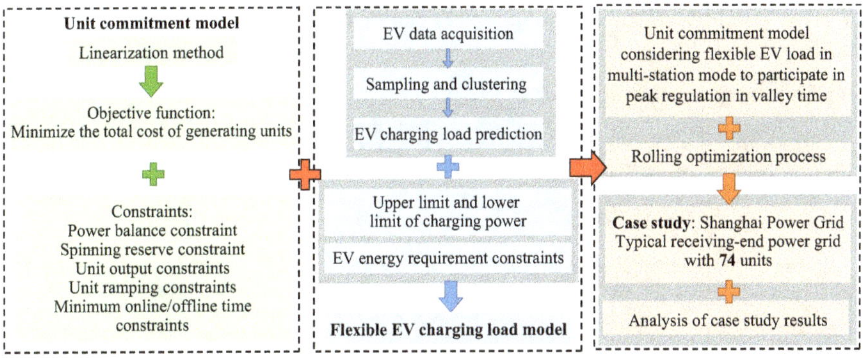

Figure 6.10 The framework of the day-ahead model

optimization window, which includes implemented intervals and unimplemented intervals. The detailed steps are as follows [17]:

(a) Measure values of variables at the initial time and predict them in the remaining time period.
(b) ISO schedules resources in an advanced optimization window through the unit commitment model and executes the optimization strategy in the implemented intervals.
(c) The state of the power system is constantly updated, waiting for the next optimization window. Repeat (a) and (b).

6.4.1.4 Charging optimization model in the day-ahead stage

(1) Objective function

$$\min \sum_{i=1}^{N_G} \sum_{t=1}^{T} (cp_{i,t} + cu_{i,t} + cd_{i,t}) + c_{EV} \sum_{t \in S_T} P_{EV,t} \tag{6.10}$$

The objective function in (6.10) is to minimize the comprehensive costs in the power grid, which is composed of fuel costs, startup costs, shutdown costs and peak regulation subsidies for EV users (Figure 6.10). In (6.10), N_G is the number of units; T is the intervals in a rolling optimization window for peak regulation; $cp_{i,t}$, $cu_{i,t}$, and $cd_{i,t}$ are the generation cost, startup cost, and shutdown cost of unit i at time t, respectively; c_{EV} is the unit peak regulation subsidy; $P_{\mathrm{EV},t}$ is the total flexible EV load:

$$cp_{i,t} = \mathrm{CA}_i p_{i,t}^2 + \mathrm{CB}_i p_{i,t} + \mathrm{CC}_i u_{i,t} \tag{6.11}$$

$$cu_{i,t} = y_{i,t}\mathrm{CStart}_i \tag{6.12}$$

$$cd_{i,t} = z_{i,t}\mathrm{CShut}_i \tag{6.13}$$

where $p_{i,t}$ is the generating power of unit i at time t; $u_{i,t}$ is 0-1 binary variable. When the unit is online/offline, $u_{i,t}$ is 1/0. CA_i, CB_i and CC_i are fuel cost parameters of unit i. $y_{i,t}/z_{i,t}$ is the binary variable. When the unit i startup/shut down at time t, $y_{i,t}/z_{i,t}$ is equal to 1. $\mathrm{CStart}_i/\mathrm{CShut}_i$ is startup cost/-shutdown cost for one time. Through the piecewise linearization method, (6.11) can be rewritten as a linear constraint for simplification [18].

(2) Commitment constraints

$$u_{i,t} - u_{i,t-1} = y_{i,t} - z_{i,t} \tag{6.14}$$

$$y_{i,t} + z_{i,t} \leq 1 \tag{6.15}$$

$$\sum_{k=\max(t-T_{\mathrm{on},i}+1,1)}^{t} y_{i,k} \leq u_{i,t} \tag{6.16}$$

$$\sum_{k=\max(t-T_{\mathrm{off},i}+1,1)}^{t} z_{i,k} \leq 1 - u_{i,t} \tag{6.17}$$

$$u_{i,t}P_i^{\min} \leq p_{i,t} \leq u_{i,t}P_i^{\max} \tag{6.18}$$

$$p_{i,t} - p_{i,t-1} \leq u_{i,t-1}P_{\mathrm{up},i} \tag{6.19}$$

$$p_{i,t-1} - p_{i,t} \leq u_{i,t-1}P_{\mathrm{down},i} \tag{6.20}$$

Constraint (6.14) ensures the consistency of start-up variables, shut down variables and state variables. In constraint (6.15), one unit is restricted to start and shut down simultaneously. Constraints (6.16) and (6.17) are the minimum startup and shutdown time constraints, where $T_{\mathrm{on},i}/T_{\mathrm{off},i}$ is the minimum startup/shutdown time. Constraint (6.18) shows the upper limit and lower limit of power generation $p_{i,t}$. Unit ramping constraints are shown in constraints (6.19) and (6.20) and $P_{\mathrm{up},i}/P_{\mathrm{down},i}$ is the maximum ramp-up/ramp-down rate of unit i.

(3) System constraints

$$\sum_{i=1}^{N_{\mathrm{G}}} p_{i,t} = P_{\mathrm{D},t} + P_{\mathrm{EV},t} + P_{\mathrm{loss},t} \tag{6.21}$$

$$\sum_{i=1}^{N_{\mathrm{G}}} u_{i,t}P_i^{\max} \geq P_{\mathrm{D},t} + P_{\mathrm{EV},t} + R_t \tag{6.22}$$

The power balance of the system is ensured in constraint (6.21), where non-EV load and inflexible EV load are included in $P_{D,t}$; $P_{EV,t}$ represents charging load involved with DR (*flexible EV load*); network losses at time t is considered as $P_{loss,t}$. The spinning reserve requirement at time t is represented in constraint (6.22), where R_t is the spinning reserve capability.

(4) Aggregator-based DR constraints

$$\eta \sum_{t \in T_{s,\pi}} p_{s,\pi,t} \Delta t = v_{s,\pi} E_{s,\pi}^{ch} \quad (s,\pi) \in S_{flexible}^{EV} \tag{6.23}$$

$$v_{s,\pi} P_{s,\pi}^{min} \leq p_{s,\pi,t} \leq v_{s,\pi} P_{s,\pi}^{max} \quad (s,\pi) \in S_{flexible}^{EV} \tag{6.24}$$

$$P_{EV,t} = \sum_{s=1}^{N_A} \sum_{\pi=1}^{\Pi_s} p_{s,\pi,t} \quad (s,\pi) \in S_{flexible}^{EV} \tag{6.25}$$

$S_{flexible}^{EV}$ is the set of charging behavior clusters of *flexible EV load*. For the sth aggregator, its πth cluster of EV charging behavior is represented as (s, π). Constraint (6.23) ensures that the energy required by EV charging behavior clusters is satisfied, where the plug-in period $T_{s,\pi}$ is determined by start charging time $t^{plug\text{-}in}$ and end charging time $t^{plug\text{-}out}$; η is the charging efficiency; $p_{s,\pi,t}$ is the total charging power of cluster (s, π) at time t; Δt is the optimization time step. Equation (6.24) shows the upper limit and lower limit of charging power. All the flexible EV load is aggregated into $P_{EV,t}$ in (6.25).

6.4.2 Research on single-station mode participating in power grid DR

EVs can participate in DR in the single-station mode. The charging station controls the charging situation of the EVs charged at the charging station to participate in DR. Stopping charging and sending power back to the grid are the common methods to provide flexibility. In the single-station mode, EV load organized by the charging station can participate in power grid response effectively.

6.4.2.1 Typical structure of charging station with battery energy station

With the decline of battery cost, it is expected that the adaptability of battery energy stations (BES) will be wider [19]. Therefore, in the single station mode, the charging station with BES can participate in emergency DR (EDR) better.

The structure diagram of a typical charging station with BES participating in EDR is shown in Figure 6.11.

Different units have corresponding tasks. First, the duty of the utility company is to provide power to the charging station and send signals of EDR requirements. Second, the charging station operator is in charge of the optimal operation of EV and BES. Under normal conditions, BES is charged during off-peak hours at night and serves as a standby generator set only during EDR events. When an EDR event occurs, the utility company will issue important parameters, like EDR event duration and participation bonus, to invoke load shedding and the charging station operator

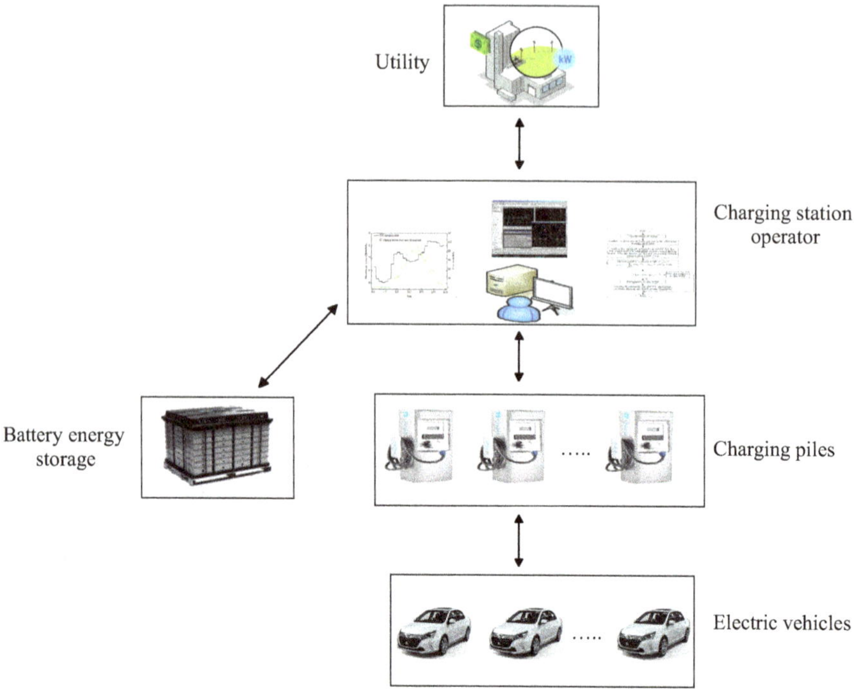

Figure 6.11 Structure diagram of a typical charging station with BES participating in EDR

must respond within a certain period [20]. In addition, the charging station operator needs to decide whether to participate in the EDR event according to the decision algorithm [10].

6.4.2.2 Charging station EV load modeling

(1) M/M/c queueing model

The $M/M/c$ queuing model can be used to estimate EV charging load [21]. In the above Kendal symbol, the first M represents the distribution of EV arrival rate. In this paper, the arrival rate of EVs should follow Poisson process to solve the randomness of an EV. The second M represents the service time distribution calculated by the EV battery charging behavior, and the c represents the number of charging piles in the charging station. The queuing model of the charging station is shown in Figure 6.12. The arrival rate of EVs follows Poisson distribution and μ indicates the number of EVs that can be serviced per unit time.

In queueing model, (6.26) shows the probability of n EVs charging simultaneously in queueing model:

$$P_t(n) = \begin{cases} \dfrac{\rho^n}{n!} P_t(0), & n < c \\ \dfrac{\rho^n}{c! c^{n-c}} P_t(0), & n \geq c \end{cases} \tag{6.26}$$

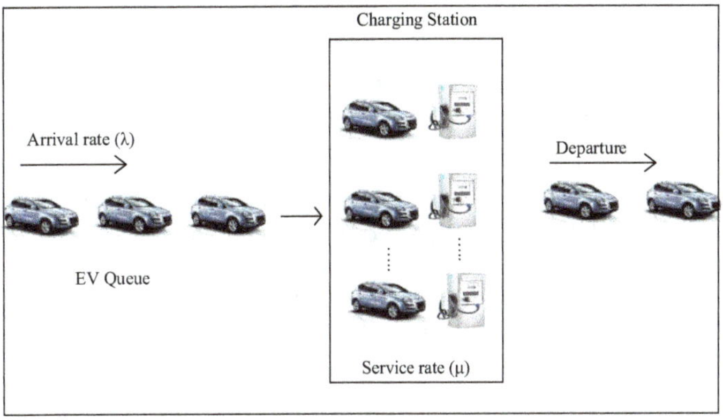

Figure 6.12 Schematic diagram of queuing theory

where $P_t(0) = \left[\sum_{n=0}^{c-1} \frac{\rho^n}{n!} + \frac{\rho^c}{c!}\frac{1}{(1-\rho/c)}\right]^{-1}$ and $\rho = \frac{\lambda_t}{\mu_t}$. ρ is the probability that there is at least one EV in the charging station. λ_t is the arrival quantity of EV per unit time and is the number of EV that can be serviced per unit time in hour t respectively. In order to ensure the rationality of the queuing system, the occupancy rate of the charging station (ρ/c) shall be less than 1.

(2) Charging behavior of EV

At the beginning of each charging phase, the state of charge (SOC) of the EV battery is determined in (6.27):

$$SOC_y = 1 - \frac{E_{C_y}}{C_{Bat_y}} \tag{6.27}$$

where E_{c_y} is the daily charging energy of EV in each charging stage, and C_{Bat_y} is the capacity of EV battery. The charging time T required for an EV is given by the formula (6.28):

$$T = \frac{SOC_y - b_l}{a_l} \tag{6.28}$$

where a_l and b_l can be obtained from piecewise linearizing of the battery charging behavior as explained in [21]. The (6.29) gives the charging current $I_{t,y}$ consumed by the EV within the charging time T in hour t:

$$I_{t,y} = \min\left(\frac{E_{C_y}}{VT}, I_{\text{Max}}\right) \tag{6.29}$$

where I_{Max} is the upper limit of charging current, and V is the charging voltage constant.

Therefore, the total charging power of n EVs is given by (6.30) by accumulating:

$$P_{EV,t} = \sum_{m=0}^{n} I_{m,t,y} V \tag{6.30}$$

For all possible n values, the total expected EV charging demand at time t is shown in (6.31):

$$E[P_{EV,t}] = \sum_{n=0}^{c} P_t(n)P_{EV,t} \tag{6.31}$$

Therefore, the total EV charging demand obtained will be used as the input of the next part of the EDR participation decision-making model.

6.4.2.3 Optimization problem for charging station with battery energy station participating in emergency DR

(1) Optimization model

When an EV charging station with BES is selected to participate in EDR, the charging station operator has the right to choose whether to participate in EDR events according to the decision algorithm. BES plays the role of improving the economy and flexibility of EV charging stations during EDR events.

The objective function of the optimal dispatching model is to maximize the operating profit of the charging station. In (6.32), the first item is the revenue from charging services. The second item is the profit from participating in EDR. The last term is the cost of purchasing electricity from the grid.

$$\max\left(\sum_{t\in S_H} p_{\mathrm{EV}}(t)P_{\mathrm{EV}}(t) + \sum_{t\in S_H} p_{\mathrm{EDR}}(t)\Delta P_{\mathrm{EDR}}(t) - \sum_{t\in S_H} p_{\mathrm{Grid}}(t)P_{\mathrm{Load}}^{\mathrm{EDR}}(t)\right) \tag{6.32}$$

where $P_{EV}(t)(=E[P_{EV,t}])$ represents the actual EV charging power in the corresponding time, and $p_{EV}(t)$ represents the price of the real-time. $\Delta P_{EDR}(t)$ corresponds to the EV that participate in EDR, and $p_{EDR}(t)$ is the incentive price for utility companies to participate in EDR. $P_{Load}^{EDR}(t)$ is the net load of the EV charging station during the EDR event, and $p_{Grid}(t)$ is the real time electricity price of the power grid. S_H is a set of discrete-time steps, where t_s is the start time, and H is the time span:

$$S_H = \{t_s, t_s + 1, \ldots, t_s + H - 1\} \tag{6.33}$$

The following (6.34)–(6.44) constitute the constraints of the optimization model:

$$\Delta P_{\mathrm{EDR}}(t) = P_{\mathrm{Load}}^{\mathrm{Fore}}(t) - P_{\mathrm{Load}}^{\mathrm{EDR}}(t) \quad t \in S_H \tag{6.34}$$

$$\Delta P_{\mathrm{EDR}}(t) \geq \Delta P_{\mathrm{EDR}}^{\mathrm{min}}(t) \quad t \in S_H \tag{6.35}$$

$$P_{\mathrm{Load}}^{\mathrm{EDR}}(t) \geq 0 \quad t \in S_H \tag{6.36}$$

$$P_{\mathrm{EV}}(t) = P_{\mathrm{Load}}^{\mathrm{EDR}}(t) + P_{\mathrm{BES}}(t) t \in S_H \tag{6.37}$$

where $P_{\mathrm{Load}}^{\mathrm{Fore}}(t)$ is the adjusted EV charging station baseline load prior; $P_{\mathrm{Load}}^{\mathrm{EDR}}(t)$ is the actual load during EDR events. $\Delta P_{\mathrm{EDR}}^{\mathrm{min}}(t)$ is the minimum load reduction signal from the utility. Constraint (6.36) restricts $P_{\mathrm{Load}}^{\mathrm{EDR}}(t)$ is nonnegative because the vehicle to grid technology is not considered in the EV charging station. Constraint (6.37) explains the relationship between $P_{\mathrm{EV}}(t)$, $P_{\mathrm{Load}}^{\mathrm{EDR}}(t)$ and the BES output $P_{\mathrm{BES}}(t)$:

$$P_{\mathrm{BES}}(t) = P_{\mathrm{BES}}^{\mathrm{dis}}(t) - P_{\mathrm{BES}}^{\mathrm{ch}}(t) \quad t \in S_H \tag{6.38}$$

$$0 \leq P_{\text{BES}}^{\text{dis}}(t) \leq P_{\text{BES}}^{\text{dis,max}} v_{\text{dis}}(t) \quad t \in S_H \tag{6.39}$$

$$0 \leq P_{\text{BES}}^{\text{ch}}(t) \leq P_{\text{BES}}^{\text{ch,max}} v_{\text{ch}}(t) \quad t \in S_H \tag{6.40}$$

$$v_{\text{dis}}(t) + v_{\text{ch}}(t) \leq 1 \quad t \in S_H \tag{6.41}$$

$$SOC(t) = SOC(t-1) - \frac{\eta^{\text{dis}} P_{\text{BES}}^{\text{dis}}}{C_{\text{rated}}} + \frac{\eta^{\text{ch}} P_{\text{BES}}^{\text{ch}}}{C_{\text{rated}}} \quad t \in S_H \tag{6.42}$$

$$SOC^{\min} \leq SOC(t) \leq SOC^{\max} \quad t \in S_H \tag{6.43}$$

$$p_{\text{EV}}(t) = K_{\text{EV}} p_{\text{Grid}}(t) t \in S_H \tag{6.44}$$

Constrains (6.38)–(6.44) are the BES constrains. Where $P_{\text{BES}}^{\text{dis}}(t)/P_{\text{BES}}^{\text{ch}}(t)$ is the discharge/charge power of the BES; $P_{\text{BES}}^{\text{dis,max}}/P_{\text{BES}}^{\text{ch,max}}$ is the BES discharge/charge power upper bound; $v_{\text{dis}}(t)/v_{\text{ch}}(t)$ is the binary discharge/charge state variable; $\eta^{\text{dis}}/\eta^{\text{ch}}$ is the discharge/charge efficiency parameter; $SOC^{\min}/SOC^{\max}$ is the lower/upper bound of $SOC(t)$. Constrain (6.41) is a logical formula to prevent the BES from being operated in charging and discharging modes simultaneously. Equation (6.42) constrains the relationship of the SOC of the BES between two adjacent discrete time steps. $p_{\text{EV}}(t)$ is the price of EV charging service. To break even the operation and maintenance cost of EV charging station along with maintaining operating profit, the constant K_{EV} is usually set higher than 1. $p_{\text{EDR}}(t)$ is the incentive price offered by the utility.

Therefore, combined with the objective function and constraints, the optimal decision optimization model for a charging station with BES participating in EDR can be formed. Because all constraints are linear constraints with binary variables, the model can be effectively solved by commercial solvers as a mixed-integer linear programming (MILP) model.

(2) Optimization results

To study the impact of incentive price of EDR, the detailed parameters of two different EDR events are shown in Table 6.3 [22]. The EDR event will be notified at 16:00 and the duration is 4 h. The difference between cases 1 and 2 is the inductive price of EDR. The response time is set to 1 h for the decision of the charging station operator. Figure 6.13 shows the profit of the charging station when the inductive price is 85 \$/MWh and 150 \$/MWh. The results show that when the incentive price is high, the charging station operator profit enough to respond to the EDR signal sent by the power company.

Table 6.3 Parameters of EDR cases

Parameter	**Case 1**	**Case 2**
EDR notification time	16:00	16:00
EDR decision making duration	1 hour	1 hour
EDR event time	17:00–20:00	17:00–20:00
$p_{\text{EDR}}(t) \quad t \in S_H$	85 \$/MWh	150 \$/MWh
$S_{\text{OC}}(t_s - 1)$	0.85	0.85

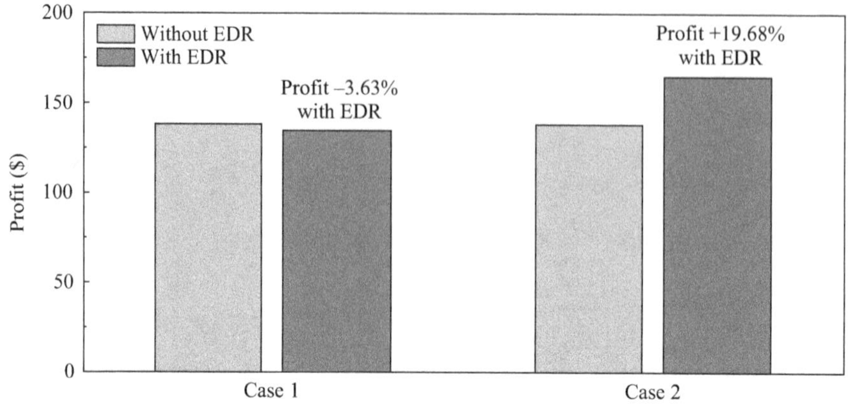

Figure 6.13 Profit comparison when the inductive price varies

6.5 Practical experience on EVs participating in DR

As the world sees the rapid growth of EVs and their tremendous charging loads, DR programs are conducted in many countries and regions to coordinate the charging and to meet the requirements of the power system. The demand information in the power system is gathered in the operation units, and the demand signals are published to potential responsible agents. EV charging loads aggregated as a responsible agent, are adjusted upward or downward.

In practice, pilot projects are executed and can provide experience in different electricity markets, while these experiments can be divided into structural pilot projects and event pilot projects. The flexibility of charging demand and the response characteristics of customers are analyzed through these projects as experimental references.

6.5.1 DR pilot projects – in structural mode

Structural pilot projects launch the DR programs periodically, e.g., in a specific period of every single day. Daily DR projects aim to shift the electricity loads in peak shaving and valley filling. The DR signals in these periodical projects are price curves in general, and the revenue settlements are implemented monthly or yearly. The price mechanism can be in three categories: time-of-use tariffs (TOU), critical peak pricing (CPP), and real-time pricing (RTP) [23].

As the DR programs are implemented in a period and the price information is openly published, structural DR projects allow responsible agents to draw up specific user plans, e.g., monthly contracts, service packages to EV groups. The structural features are reflected in categorized customers and their administrable response behaviors.

6.5.1.1 Projects organized by the State Grid Corporation of China

As the world's largest utility, State Grid Corporation of China (SGCC) supplies power to a 1.1 billion population with a service area covering 88% of Chinese territory. State Grid EV Service Co., Ltd (SGEV) is one of the subsidiaries directly managed by State Grid and conducts business in 14 provinces in China [24]. The services of SGEV include EV charging management, green electricity trade and interaction between EV aggregators and power systems. Routine DR projects setting specific periods in one day as DR time are one of the methods to improve its service quality and reduce operation cost.

- DR area: Zhejiang Province, Shanxi Province, Tianjin City, in China.
- DR period: 00:30–07:00, 12:30–16:00, each day.
- DR parts: (A) SGEV, as a schedulable branch of grid operator; (B) EV aggregator; (C) individual EV user (provided charging service by A or B).
- DR policy: (1) a receiving dispatch command (to regulation total charging power) and reporting total adjustable power; (2) setting DR time and non-DR time, and making a contract between A and B/C, which permits A to provide peak load regulation service; (3.1) adopting lower electricity tariffs to encourage B/C to charge in DR time; (3.2) noticing real-time charging tariff discounts to B/C according to the dispatch command in DR time (in Shanxi Province); (4) B organizing the charging of aggregated EVs, and reporting its adjustable power to A in line with the contract. The full view of the policy is concluded in Figure 6.14.
- DR performance: Providing peak load regulation service to the power system. Average DR participating electricity per month in 2021: Tianjin city – 43,000 kWh; Shanxi Province – 450,000 kWh; Zhejiang Province – 1,364 kWh.

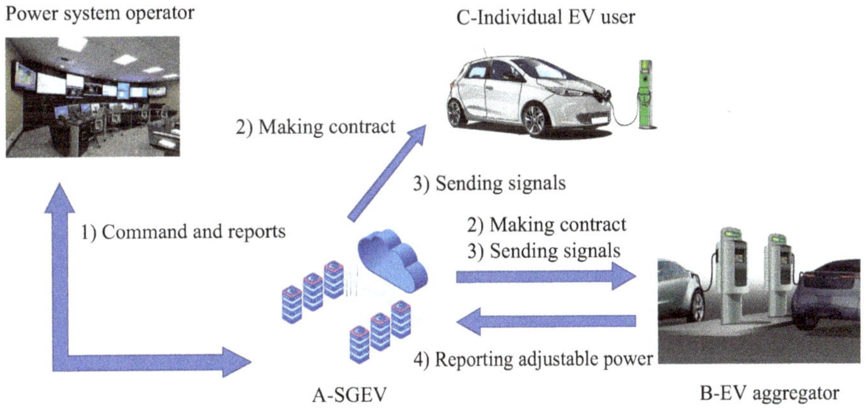

Figure 6.14 DR policy in projects organized by SGCC

6.5.1.2 Project in north China – V2G in peak load regulation market

Vehicle-to-Grid (V2G) technology enables bi-directional energy transfer between EVs and power grids. In April 2020, China officially included V2G charging pile resources into the north china ancillary service market, which enables EVs to be aggregated and discharged according to price signals [25].

- DR area: North China (Beijing-Tianjin-Tangshan area).
- DR period: All day.
- DR parts: (A) North China ancillary service market; (B) EV aggregator (able to provide no less than 10 MW/30 MWh regulation capacity); (C) Individual EV user.
- DR policy: (1) A announcing the peak load regulation tariff and the regulation demand after real-time dispatch; (2) A and B communicating via a coordinated dispatching platform; (3) B aggregating C with user agreements, which permits the supply of electricity to power grids.
- DR performance: 2430 charging stations/battery swapping stations being involved in the project, including around 24,000 charging piles. The potential EV participating quantity is up to 400,000 [25].

6.5.1.3 ChargeForward program in the USA

The ChargeForward program is open to BMW EV/PHEV owners who is also customers of Pacific Gas and Electric Company (PG&E) customers, which has kicked off its third phase. This collaboration launched in 2015 has been one of the longest-running partnerships between an electric utility and an automaker. This project aims to test the ability of EVs to support the electric grid and provide benefits to customers through vehicle-grid-integration applications that enable smart charging and DR [26].

- DR area: Northern California, USA.
- DR period: All day. (last for 24 months in phase 3).
- DR parts: (A) Utility – PG&E; (B) EV aggregator – BMW; (C) BMW EV owner.
- DR policy: (1) A sending an alert to B, indicating the quantity and duration to curb load when cutting demand emerges at any time of day; (2) B going to signal the telemetry equipment in each participating vehicle, and going to halt the charging for the duration of the event; (3) C going to get compensations or incentives through its participation, amounting to $150 at sign-up and an annual price of $250.
- DR performance: Around 3,000 EVs will participate in the DR project (100 in phase 1, and 400 in phase 2). Renewable energy usage can be more than double through the DR project, and over 1 million miles were powered by 100% renewable energy charging during a one-year period [26].

6.5.2 DR pilot projects – in event mode

Event pilot projects launch the DR programs in a specific period separately according to the notice of the power system operator. Compared to the structural DR projects, event projects see a lower implementation frequency, and often show seasonal characteristics and the characteristics of the electricity load in holidays. In contrast, this kind of DR projects generally cover a longer duration, e.g. tens or hundreds of hours. Hence, event DR projects are able to provide various services, including but not limited to peak-load regulation and emergency support.

As an independent program, power system operators offer incentives or compensations to responsible agents in event DR projects, other than the general electricity tariffs. This type of DR is open to various participants, while EV aggregators are able to self-organize the charging loads based on the DR signal. The policy and backgrounds of event DR projects are different, and there exists no fixed mechanism.

6.5.2.1 Serial pilot projects in Shanghai

As an economically developed metropolis in China, Shanghai sees a huge peak-valley difference in electricity consumption, while the city relies on external electricity and renewable energy in the power supply side. DR is one of the effective countermeasures to solve these problems, and Shanghai is the first city to implement DR projects nationwide. Current DR programs, including peak-shaving and valley-filling services, are assessed in different time scales, i.e. medium and long term mode, intra-day mode, and fast response mode. During 2015–18, Shanghai conducted a series of eight short-team DR projects, and involves increasing areas and participants.

EVs were first included in the DR project in 2019, as Shanghai saw an EV population of 300,000 in this year. An aggregator-based structure was designed, where invited EV aggregators are responsible for organizing users under the coordination of the municipal control center. The organization structure is demonstrated in Figure 6.15.

Three DR projects in 2019 with the participation of EVs were conducted at 2 am–5 am, 7 June, 12 am–2 pm, 9 August, and 10 am–11 am, 5 December, respectively [27]. In these projects, EVs responded to the power grid signals collaborating with other flexible loads, e.g., energy storage, air conditioner. Although the duration of projects was not long, various characteristics of EV participating in DR were evaluated in the serial projects, e.g., the advantages compared with other loads, the characteristics of different kinds of charging methods in DR, and the performance comparison in peak-shaving and valley-filling. Comparative programs were executed in 2021, which

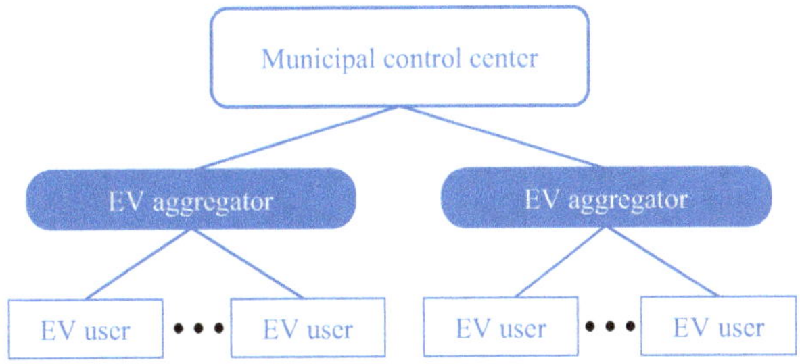

Figure 6.15 Organization structure of DR projects in Shanghai

retains the organizational structure of those in 2019. The compensation standard was changed, and users' response was analyzed.

- DR area: Shanghai, China.
- DR period: Several hours.
- DR parts: (A) virtual power plant platform; (B) EV aggregator (no capacity limit, least participation duration 1 h), including private charging piles, specific charging piles and battery swapping stations; (C) EV users.
- DR policy: (1) Registration: A sending the project plan to B in advance, while B inviting C to sign up for the DR project; (2) preliminary report: B predicting the DR capacity curves with the sign-up information, and reporting it to A at preliminary stage (including medium-term curves and day-ahead curves); (3) online report: B sending guiding signals to C based on the charging plan, collecting the real-time charging information, and reporting it to A in every 15 min; (4) Settlement: A paying B in line with the incentive standard and the DR performance, while B compensating C based on the sign-up agreement.
- DR incentive standard: The DR performance was evaluated with the baseline of loads and the real consumption, while the DR incentive was calibrated with impact factors. Impact factors include electricity quantity in DR, response duration in DR, response efficiency in DR and participation ratio. The reference value of the incentive is 30 CNY/kW for peak-shaving and 12 CNY/kW for valley-filling.
- DR performance: (1) Private charging piles are fit for providing valley-filling service, as the charging electricity increased to 7.8 times of base value in the project. The profit of single participation was more than 14 CNY. (2) Specific charging piles are fit for providing peak-shaving service, as the charging electricity decreased to 25% of the base value in the project. The profit of single participation was more than 23.5 CNY. (3) Battery swapping stations are fit for providing peak-shaving service, as the charging electricity was reduced by 81.2% at maximum in the project. And battery swapping stations showed advantages in response efficiency. The profit of a single station for one participation was more than 500 CNY [27].

6.5.2.2 Project helping release California's rolling blackouts

Extreme weather continues to intensify due to climate change, e.g., continuing extreme heat. As power grids decarbonize, a greater volume and faster deployment of these resources are required to keep the power supply at nights. Charging service providers participate in utility and grid-sponsored DR programs to monetize the flexibility of aggregated EV charging loads. In the summer of 2020, record heat in California led the state's grid operator to call on residents to reduce electricity usage by issuing a "flex alert." In consequence, California experienced rolling blackouts for the first time in 19 years, with nearly a million residents affected. The smart charging piles were activated automatically and responded to the emergency demand.

- DR area: California, USA.
- DR period: 3.6 h/day in total, standby all day [28].

- DR parts: (A) Power system operator – CAISO; (B) EV aggregator – Enel X; (C) EV user.
- DR policy: (1) C installing application software "JuiceNet" and agreeing to hand over the charging control right to B; (2) B able to aggregate charging piles into a virtual power plant with capacity of automatic DR and to bid in the spot market. (3) A announcing DR signals (2,750 times per month on average) and issuing a "flex alert" in an emergency.
- DR performance: (1) The JuiceNet customers participate in the DR project around 6,000 times per month in the common market; (2) during the rolling blackouts, the average dispatch event duration to JuiceNet customers was just over 3.6 h per day, with event participation rates above 90% [28].

6.5.3 Practical experience

The major contribution of EV in current DR programs is peak load regulation, while various types of services, e.g., frequency regulation, emergency support, can be provided in the future. V2G technology can improve the performance in DR.

With a high-performance and small-capacity battery, EVs are generally aggregated by a charging service provider to participate in DR programs, while individual EV users could also participate in the programs directly. Diverse charging infrastructures and aggregators are expected to be involved in DR programs. Different subjects show advantages in different kinds of grid services. The carbon emission reduction effect has been tested in pilot projects, and EVs will play an important role in participating in DR jointly with renewable energy.

The market mechanism of the DR program should be transparent, and an optimal incentive mechanism or a price system needs to be studied. Specified DR policy is recommended to be designed for different scenarios, different participants and different regulation targets. The research on users' behaviors and the game model among interest bodies will be an important preliminary study. Intelligent equipment and user education are crucial for large-scale EVs participating in DR programs. In terms of hardware, intelligent equipment provides users a convenient participation approach, which has a great influence on users' participation will. It also provides aggregators timely information interaction approach and data for accurate predictions. In terms of user education, the advertising effect can expand the scale of imitators and promote DR programs.

6.6 Summary and prospect

6.6.1 Summary

The focus of this chapter, EV massive resources mining and DR, is illustrated from four aspects in the above sections. The DR technology of massive EVs is thoroughly discussed, according to the basic conditions for its development, the technological capability and potential evaluation, the implementation mode, and the practice experience of this technology in real scenarios.

On top of this analysis, the feasibility of this technology is fully demonstrated. The charging facilities of massive EVs have great potential to play a crucial role in the future power system, as the largest distributed energy storage in the whole society. Thus, current work should be concentrated on technology iteration, optimization, further deep application and improvement of relevant mechanism construction and industry supervision.

6.6.2 Prospect

According to some cases above, some future development directions for EVs to participate in DR are put forward.

6.6.2.1 Technology breakthrough of EVs

There are many cases abroad to promote EVs as a flexible resource of the power system. The implementation mode and scale of cases in different regions are different. Generally speaking, V1G programs are essentially a type of demand-side management strategy, which many utilities compensate customers for their participation. By agreeing to shift their energy use to the less in-demand hours, utilities can offer financial incentives, cheaper energy prices, and other benefits. V2G or V2X means that the energy between vehicle and network or vehicle and X can flow in both directions. The technical requirements and business model of V1G are relatively simple, and there have been many successful cases, but the sustainable business model and user participation are still the problems to be solved. For V2G, most of the cases are still in the trial stage. So the technology of V2G should be broken through.

6.6.2.2 Development of charging facilities

The development of charging facilities is an indispensable basis for the interaction between vehicles and networks. For example, private charging piles in many single storey houses in the United States are easy to mobilize participation, but the development of public charging infrastructure, especially rapid charging infrastructure, is uncertain, and its utilization scenario prediction is difficult. So in order to develop the DR, the location and capacity of the charging facilities should be planned reasonably. And the technology of charging power control level should be improved. More precise control in a larger power range can help EVs participate in DR effectively.

6.6.2.3 Market mechanism on the user side

In addition to the technological breakthrough of EVs and charging facilities, the market mechanism on the user side is also one of the key factors. In order to answer the questions of how to effectively utilize and manage the charging load of EVs and which types of EVs, analyzing the charging load characteristics of users is necessary. By considering user response characteristics to electricity prices or control signals, user wishes and other factors make a series of subsidy policies and establish a precision DR model to help EVs participate in DR effectively.

6.6.2.4 Exploration of more possibilities for EVs in DR

By 2030, China promises to reduce carbon dioxide emissions per unit of GDP by 60–65% compared with that in 2005, and increase the proportion of non-fossil energy in primary energy consumption to 20%, reaching the goal of "peak carbon." By 2060, more than half of all energy sources will be produced from non-fossil fuels, while being "carbon neutral." With the "double carbon" targets, China, as the largest EV market in the world, needs to rely on the flexibility of the massive EVs charging/discharging to establish a green, stable and safe energy system. By promoting efficient collaboration between EVs and renewable energy with the "double carbon" targets, it is helpful for wind-PV accommodation. And PV-EES-EV charging stations are good combinations of EVs and renewable energy. Through carrying out more V2G pilot programs, more experience can be concluded and the experience can be used to help EVs participate in more types of auxiliary services.

References

[1] Raghavan. S., Tal G.: 'Plug-in hybrid electric vehicle observed utility factor: Why the observed electrification performance differ from expectations'. *International Journal of Sustainable Transportation*. 2020;**4**:1–46.

[2] Mark Kane.: *World's Top 5 EV Automotive Groups Ranked by Sales: Q1–Q4 2020* [online]. 2021. Available from https://insideevs.com/news/486325/world-top-ev-automotive-groups-2020/ [Accessed Aug 2021].

[3] Xinhua.: *The Ministry of public security released the data of national motor vehicles and drivers in the first half of 2021* [online]. 2021. Available from http://www.gov.cn/xinwen/2021-07/06/content_5622875.htm [Accessed Aug 2021].

[4] EVCIPA.: *Operation status of national electric vehicle charging infrastructure in April 2021* [online]. 2021. Available from http://www.evcipa.org.cn/ [Accessed Aug 2021].

[5] *Expanding private charging infrastructure is key support to ramping-up electric mobility* [online]. 2020. Available from https://www.vda.de/en/press/press-releases/200708-Expanding-private-charging-infrastructure-is-key-support-to-ramp-ing-up-electric-mobility.html [Accessed Aug 2021].

[6] General Office of the State Council.: *Guidance on accelerating construction of the electric vehicle charging infrastructure* [online]. 2015. Available from http://www.gov.cn/zhengce/content/2015-10/09/content_10214.htm [Accessed Aug 2021].

[7] Xinhua.: *China unveils plan for new energy vehicle industry* [online]. 2020. Available from https://www.chinadailyhk.com/article/148125 [Accessed Aug 2021].

[8] Liu Z., Xie Y., Feng D., Zhou Y., Shi S., Fang C.: 'Load forecasting model and day-ahead operation strategy for city-located EV quick charge stations'. *8th Renewable Power Generation Conference (RPG 2019)*; Shanghai, China, Oct 2019, pp. 1–6.

[9] Wang M., Mu Y., Shi Q., Jia H., and Li F.: 'Electric vehicle aggregator modeling and control for frequency regulation considering progressive state recovery'. *IEEE Transactions on Smart Grid*. 2020;**11**(5):4176–89.

[10] Upadhaya B., Feng D., Zhou Y., Gui Q., Zhao X., and Wu D.: 'Optimal decision making model of battery energy storage-assisted electric vehicle charging station considering incentive demand response'. *8th Renewable Power Generation Conference (RPG 2019)*; Shanghai, China, Oct 2019, pp. 1–8.

[11] Qian K., Zhou C., Allan M., and Yuan Y.: 'Modeling of load demand due to EV battery charging in distribution systems'. *IEEE Transactions on Power Systems*. 2011;**26**(2):802–10.

[12] Zhao X., Xu R., Zhou Y., *et al*: 'Aggregator-based demand response mechanism for electric vehicles participating in peak regulation in valley time of receiving-end power grid'. *2020 IEEE/IAS Industrial and Commercial Power System Asia (I&CPS Asia)*; Weihai, China, July 2020, pp. 187–93.

[13] Chen X., Leung K., Lam A., and Hill D.: 'Online scheduling for hierarchical vehicle-to-grid system: design, formulation, and algorithm'. *IEEE Transactions on Vehicular Technology*. 2019;**68**(2):1302–17.

[14] Xu N. and Chung C.: 'Reliability evaluation of distribution systems including vehicle-to-home and vehicle-to-grid'. *IEEE Transactions on Power Systems*. 2016;**31**(1):759–68.

[15] Bessa R., Matos M.: 'Global against divided optimization for the participation of an EV aggregator in the day-ahead electricity market. Part I: Theory'. *Electric Power Systems Research*. 2013;**95**:309–18.

[16] Fang C., Zhao X., Xu Q., Feng D., Wang H., and Zhou Y.: 'Aggregator-based demand response mechanism for electric vehicles participating in peak regulation in valley time of receiving-end power grid'. *Global Energy Interconnection*. 2020;**3**(5):453–63.

[17] Bakirtzis E., Marneris I., Vagropoulos S., Biskas P., and Bakirtzis A.: 'Demand response management by rolling unit commitment for high renewable energy penetration'. *2018 International Conference on Smart Energy Systems and Technologies (SEST)*; Sevilla, Spain, Sep 2018, pp. 1–6.

[18] Carrion M., Arroyo J. M.: 'A computationally efficient mixed-integer linear formulation for the thermal unit commitment problem'. *IEEE Transactions on Power Systems*. 2006;**21**(3):1371–78.

[19] B. N. E. Finance, *New Energy Outlook 2019* [online]. 2019. Available from https://about.bnef. com/new-energy-outlook [Accessed Aug 2021].

[20] *Emergency Demand Response Manual* [online]. 2020. Available from https://www.nyiso.com/documents/20142/2923301/edrp_mnl.pdf/8f34b03 [Accessed Aug 2021].

[21] Shukla A., Verma K., and Kumar R.: 'Voltage-dependent modelling of fast charging electric vehicle load considering battery characteristics'. *IET Electrical Systems Transportation*, 2018;**8**(4):221–30.

[22] Wang J., Wu D., Zhao W., Shi S., Upadhaya B., and Shi Y.: 'Queueing theory-based optimal decision-making model of battery energy storage-assisted fast charging station participating in emergency demand response'. *2020 IEEE*

Sustainable Power and Energy Conference (iSPEC); Chengdu, China, Nov 202:2110–5.

[23] Paterakis N., Erdinç O., and Catalão J.: 'An overview of demand response: key-elements and international experience.' *Renewable and Sustainable Energy Reviews*. 2017;**69**: 871–91.

[24] State Grid Corporation of China. *Corporate profile* [online]. 2021. Available from http://www.sgcc.com.cn/html/sgcc_main_en/col2017112307/column_2017112307_1.shtml [Accessed Aug 2021].

[25] North China Grid Corporation.: *V2G charging piles were formally brought into auxiliary service market and settled* [online]. 2020. Available from http://chuneng.bjx.com.cn/news/20200421/1065131.shtml [Accessed Aug 2021].

[26] Bussiness Wire.: *PG&E and BMW Group Taking Next Step in Powering Electric Vehicles with Renewable Energy and Supporting Grid Reliability* [online]. 2021. Available from https://www.businesswire.com/news/home/20210322005580/en/PGE-and-BMW-Group-Taking-Next-Step-in-Powering-Electric-Vehicles-with-Renewable-Energy-and-Supporting-Grid-Reliability [Accessed Aug 2021].

[27] NRDC.: *Commercial prospects for electric vehicles interacting with the grid in Shanghai* [online]. 2020. Available from http://www.nrdc.cn/information/informationinfo?id=250&cid=49&cook=2 [Accessed Aug 2021].

[28] Moloughney T.: *A peek into the future of how the grid will benefit from smart EV charging* [online]. 2020. Available from https://insideevs.com/news/440433/enelx-smart-chargers-help-california-grid/ [Accessed Aug 2021].

Chapter 7

Demand response measurement and verification approaches: analyses and guidelines

Hamidreza Arasteh[1]*, Niki Moslemi*[1] *and Seyed Mohsen Hashemi*[1]

Demand response (DR) programs are defined as the ability of customers to change their consumption pattern in response to market/system signals. Nowadays, DR programs are interested worldwide as an essential part of the future power system and also considered as virtual generation resources. However, an accurate measurement and verification (M&V) approach is needed to implement these programs successfully. Indeed, the evaluation of the real potential of a DR program that is enabled during a DR event is depended on an evaluation method that should be employed to estimate the consumption behavior of the customers if they have not participated in DR. In this regard, customer base-load (CBL) estimation is defined as the approach to estimate the customers' load levels if they have not received DR calls. Then, by computing the difference among the estimated baseline and measured load data, the real potential of DR would be calculated. So, the determination of the real potential of DR is dependent on the difference between the estimated baseline and measured load data. Since various factors (such as load type, weather condition, day of a week, etc.) could affect the CBL, it is a challenging and complex task to provide an accurate estimation of the CBLs.

Generally, three steps could be considered to estimate the CBLs:

- Data selection: refers to the task to determine which data should be used for the estimation procedure. For instance, if the event day is a weekday, the historical data from the weekdays should be used.
- Computation: refers to the approach that is employed to estimate the CBLs.
- Adjustment: refers to the modifications that could be applied to the initially estimated baseline to enhance the accuracy.

In addition to the accuracy of the estimation, other aspects are also crucial to select the appropriate approach. Indeed, selecting the suitable method is based on four pillars: accuracy, simplicity, integrity, and alignment.

[1]Power Systems Operation and Planning Research Department, Niroo Research Institute, Tehran, Iran

The aims of this chapter are to introduce the concepts and characteristics of M&V approaches, introduce different classes of CBL estimation methods, compare the features of different techniques, present some practical results, and finally provide a guideline and suggestions to help the decision-makers to select the appropriate approach.

Keywords: Demand response, Customer baseline estimation, DR measurement and verification, Guideline, Practical results

7.1 Introduction

7.1.1 Concepts

Demand response (DR) is defined as the ability of customers to modify the consumption pattern to help the operators to provide the required energy demand in a cost-effective and reliable manner [1,2]. DR programs could be categorized as time-based and incentive-based programs [3]. In the broader view, using the demand side potentials can improve the power system operation capabilities. For example, in [4–7], the system security is increased using the load control mechanisms. Also, the demand side resources can be employed by the EMS of the MGs [8] to achieve the economic goals. There should be precise mechanisms to measure and verify the performance of the responsive loads during the DR events. Different methods have been employed to evaluate the baselines/customer base-loads (CBLs). These approaches could be distinguished based on their features. As an instance, baseline estimation approaches may be dynamic or static, they may use historical data or utilize statistical sampling, or they may use hourly or daily data [9–11].

One of the essential factors for the successful implementation of DR programs is to perform an accurate measurement and verification (M&V) [12]. The determination of the load reduction due to the performance of DR depends on the difference between the estimated baseline and the measured load levels. CBLs are not measurable and should be computed based on the available data. Several factors such as load types, weather conditions, day of the week, etc. could affect the baseline calculations. Therefore, the accurate determination of CBLs is a complex and challenging procedure [13].

The implementation of DR needs to identify the baseline to evaluate whether they have been enabled successfully or not. In other words, in order to measure and verify the success level of energy service companies, it is needed to employ the accurate M&V calculations. Hence, CBL estimation plays a pivotal role in DR evaluation. As explained, baseline estimation is the task of forecasting the consumption patterns, if DR is not called. Indeed, by using the baseline estimation, the actual load reduction resulting from the DR is calculated. However, the accurate determination of the CBLs is challenging. CBLs could be diverse due to some phenomena that are not necessarily related to DR. Independent to the price changes or DR incentives, factors such as weather conditions, programs of the broadcasting organization, seasonal variations, and holidays are some of the factors that could affect the consumption patterns [14].

In baseline estimations, such factors should be taken into account. However, accuracy is not the only criterion to evaluate the performance of the baseline estimation methods. A suitable method is simple enough so that all the DR beneficiaries could easily comprehend and compute the baselines. The employment of simple approaches could also decrease the managerial costs and would make it enjoyable. Moreover, DR providers may be able to employ some strategic behaviors to manipulate the baseline calculation approaches to enhance the forecasted load levels and receive higher incentives [15]. Therefore, a baseline estimation method should be able to handle these behaviors to ensure the proper implementation of DR. In addition, the participants should be incentivized well to be motivated to keep their participation in DR events. Consequently, it is a complex and challenging task to preserve all these goals, since they may be conflicting with each other. As an instance, a complicated method could preserve the baseline estimation approach from the manipulations. However, it may lose the simplicity of the calculations and would not be interesting for all the DR players. Similarly, a simple method may not model various factors (such as calendar data or weather conditions) and so will lose the accuracy criteria. Due to such problems, several methods have been developed, and the most appropriate approach should be selected based on various factors such as the types and aims of DR programs, frequency and duration of the events, and notification periods [16, 17].

Baseline estimation aims to estimate the amount of loads that would be needed in the hour and day of the event if DR programs are not enabled. As shown in Figure 7.1, the difference between the baseline and measured load levels is the actual enabled capacity of DR. Figure 7.1 shows the measured consumption data after the implementation of DR against the estimated CBL [18,19].

Some important definitions and concepts regarding the CBL studies are presented in the following:

- DR event: request for load reduction during a time interval of an event day by participating in one of the DR programs is called a DR event.

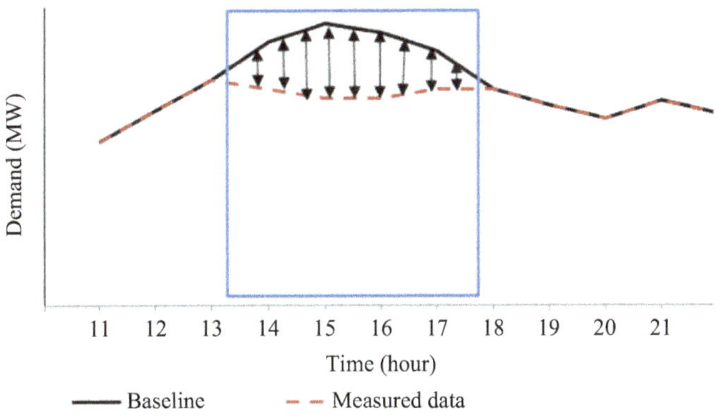

Figure 7.1 Measured consumption data after the implementation of DR vs. the estimated CBL

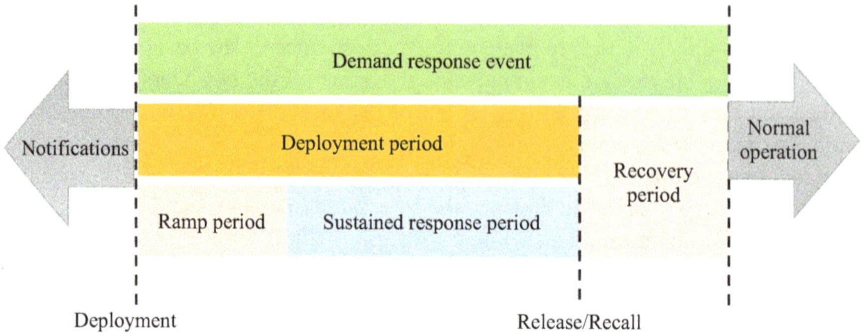

Figure 7.2 Timing of DR events [16]

- Timing of DR event: it denotes all the time-intervals to implement DR, which includes:
  - Deployment;
  - Ramp period;
  - Sustained response period;
  - Release/recall;
  - Recovery period.
- Deployment period: The time interval from "Deployment" to "Release/Recall" is called the "deployment period."

The timing of DR events and the definitions mentioned above have been illustrated in Figure 7.2 [16].

7.1.2 Literature review

Several studies have investigated the M&V of DR programs. Generally, such researches aim to find an appropriate approach to estimate the CBLs. The authors of [20] used the neural network method to calculate the CBL of the industrial loads. They have concluded that this method has better performance in comparison to the linear and polynomial regression methods. Gabaldón *et al.* [17] proposed a simple method to calculate the CBL that uses some of the adjustment factors based on the physically-based models. Reference [21] proposed a method to calculate the CBLs of the residential customers, who their daily loads vary randomly. They use a virtual control group for each DR event and combine it with difference-in-differences to form the V-CBL method. Ziras *et al.* [22] analyzed the local flexibility markets (LFMs), as a solution to appropriately react against the intermittency of the DERs considering the customers' activities. They used the CBL method to monitor and analyze the performance of the consumers in the provision of the flexibility services. In general, the baseline methods can be categorized as Figure 7.3 [22].

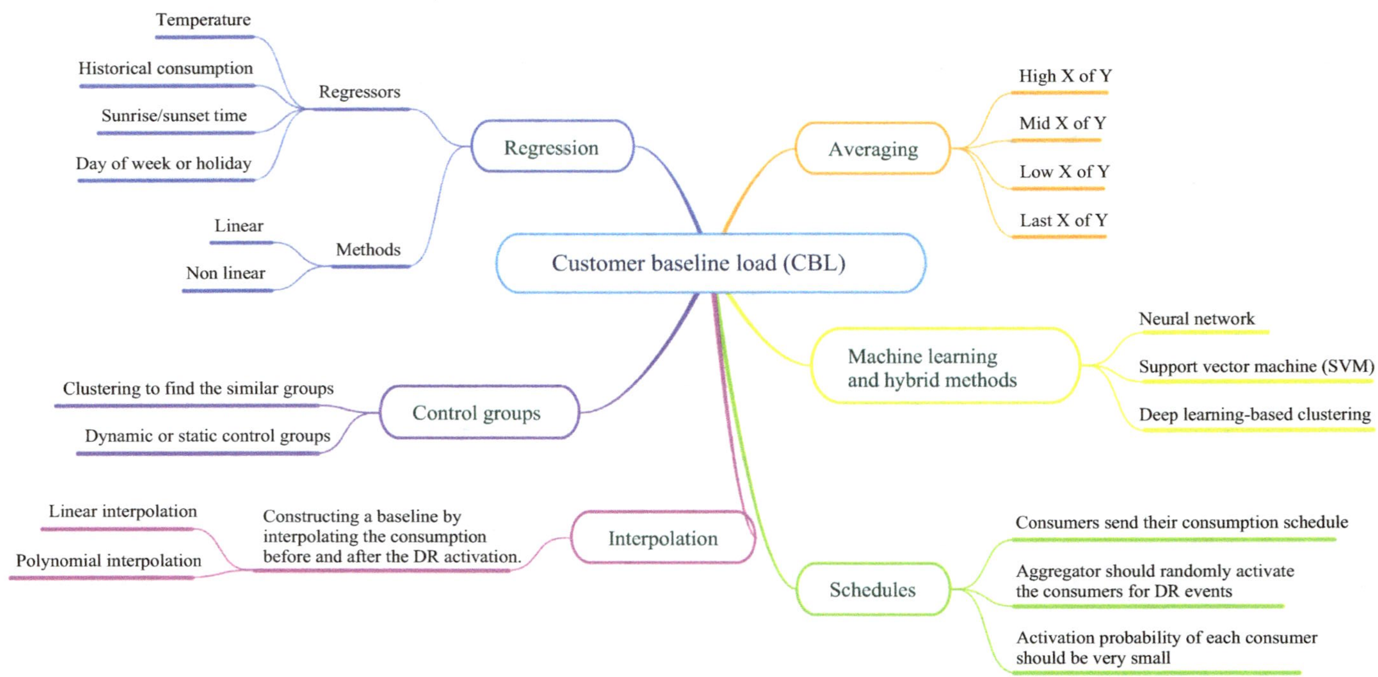

Figure 7.3 Overall categories of the baseline calculation methods [22]

Averaging methods are other types of approaches to estimate the CBLs. In these methods, the baselines are calculated based on the average consumption of the consumers in X days from the Y non-DR days [23, 24]. In some researches, the averaging methods have been adjusted by some influential factors, such as the weather [14]. The regression methods use the historical data to determine the dependence of the customer's load to the different regressors of the temperature, sunrise time, sunset time [23], etc. The authors of [25] used several regression methods, including different seasonal and daily time scales in which different regressors such as the temperature are considered. In [26], regressors such as days of weeks or holidays are considered. The relationship between the customer's load and the hour of the day has been analyzed in [27] using a cubic regression method. According to [28], grouping the customers will improve the accuracy of the CBL calculation in the regression method.

In the control group method, the customers participating in DR events have been compared with a similar load group to evaluate their performance. Müller *et al.* [29] divided the aggregators' customers into two parts for calculating the CBLs. One part was for the control group and another part was the DR participants. This method may require reserving a large share of the consumers as the control group to determine the performance of other groups. In order to deal with this drawback, the authors of [21] have proposed a dynamic control group in which different control groups are made for various DR events. Reference [30] used the k-means clustering method to find the most similar control group for each load. For this purpose, they propose a clustering technique to determine typical load patterns for the consumers. The authors of [11] determined the optimal control groups based on the load curves of each customer. They use a constrained regression method. Zhang *et al.* [31] assessed the correlation of the load patterns of the customers in the non-event days. By this method, they split the loads to the DR participants and the control groups.

Machine learning is another popular method to calculate the CBL. The authors of [9] used the support vector machine (SVM) to estimate the base-load based on the smart meters and the weather data. The machine learning-based methods may be combined with other approaches to perform the hybrid procedures. The authors of [32] compared the accuracy of the SVM with the methods of averaging and the exponential moving average. They concluded that the SVM has better performance for residential consumers. The authors of [33] used a combination of the methods to improve the performance of the CBL calculation procedure. They disaggregate the PV generation and the residential load using the SVM method. This data disaggregation enhances the performance of the averaging and regression methods. Sun *et al.* [34] extracted a set of the daily load profiles by applying the deep-learning method for the non-participating consumers. Also, they create the training and testing data sets by assessing the energy consumption of the loads before and after the DR events. A machine learning approach based on the Gaussian process regression is used in [35]. Also, the k-means clustering method and a self-organizing map are utilized in [36] to estimate the load during a DR event. The authors of [37] and [24] used the linear and polynomial interpolation methods to determine the CBLs, respectively, based on the load's energy consumption before and after the DR events. In some methods, the load forecast of the residential consumers is considered as the baseline load. These scheduling-based methods may be performed based on the aggregators [38], consumers [39], or even the individual appliances [40,41]. The authors of [42] proposed the usage of the CBLs. In this structure, the consumers are incentivized to be truthful. According to [43], the mentioned structure has better performance when the activation probability of the loads is very low.

7.1.3 Classification of CBL estimation methods

As explained, the goal of baseline estimation is to compute the DR performance to reduce the load levels and show how DR implementation is successful. Indeed, the baseline could be considered as an index of the expected value of the load. Since baseline estimation methods should be accurate and simple, averaging and regression are two widely used approaches to compute CBLs. North American Energy Standards Board (NAESB) categorizes the baseline calculation methods into five groups [16]:

- Baseline type I (BT-I): This class of baseline uses the historical data (usually, using the 1-h time intervals) and also may utilize weather and calendar data (such as holidays). In other words, this is a performance evaluation method based on the customers' historical data that may include other factors, e.g. weather and calendar. Nowadays, BT-I is the most prominent approach to calculate CBLs. Averaging method, regression, rolling average, and comparable day techniques are some of the approaches in the BT-I group.
- Maximum base load (MBL): In this method, the system's data during a period (e.g. the last season) is used to create a constant level of demand. This is a performance evaluation method based on the ability of the customers to reduce their loads to the specific levels. MBL is also named as firm service level in PJM.
- Meter before – meter after (MBMA): In this approach, the baseline is created by using the load data immediately before and after an event. In other words, in this method, electricity demand data before an event will be compared with similar measurements after the event.
- Baseline type II (BT-II): In this approach, statistical sampling is employed to estimate the baselines for the customers who are not equipped with individual metering devices. Hence, BT-II is a performance evaluation approach that uses the statistical sampling technique to estimate the customers' load levels.
- Generation: This type of baseline is applicable for facilities with on-site generation and is not discussed in this chapter.

Baseline estimation methods are different based on the load shape, required data, time intervals of historical data, aims, and design of the programs. These approaches are presented to forecast the customers' load levels if customers are not called by DR to calculate the actual participation capacities in DR.

7.1.4 Features of CBL estimation methods

A proper baseline should not incentivize or penalize the customers because of the regular operation. Moreover, it should consider the changes of the activities that are

Table 7.1 Central concepts in baseline estimation studies [12]

Measurement granularity	Load measurement time intervals (e.g. 5-min intervals)
Profile baseline (from now on, for the sake of simplicity, "baseline" is used instead of "profile baseline")	It indicates an approach in which the baseline is estimated using the historical data of the past similar days through a granular measurement (e.g. hourly) of the customers' load data. Therefore, customers' load patterns are evaluated dynamically during a 24-h horizon time.
Static baseline	Despite the hour or day of a week, a static baseline uses the average load during a long period (like a season) to estimate the baseline.
Individual baseline	Individual baseline refers to an approach that calculates the baseline for a particular customer or a region and then integrates all the individual baselines to create the baseline of whole the system.
Portfolio baseline	Portfolio baseline is the calculation of the baseline for all the systems under study, which is involved in DR programs.
Baseline window or look-back window	It refers to a time interval (like, the number of days) in which all the required data are extracted from this interval to calculate the baseline.
Exclusion rules	The rules that would be applied to the data of the look-back window to remove the undesired data, such as the data from the event days or holidays.
Baseline adjustments	When a baseline is initially created based on the historical data, a specific value may be added to enhance the accuracy of the calculations. To this end, usually, data from 2 to 4 h before the DR deployments on an event day are being used. In this regard, the averages of the differences between the estimated baselines and real measured data during the adjustment windows (i.e. time intervals that their data are used to calculate the values of adjustments) are used to modify the initially estimated baseline during the DR event periods.
Adjustment cap	Adjustments may be capped or uncapped. Capped adjustments indicate that the amounts of increasing/decreasing adjustments should not exceed a maximum value, while the amount of the adjustments is not limited in the uncapped manners.

not related to DR (such as the development of new businesses or normal commercial activities).

The main concepts in baseline estimation studies are presented in Table 7.1 [12]. It should be mentioned that comprehensive explanations are provided in the following sections.

In order to evaluate the performance of the baseline estimation approaches, some principles should be considered as the pillars of the baseline estimations that are illustrated in Figure 7.4 [44,45].

Based on Figure 7.4, a baseline estimation approach should be accurate and simple enough, not be easily manipulatable, and also be aligned with the DR goals. These features are explained in detail as follows.

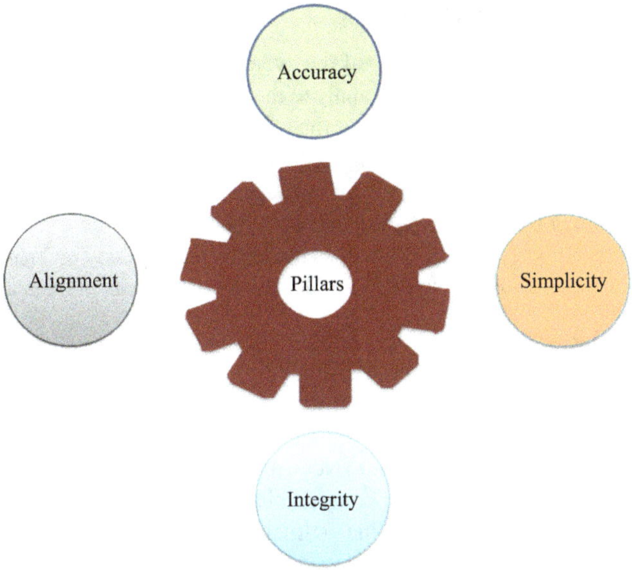

Figure 7.4 Pillars of CBL estimation approaches

- *Accuracy:* it means how the estimated baseline is close to the actual load data. Responsive customers should not gain a credit less or more than their actual participation.
- *Integrity:* Baseline calculation approaches should prevent the customers from manipulating the baseline by changing their consumption behaviors. In other words, customers should not be able to change their consumption pattern in a way to affect the computation results.
- *Simplicity:* Baseline calculation methods should be simple enough to be intelligible by all the beneficiaries, e.g. electric customers.
- Alignment: The baseline estimation methods should be aligned with the DR implementation goals, i.e. peak reduction, in a way that all the load management attempts result in the proper incentives for responsive customers.

These features may be conflicting with each other. As an instance, preventing the baseline manipulation needs to employ more complex tools that conflicts with the feature of simplicity. All in all, the baseline estimation approaches should keep the compromises among these features.

7.2 An overview of different CBL estimation approaches

Different CBL estimation approaches are investigated and analyzed in this section.

7.2.1 Averaging method

One of the most essential advantages of the averaging methods is simplicity. Hence, all the beneficiaries would be able to analyze the data that will enhance the clarity of the procedure. By using other features of the baselines, it may be possible to improve the accuracy and integrity of the averaging methods.

Averaging methods are the most usable approaches among the BT-I methods that utilize the recent historical data to estimate the CBLs during the specified time intervals. These methods are also known as the representative day or X of Y methods. This approach is based on the selection of X days within Y recent days and could be classified as [16]:

- High X of Y: In this approach, a Y-day baseline window proceeding the event days is considered. The data from X days with the highest load average within this baseline window is used to calculate the baseline. As an instance, a 5 of 10 methods has been used in NYISO (New York Independent System Operator) in which five days with the highest consumptions within a 10-day baseline window are selected to perform the baseline estimations.
- Low X of Y: In this approach (employed in New England ISO), a Y-day baseline window proceeding the event days is considered, and data from X nonholiday days with the lowest load average within this baseline window is used to calculate the baseline.
- Middle X of Y: In this approach, some days with the highest and lowest consumptions are removed from the Y-day baseline window so that the data of the remaining X days are utilized to perform baseline calculations.

Moreover, in X of Y approaches, weekdays and holidays should be investigated separately, where:

- Weekdays: In this approach, the averaging procedure is employed based on the data from the most recent weekdays, in which, all the weekends, holidays, and event days should be excluded from the baseline window.
- Holidays: In this approach, data from the weekends and holidays could be used as the input data for the baseline calculations.

From now on, for the sake of simplicity, the terms of X of Y or X/Y are used to indicate the high X of Y method. The following considerations could be taken into account to determine the numbers of X and Y [16]:

- Baseline window: As explained, the baseline window is a Y-day interval that its data is used to perform the baseline estimation calculations. Considering the limitations on the baseline window could help to avoid the utilization of the outdated data that may not be representative of the load levels of the event days.
- Exclusion rules: In the baseline estimation procedure, some days would be excluded since the consumption patterns of these days are substantially different from the event days. Indeed, these days could not be well-representative of the event days. Usually, holidays, weekends, and also the event days in which

DR programs have been previously enabled are not considered in the calculations. North American Electric Reliability Corporation (NERC) has introduced holidays as the "Off-peak days," and this standard is utilized by many numbers of programs. In addition to these fundamental cases, thresholds and scheduled shutdowns have been considered. In 2008, PJM investigated the consideration of the thresholds in the selection of the baseline window. As an instance, a 10% threshold means that if the average of the loads during a specified time interval in a day is less than 10% of the average of loads in the similar time interval during all the Y days of the baseline window, that day should be excluded from the calculations. Also, large swings in energy and curtailments could be considered as the exclusion rules [25].

- Relationship between X and Y: After selecting the Y days as the baseline window, these days would be narrowed down to a sub-set of X days. This sub-set should be formed based on the nature of the programs. For instance, it is expected that a DR event day in summer would be occurred in a day with extreme weather conditions when the load levels are high. However, all the Y days of the baseline window are not the days with high-load levels. Hence, if a baseline estimation approach uses the data from all the Y days, it will also consider the days in which the load levels are not high and are not similar to the event days. So, such an unadjusted approach will understate the customers' baselines, and reduce the incentives. In order to avoid such challenges, days with the lowest load levels could be removed from the calculations. Similarly, it may be observed that some of the DR programs are not always enabled during high load days. In such conditions, selecting the X days based on the "Middle X of Y" method may be better.
- Time intervals: Frequent intervals of data could be utilized by many DR programs. Hourly data are often used by many baseline estimation approaches, since data processing for a huge number of customers using data of shorter time intervals may be an unnecessary logistical strain.
- Baseline adjustments: As explained, the set of X days is selected from the look-back window in a way to obtain the most similar days to the event day. However, usually, the event day is not entirely the same as the selected days (especially for the consumers who are sensitive to the weather conditions). Programs associated with the peak load periods or the emergency conditions (because of the outages of the power plants) are usually required concurrently with the severe weather conditions. Therefore, some adjustments are needed to be applied to the initially calculated baselines. Baseline adjustments also could be called daily adjustments since these modifications are being computed using the data from the event days. Baseline adjustments could help to better reflect the load conditions in the event days and enhance the accuracy of the calculations for any type of DR program.
- Timing & duration: Data from 2 to 4 h period preceding the DR events is usually used for baseline adjustments. On the one hand, this period should not be too short (in order to take into account the load differences in different hours). On the other hand, this adjustment period should not be too long (typically, if it is more than 4 h, data that are not close enough to the event time may reduce the accuracy of the adjustments to consider the conditions during the events). To calculate the

adjustments, actual load data from the adjustment windows should be compared with the initial estimated data. It should be noted that the ramp period data should not be used to calculate the adjustments. Indeed, if the data from the ramp period is used, a customer who has decreased its load level earlier than the event time would lose the DR incentives and may be penalized. In addition, the consumer may be able to game the system by increasing their load levels during these periods. Such problems would negatively affect the integrity and accuracy of the estimations. So, by utilizing the data from the period before the notice of the event, it would be possible to prevent such problems.

The features of the baseline adjustments are:

a. Adjustments could be performed using additive or scalar factors:
Adjustments could be performed through additive or scalar manners. In an additive adjustment, the average of the load differences in adjustment windows would be added to the estimated baselines during the DR event hours. These additive values could be positive or negative. In a scalar approach, the average percentages of the differences during the adjustment windows would be multiplied with the estimated data during DR event periods. These scalar values also could be positive or negative.

b. Adjustments could be performed symmetrically or asymmetrically:
Baseline adjustments could be symmetrical or asymmetrical. The symmetrical approach means that the adjustments could be positive or negative. So, usually, the accuracy of the symmetrical adjustments is better than the asymmetric one. However, symmetrical adjustments may have damaging unintended consequences. For instance, if a customer decides to shut down a product line before the deployment of the DR and the meter data from after the production line has been shut off are used for the adjustments, its baseline will be decreased significantly. Under such situations, it should be noted that the time interval of such actions should not have any overlap with the adjustment windows [16]. Taking into account the alignment feature, negative adjustments may have some concerns, because, if customers typically have low load levels before the DR deployments, they will not be motivated to keep their demand at lower levels. As another instance, if in response to DR calls, an end-user decides to shut down some parts or all of its devices to perform maintenance plans, if morning adjustments are considered in the baseline calculations, symmetrical adjustments would significantly decrease its baseline. Also, if customers are aware of such influences, they may increase their demand during the adjustment windows to increase their baselines. From the administrative point of view, it is not desirable that customers keep their load levels at high levels during the adjustment window [12].

Asymmetrical increasing adjustments may also be faced with some challenges, since they may result in an overestimation of the baselines due to some problems, such as gaming. In addition, the baselines of the customers who, unusually, have high-load levels during the adjustment windows may be overestimated. It is noteworthy that, in these cases the problem is with the accuracy of the approaches, not the alignment. Under such situations, the customers may be received higher incentives, and they will be motivated to have participation. So, there is an alignment between the goals of the programs and the incentives [12]. Also, it is essential to note that if the increasing adjustments are not applied, it is possible that customers are not motivated to participate in DR during the days with high-load levels. So, harmful consequences may be occurred [12].

c. Adjustments could be capped or uncapped:
In order to limit the amounts of adjustments, a maximum range may be considered. Based on [16], the capped adjustments may penalize the consumers if the load levels are abnormally high in the event days.

In the symmetrical adjustments, considering a cap for the adjustments (especially for the decreasing adjustments) could help to keep the alignment characteristic, as well as the customers' satisfaction. However, considering a cap for the incremental adjustments may be resulting in some undesirable consequences. For example, if the event day has a severe weather condition with a very high temperature, the demand levels would normally be much more than the previous days through the baseline window. So, if the adjustments are capped, the customers may be penalized even if they are responsive to the DR calls.

Therefore, it should be tried to consider different consequences of the decisions regarding the characteristics of the baseline calculation methods.

7.2.2 Regression method

The regression method uses several variables such as the hour, day type (in a week), sunrise, and sunset to calculate the CBL. The simultaneous employment of the regression method and the ones mentioned above can improve computational accuracy. In other words, as the regression method uses a large number of variables, it is one of the most accurate methods. During the past 10 years, several researchers have compared the regression method and X/Y methods. According to LBNL, the X/Y methods have better performance than the regression method for the consumers with high load variability. According to [46], the performance of the X/Y methods is relatively similar to the weather regression models. According to [47], the regression method has better performance than the X/Y methods. However, the regression method may not be appropriate for real-time applications [48]. Some of the explanatory variables of the regression method used in these studies are the average load, cooling and/or heating degree days, day type indicators, and the day of average load [16].

The calculation of the regression-based CBLs is complex as needing the data of the load, weather, and day type. This method may require all the summer data for the estimation of the load levels in the event days. In such a case, it is impossible to calculate real-time CBL during an event period. It is worth noting that the real-time reporting of the CBL is required by the consumers and aggregators, as the feedback to determine their performance. Furthermore, the regression method defers the post-event performance evaluation of the consumers that may reduce their satisfaction level. Although the regression method is accurate, it is complex and is not recommended for the measurement and validation of the DR programs [16]. According to the high complexity of the regression-based methods, they cannot be manipulated easily, which means the improvement of the integrity. In return, they cannot provide a proper understanding for the DR players about the exact relationship between the load reduction and the incentives [12]. Accuracy improvements against the loss of simplicity may face the entities with some challenges since incentives would not be clear enough [12].

7.2.3 Other CBL calculation methods

This section introduces some other types of CBL calculation methods as the followings:

- *Comparable day method:* In this method, the CBLs of the event days are calculated using the data from their similar days. Unlike the averaging methods, this approach only uses the data of a single day. The comparable day method is faced with two main challenges [16]:
  - The CBL cannot be calculated during the event period.
  - There is no clear criterion for the selection of a similar day that would make it difficult to evaluate how a day is appropriate to be selected as a similar day.
- *Rolling average method:* These methods use historical data and give greater weight to the recent days. It is more accurate for the consumers with approximately constant load patterns. If a customer is active only in winter since the rolling average method will also consider the data of other seasons, the accuracy of the calculations to estimate its winter load pattern would be decreased [16].
- *Virtual control group method:* In this method, the customers are divided into two groups and one of them controls the behavior of another one. These groups should be very similar in different aspects such as the load profile, types of electrical devices, and energy consumption levels. In general, randomized group control (RGC) and non-experimental group control methods are utilized to choose the control groups. Based on the RGC method, the selection of the control group and the responsive group should be made in a way that both of them have similar characteristics and conditions. The RGC method is used in different companies such as Statewide Pricing Pilot [49], Anaheim Critical-Peak Pricing Experiment [50], Olympic Peninsula Project [51], Kyushu Electricity Pricing Experiment [52], and Energy Demand Research Project [53]. The main drawback of the RGC method is its implementation cost, as it requires that all the consumers be equipped with smart measuring infrastructures. Moreover, in the systems with high penetration of the smart metering equipment, most of the consumers receive the load reduction signals. So, it is a difficult task to find a control group that is not receiving the DR signals. Under these conditions, the non-experimental methods could be used to determine the control groups, in which, the utility companies create a database containing load data of the consumers such as the load profile, annual or daily energy consumption, building infrastructure, demand levels, etc. By using this approach, control groups are selected based on the customers' characteristics using the available data of this database. Although this approach is cheaper than another one, it has some drawbacks. It requires a large amount

of data to determine the various energy consumptions, and it is costly to gather such a volume of data. Another drawback of this method is that the stored load data are static and cannot determine the dynamic behavior of the loads.

- *Machine learning method:* This method employs neural networks in which different effective factors (such as the average temperature of a day) are used as the learning coefficients to calculate the CBLs [30].
- *Polynomial interpolation method:* This method is based on the actual data of the day and uses a polynomial function (usually a fourth-degree function) to estimate the CBL by using data from the period before or after the event hours.
- *Cluster-based baseline calculation:* In these novel approaches, the actual data of the event day are used rather than the historical data [31, 54]. This method considers different effective factors such as weather data. Also, the following clustering methods are being used by this approach to cluster the customers:
  - Based on the historical load data.
  - Based on the decision tree and using the demographic information.

Each cluster is divided into two subsets: participants and control groups. The average of the load levels of a control group is used to calculate the baseline of the participated consumers. It is worth noting that the weighted regression methods can be used instead of the averaging methods to consider more factors, such as weather conditions.

7.3 Comparison of different baseline estimation methods

As mentioned, the performance of the baseline estimation methods could be evaluated considering four criteria, i.e., simplicity, accuracy, integrity, and alignment. The performances of some of the baseline estimation methods are compared in the following [12,16].

- *X of Y:* This approach is simple enough and makes it easy to have proper communication with customers. Since the temperature data is not the input of these methods, they may not have high accuracy, especially if the weather condition is approximately constant during a long period and it varies significantly in the event days. Also, customers may be able to manipulate their consumption patterns through the X days in order to enhance their baselines and receive more incentives. However, it should be noted that these drawbacks could be eliminated by employing some actions such as adjustments (as an instance, if suitable adjustments are applied, they could significantly reflect the effects of weather conditions and improve the accuracy. Also, the appropriate selection of the look-back window could improve the integrity).
- *Regression:* Since different variables such as weather temperatures are used by this approach, it has high accuracy. However, its complexity makes it difficult to have proper communications with customers. From the integrity point of view, it is not easily possible to manipulate the consumption patterns, since this approach is based on data from a long period, and changing data in some limited hours

would not significantly affect the results. The disadvantage of this approach is its weakness to identify the substantial changes in the consumption patterns (such as installing the storage devices).

- *Machine learning:* Since this approach can consider the temperature data, it is a good choice to estimate the residential baselines. The machine learning method could identify the relationships between the inputs and outputs. However, this approach is more complex than the previous ones and would make it difficult to have effective communication with the customers. From the accuracy viewpoint, it is more accurate than others. In addition, it is based on enormous data, and so the possibility to manipulate its results is near to zero.
- *Polynomial interpolation method:* This approach is based on the interpolation of the data of limited hours. Therefore, it is simple and does not require a high amount of input data. Also, the accuracy of this approach is acceptable since it uses data from an interval that is close to the event hours. However, variables such as temperature data are not the inputs of this method. Moreover, it is under a high risk of manipulation, because it is based on data from a short period.
- *Virtual control group method:* This method has a high accuracy in estimating the baselines. It is not manipulatable because the customers are not compared with their own historical data. Also, one of the important features of this method is its scalability which could cover a large number of customers. In this approach, the various data of several consumption patterns are available for the distribution companies or aggregators (such as the types of the devices, number of family members, and the area of houses) that could be used to estimate the baselines. The main drawback of this method is that a massive number of factors should be considered to be able to correctly select the control group, which most of them are not available.
- *Maximum base-load (MBL):* This method is based on the ability of a responsive load to reduce its load to a specified load level [16] which is calculated by subtracting the contracted load reduction amount from the maximum expected load level. This method is sometimes called as "drop to" method, as the consumer should reduce its load to a specified load level. In return, most BT-I methods are called "drop by" methods as the consumer is aware of the load reduction amount while the load level is not a specific value. Unlike the BT-I methods, which create a dynamic load profile, the MBL is a static baseline calculation method. It is worth noting that in the MBL method, while the normal loads are lower than the specified load levels, consumers could be considered as the responsive participants without the need for any load reduction.

The Average Coincident Load (ACL) and Peak Load Contribution (PLC) are two popular methods based on the MBL. The ACL method has been employed by NYISO, and the PLC method has been used by the PJM in the Emergency Load Response program [16]. In both of these methods, the peak hours of the previous year are used to calculate the average values of the consumers' peak loads, which would be considered as their baselines during all hours. These two methods are different from each other in the process of identifying the peak hours.

MBL is of two types: coincident and non-coincident. In the coincident type, the system peak hours are being considered as the peak hours of this approach. In return, in the non-coincident type, the peak hours of different individual consumers are being used instead of the system peak hours.

- *Meter Before – Meter After (MBMA):* "A performance evaluation methodology where electricity consumption or demand over a prescribed period before deployment is compared to similar readings during the Sustained Response Period" [16]. This method is appropriate for the ancillary service events due to the minimal notice and reduced event durations. In general, the ancillary service programs aim to reduce the load in a short period of time.
  The properties of the MBMA are as the following:
  - The CBL is static.
  - It uses the historical data of each individual consumer.
  - It is based on the historical data during a short time.

The previously analyzed DR programs and their related CBLs focused on the emergency and energy services in which the consumers participate for around 4–8 h. In return, in the ancillary service programs, the participation period is short (generally around 10 min–2 h). So, the ancillary service programs usually use the MBMA method for calculating the CBL.

- *Baseline type II (BT-II):* This is a performance assessment method that uses statistical sampling to estimate the energy consumption of the aggregated demand resource, where interval metering is not possible for all consumers. Often, historical data are used to calculate CBLs. However, sometimes data are not separately available for each consumer, and the existing data is related to a group of customers. Under these conditions, CBLs could be calculated for the groups of loads, and then a method should be used to determine the share of each consumer. As the industrial and commercial loads are equipped with metering devices, the BT-II method is not commonly used for them. In return, this method is appropriate for the residential consumers that installing metering devices for them is costly. By increasing the number of metering devices for these consumers, the need to employ the BT-II method would be decreased.

7.4 Accuracy evaluation indexes

To determine the accuracy of the methods, the calculated CBL could be compared with the actual loads during the periods without DR. Here, some criteria are introduced to assess the accuracy of the methods.

7.4.1 RMSE and RRMSE

Root mean squared error (RMSE) is calculated using the following equation [55]:

$$\mathrm{RMSE} = \sqrt{\frac{1}{n}\sum\nolimits_{i=1}^{n} (\hat{y}_i - y_i)^2} \tag{7.1}$$

In this equation, n is the number of data samples, i is the indicator for the number of data, y_i is the actual value of the ith sample, and $\hat{y}_i$ is the estimated value of the ith sample.

The relative root mean squared error (RRMSE) [56] can be calculated using (7.2), in which, $\bar{y}$ is the average of the actual data.

$$\text{RRMSE} = \frac{\text{RMSE}}{\bar{y}} = \frac{\sqrt{\frac{1}{n}\sum_{i=1}^{n}(\hat{y}_i - y_i)^2}}{\frac{1}{n}\sum_{i=1}^{n} y_i} \tag{7.2}$$

7.4.2 MAPE and MAE [57, 58]

The mean absolute percentage error (MAPE) is calculated using (7.3) [55]:

$$\text{MAPE} = \frac{100}{n}\sum_{i=1}^{n}\left|\frac{y_i - \hat{y}_i}{y_i}\right| \tag{7.3}$$

Also, the mean absolute error (MAE) is calculated as the following [21]:

$$\text{MAE} = \frac{1}{n}\sum_{i=1}^{n}\left|y_i - \hat{y}_i\right| \tag{7.4}$$

It should be noted that, regardless of whether the estimated values are more or less than the actual values, the differences among these values are used by these indexes. Therefore, it is suggested here to take into account the positivity or negativity of the differences, since it shows that the estimation methods are over/under-estimating the load behaviors.

7.5 Guidelines and suggestions to select a proper baseline estimation method

Based on the literature, some guidelines and suggestions are summarized in this section. To enhance the accuracy of the methods, it would be beneficial to utilize some approaches, such as data-mining or clustering-based methods. In addition, the integrity criterion denoted that an appropriate method should not be vulnerable to manipulations. Moreover, the CBL estimation method should be designed in a way to incentivize the customers to participate in DR programs (that is defined as alignment). Simplicity indicates that the calculation of the baseline should be simple enough to be computable using the simple available tools and also be easily understandable by all the players. However, it should be noted that if an approach is too simple, it may lose the accuracy, integrity, or alignment.

Based on the literature, in the averaging methods, the following guidelines should be taken into account [12,16]:

- If X approaches Y, an unadjusted baseline would increasingly understate the value of CBLs;
- If adjustment is used, the X to Y proportions are less important;

- Usually, the utilization of the uncapped adjustments is better than the capped adjustments from the accuracy point of view. Hence, it is suggested to employ uncapped adjustments.
- Based on the experiences, usually if $0.4<X/Y<0.8$, the error is minimum. Moreover, if $0.4 <X/Y<0.8$, an increase in X will probably decrease the errors.
- The utilization of the increasing/decreasing adjustments is suggested, since, usually the utilization of load data immediately preceding the events could improve the accuracy of the estimation.
- X of Y methods is generally able to provide a satisfying tradeoff between the simplicity and accuracy. Based on [59], the accuracy of the X of Y and regression methods are close to each other, while the regression methods are more complex. Therefore, based on [59], X of Y methods are preferred to regression-based methods due to less managerial costs.
- Symmetrical adjustments will be able to increase or decrease the initially estimated baseline. If the uncapped adjustments are employed, the load data during the preceding hours of the events is utilized nicely to improve the accuracy of the calculations. However, the utilization of the asymmetrical adjustments would be more interesting for responsive customers.
- The symmetric adjustments are appropriate for the programs that are not designed for the extreme load conditions. In these programs, the load levels of different days are similar, and there are the same probabilities for the unadjusted baselines to be higher or lower than the actual loads.
- When DR events are called on-peak periods, by the symmetric adjustments, the consumers may need to take some actions in the baseline adjustment period to prevent the baseline understatements. While, given an asymmetrically adjusted baseline, they need no special action. The asymmetric adjustments are effective solutions to reduce the negative consequences of the early curtailment actions. These methods have no negative effects on the logical behaviors of the energy users to reduce their load at the time of grid stress.
- Adjustment windows should not have any overlap with the ramp period.
- If $X < Y$, since X days with the highest energy demand for each individual customer may not be the same with other customers, if the combined load data of all the customers is used, the estimated baseline may be less than the way in which the baseline is individually calculated for each customer. Hence, usually, it is better to calculate the baseline separately for each individual customer.
- A look-back window with ten working days could have satisfying results since it is not a too short or too long period.
- For the industrial and commercial customers, a 10 of 10 approaches could be a proper choice, since these types of customers typically have constant consumption patterns and are not highly dependent on weather. If the consumption pattern is intermittent, this method may not be suitable enough [60]. For such loads, the usage of the control group method or the utilization of the periods before or after an event with similar weather might be proper. The identification of the suitable methods to evaluate the baselines for the customers with the intermittent load patterns is a challenging task and needs more investigations

and analysis and is highly dependent on the specifications of the system under study.

- Although the utilization of the MBL methods is simple, they may significantly overestimate the baselines and be resulted in high errors. So, usually, X of Y methods are preferred to the MBL methods.

As it could be observed from the literature, none of the approaches are necessarily better than others. It is vital to implement different approaches in the system under study and analyze the results. The appropriate method should be identified based on the analysis of the load specifications in the system during the DR implementation period.

7.6 Practical results

In this section, a sample of the calculation results for the baseline estimation is presented using the actual data in 2020 (spring, summer, and fall). To this end, a calculation tool is developed and 44 averaging approaches are simulated and their performances are examined. All these approaches are based on the X of Y methods. According to these models, the historical load data of X-days with the maximum average load levels from a Y-day look-back window is utilized. It should be noted that holidays and weekends are excluded from the computations. It is noteworthy that the days with entirely different consumption patterns should also be excluded.

In these calculations, hours 12–17 are the DR event times when the DR aggregators should enable the DR potential. Adjustments should be applied using the data from the period when the customers have not changed their behaviors in response to DR signals. Here, it is assumed that DR deployment time is 12 PM, and customers may react to DR signals since 11 AM. Therefore, 9 AM and 10 AM are selected to calculate the adjustments. Table 7.2 shows the calculation types that are employed here.

The accuracies of the calculation methods are evaluated using the RMSE, EEMSE, MAE, and MAPE indexes.

The simulation results show that in all cases, the utilization of the scalar adjustments leads to better performances compared to the additive ones. Moreover, generally, during the summer and fall, in which load levels are not usually increasing/decreasing and show some oscillations around some values, symmetrical adjustments are better than asymmetric ones. Based on the results, it is not possible to explicitly select an approach as the best one. Such achievements were expectable based on the available experimental results in the literature. Indeed, based on the existing studies and practical experiments, none of the approaches are always better than others. The selection of the best method is wholly based on the features of the system under study, policies, and the period that DR is enabled.

Tables 7.3–7.5 show four methods with better results during spring, fall and summer.

As it could be observed, in spring, symmetrical and asymmetrical two of three methods and asymmetrical three of three and four of five methods, all of them using

Table 7.2 Employed methods

Method			**Adjustment**				
No.	**X/Y**	**Symmetrical increasing/decreasing**	**Asymmetrical increasing/decreasing**	**Additive**	**Scalar**	**Capped**	**Uncapped**
1	3/3	✓		✓			✓
2	3/3	✓			✓		✓
3	3/3		✓	✓			✓
4	3/3		✓		✓		✓
5	2/3	✓		✓			✓
6	2/3	✓			✓		✓
7	2/3		✓	✓			✓
8	2/3		✓		✓		✓
9	10/10	✓		✓			✓
10	10/10	✓			✓		✓
11	10/10		✓	✓			✓
12	10/10		✓		✓		✓
13	4/5	✓		✓			✓
14	4/5	✓			✓		✓
15	4/5		✓	✓			✓
16	4/5		✓		✓		✓
17	6/11	✓		✓			✓
18	6/11	✓			✓		✓
19	6/11		✓	✓			✓
20	6/11		✓		✓		✓
21	10/11	✓		✓			✓
22	10/11	✓			✓		✓
23	10/11		✓	✓			✓
24	10/11		✓		✓		✓
25	10/15	✓		✓			✓
26	10/15	✓			✓		✓
27	10/15		✓	✓			✓
28	10/15		✓		✓		✓
29	15/15	✓		✓			✓
30	15/15	✓			✓		✓
31	15/15		✓	✓			✓
32	15/15		✓		✓		✓
33	5/10	✓		✓			✓
34	5/10	✓			✓		✓
35	5/10		✓	✓			✓
36	5/10		✓		✓		✓
37	8/10	✓		✓			✓
38	8/10	✓			✓		✓
39	8/10		✓	✓			✓
40	8/10		✓		✓		✓
41	5/6	✓		✓			✓
42	5/6	✓			✓		✓
43	5/6		✓	✓			✓
44	5/6		✓		✓		✓

Table 7.3 *Four baseline estimation methods with better results during spring*

Method	Accuracy evaluation index	5/6	8/10	5/10	15/15	10/15	10/11	6/11	4/5	10/10	2/3	3/3
Symmetrical using additive adjustments	RRMSE											
	RMSE											
	MAE											
	MAPE											
Symmetrical using scalar adjustments	RRMSE										2	
	RMSE										2	
	MAE										3	
	MAPE										3	
Asymmetrical using additive adjustments	RRMSE											
	RMSE											
	MAE											
	MAPE											
Asymmetrical using scalar adjustments	RRMSE								3		1	4
	RMSE								4		1	3
	MAE								4		1	2
	MAPE								4		1	2

Table 7.4 *Four baseline estimation methods with better results during summer*

Method	Accuracy evaluation index	5/6	8/10	5/10	15/15	10/15	10/11	6/11	4/5	10/10	2/3	3/3
Symmetrical using additive adjustments	RRMSE											
	RMSE											
	MAE											
	MAPE											
Symmetrical using scalar adjustments	RRMSE		4		2		3			1		
	RMSE		4		2		3			1		
	MAE		4		1		3			2		
	MAPE		4		2		3			1		
Asymmetrical using additive adjustments	RRMSE											
	RMSE											
	MAE											
	MAPE											
Asymmetrical using scalar adjustments	RRMSE											
	RMSE											
	MAE											
	MAPE											

Table 7.5 Four baseline estimation methods with better results during the fall

Method	**Accuracy evaluation index**	**5/6**	**8/10**	**5/10**	**15/15**	**10/15**	**10/11**	**6/11**	**4/5**	**10/10**	**2/3**	**3/3**
Symmetrical using additive adjustments	RRMSE											
	RMSE											
	MAE											
	MAPE											
Symmetrical using scalar adjustments	RRMSE	1	3						2	4		
	RMSE	1	3						2	4		
	MAE	1	3						4	2		
	MAPE	1	4						2	3		
Asymmetrical using additive adjustments	RRMSE											
	RMSE											
	MAE											
	MAPE											
Asymmetrical using scalar adjustments	RRMSE											
	RMSE											
	MAE											
	MAPE											

the uncapped and scalar adjustments are the most appropriate approaches. However, the differences are not significant. Against the results in spring, in summer, the symmetrical 10 of 10, 10 of 11, 15 of 15, and 8 of 10 using the scalar and uncapped adjustments show better performances. Also, during the fall, the symmetrical 10 of 10, 8 of 10, 4 of 5, and 5 of 6 methods using the scalar and uncapped adjustments are the most proper approaches.

It is noteworthy that the load trend in spring is different from summer and fall. Therefore, the results presented in Table 7.3 (that belongs to spring) are considerably different from the results represented in Tables 7.4 and 7.5 that belong to fall and summer, respectively. During spring, the average of load levels is increasing from the beginning until the end of the period, while in fall and summer seasons, often, the behavior of the load levels is in a way that it is growing until some specific levels and then violate around these levels. Based on the investigations, the 10 of 10 approaches are usually an appropriate method when the consumption patterns are not highly intermittent.

Based on the literature, it could be concluded that the 8 of 10 using the symmetrical and uncapped adjustments could be a proper approach, due to the following criteria:

- X increments through the $0.4 < x/y < 0.8$ interval could decrease the error [16];
- A 10-day look-back window containing the working days could be a proper choice [12];
- Symmetrical adjustments could improve the accuracy of the calculations [12,16];
- Uncapped adjustments are usually better than the capped ones [16].

As the final comparison, based on the practical results and also what could be found from the literature, two approaches are selected to be compared with each other: 10 of 10 vs. 8 of 10 (both of them using the symmetrical, scalar, and uncapped adjustments). The RRMSE and MAPE indexes are used for the comparison. Based on the comparison results, it seems that the symmetrical 10 of 10 approaches using the scalar and uncapped adjustments is the most appropriate tool to estimate the CBLs in the test case of this chapter during the fall and spring seasons.

7.7 Concluding remarks and outlook

In this chapter, different baseline estimation methods are introduced, and their specifications are investigated. To determine the accuracy of the approaches, four indexes have been used: RMSE, RRMSE, MAPE, and MAE. It should be noted that based on these indexes, only the differences between the actual and estimated values are considered in the calculations, and the negativity or positivity of the errors are not considered (that denotes the underestimation or overestimation of the calculations). Hence, it is suggested that not only the accuracy indexes are used but also the negativity or positivity of the errors be considered to avoid severe underestimating and ensure the alignment to enhance the participation of the customers in DR.

As discussed in this chapter, a suitable baseline estimation method should provide a tradeoff between different features, i.e. accuracy, simplicity, integrity, and alignment. Quantitative studies are needed to evaluate the accuracy of the methods in the system under study.

On the one hand, based on our experience, in most cases in which the consumption patterns do not show sudden changes, the 10 of 10 approaches could have satisfied performance. On the other hand, based on the literature, it seems that some approaches, such as 8 of 10, may also be able to show good performances. As mentioned, the selection of the appropriate method is wholly based on the specifications of the system under study, and it is required to implement and analyze different approaches.

In this chapter, the concepts of the baseline estimation are introduced, and different evaluation methods and their advantages and disadvantages are presented. Then, comparisons have been provided between different approaches. In addition, a guideline is provided that could be used by all the beneficiaries of DR programs for selecting the appropriate baseline estimation approaches. Moreover, some practical results and experiences have been presented, and the relevant analyses have been explained. It should be noted that the provided guideline is a set of scientific and experimental notes that could help different players in making the DR-oriented decisions. However, the final decision about the appropriate method should be made after the investigation and implementation of different methods in the system under study. Indeed, the selection of the best approach is completely based on the system's specifications during the DR implementation period and is not clear before such studies. Our experiences based on the specifications of the system show that during the fall and summer, symmetrical 10 of 10, 8 of 10, 15 of 15, and 10 of 11 using the scalar and uncapped adjustments are suitable. However, during spring, when the consumption pattern is utterly different with fall and summer (it shows smoothly increasing behavior from the beginning to the end of the spring), the symmetrical and asymmetrical 3 of 4 methods and also the asymmetrical 3 of 3 and 4 of 5 approaches using the scalar and uncapped adjustments show better performances. These decisions are made based on the values of the RMSE, RRMSE, MAE, and MAPE indexes. It is suggested to select the approaches that firstly have the minimum error and secondly do not underestimate the baselines (in order to support the activities of the startup companies and increase the participation rates in DR programs). Also, it is suggested that the days with typically lower load levels (like holidays and weekends) be excluded from the look-back window to avoid the underestimating the baselines. Finally, it should be emphasized that the selection of the appropriate baseline is wholly based on the load specifications in the system under study during the DR implementation period and also should be done based on the expectations and goals of the decision-makers taking into account the accuracy, simplicity, integrity, and alignment, as the pillars of the baseline estimations. The notes that have been presented in this chapter are a guideline regarding the selection of the appropriate approach based on the scientific findings and practical experiences.

Acknowledgments

This work is supported by the Niroo Research Institute (NRI) under Contract No. 169001. Financial supports granted by TAVANIR and NRI are gratefully acknowledged.

References

[1] Arasteh H., Vahidinasab V., Sepasian M.S., and Aghaei J., "Stochastic system of systems architecture for adaptive expansion of smart distribution grids," *IEEE Trans. Ind. Inform.*, 2018, vol. 15, no. 1, pp. 377–89.

[2] Arasteh H., Bahramara S., Kaheh Z., *et al.*, "A system-of-systems planning platform for enabling flexibility provision at distribution level," in *Flexibility in Electric Power Distribution Networks*, CRC Press, pp. 41–65.

[3] Arasteh H.R., Moghaddam M.P., Sheikh-El-Eslami M.K., and A. Abdollahi, "Integrating commercial demand response resources with unit commitment," *Int. J. Electr. Power Energy Syst.*, 2013, vol. 51, pp. 153–61.

[4] Hashemi S.M., Vahidinasab V., Ghazizadeh M., and Aghaei J., "Load control mechanism for operation of microgrids in contingency state," *IET Gener. Transm. Distrib.*, 2020, vol. 14, no. 23, pp. 5407–17.

[5] Hashemi S.M., Vahidinasab V., Ghazizadeh M.S., and Aghaei J., "Valuing consumer participation in security enhancement of microgrids," *IET Gener. Transm. Distrib.*, 2018.

[6] Hashemi S.M., Vahidinasab V., Ghazizadeh M.S., and Aghaei J.: "Reliability-oriented DG allocation in radial microgrids equipped with smart consumer

switching capability," in *2019 International Conference on Smart Energy Systems and Technologies (SEST)*, 2019, pp. 1–6.

[7] Hashemi S.M., Vahidinasab V., Ghazizadeh M.S., and Aghaei J.: "Security constrained operation of radial micro grids based on the loads' Vulnerabilities and Flexibilities," *TABRIZ J. Electr. Eng.*, 2020, vol. 50, no. 3, pp. 1441–53.

[8] Hashemi S.M. and Vahidinasab V.: "Energy Management Systems for Microgrids," *Microgrids Adv. Oper. Control Prot.*, 2021, pp. 61–95.

[9] Chen Y., *et al.*: "Short-term electrical load forecasting using the support vector regression (SVR) model to calculate the demand response baseline for office buildings," *Appl. Energy*, 2017, vol. 195, pp. 659–70.

[10] Coughlin K., Piette M.A., Goldman C., and Kiliccote S.: "Statistical analysis of baseline load models for non-residential buildings," *Energy Build.*, 2009, vol. 41, no. 4, pp. 374–81.

[11] Hatton L., Charpentier P., and Matzner-Løber E.: "Statistical estimation of the residential baseline," *IEEE Trans. Power Syst.*, 2015, vol. 31, no. 3, pp. 1752–59.

[12] "The Demand Response Baseline," ENERNOC, 2009.

[13] Chan S.-C., Tsui K.M., Wu H.C., Hou Y., Wu Y.-C., and Wu F.F.: "Load/price forecasting and managing demand response for smart grids: methodologies and challenges," *IEEE Signal Process. Mag.*, 2012, vol. 29, no. 5, pp. 68–85,.

[14] Wi Y.-M., Kim J.-H., Joo S.-K., Park J.-B., and Oh J.-C., "Customer baseline load (CBL) calculation using exponential smoothing model with weather adjustment," in *2009 Transmission & Distribution Conference & Exposition: Asia and Pacific*, 2009, pp. 1–4.

[15] Wang X. and Tang W.: "To Overconsume or Underconsume: Baseline Manipulation in Demand Response Programs," in *2018 North American Power Symposium (NAPS)*, 2018, pp. 1–6.

[16] "The Demand Response Baseline," ENERNOC, 2011.

[17] Gabaldón A., García-Garre A., Ruiz-Abellón M.C., Guillamón A., Álvarez-Bel C., and Fernandez-Jimenez L. A.: "Improvement of customer baselines for the evaluation of demand response through the use of physically-based load models," *Util. Policy*, 2021, vol. 70, p. 101213.

[18] Song T., *et al.*: "A cluster-based baseline load calculation approach for individual industrial and commercial customer," *Energies*, 2019, vol. 12, no. 1, p. 64.

[19] Rossetto N.: "Measuring the intangible: an overview of the methodologies for calculating customer baseline load in PJM," 2018.

[20] Arunaun A. and Pora W.: "Baseline calculation of industrial factories for demand response application," in *2018 IEEE International Conference on Consumer Electronics-Asia (ICCE-Asia)*, 2018, pp. 206–12.

[21] Lee E., Lee K., Lee H., Kim E., and Rhee W.: "Defining virtual control group to improve customer baseline load calculation of residential demand response," *Appl. Energy*, 2019 vol. 250, pp. 946–958.

[22] Ziras C., Heinrich C., and Bindner H.W.: "Why baselines are not suited for local flexibility markets," *Renew. Sustain. Energy Rev.*, 2021 vol. 135, p. 110357.

[23] Wijaya T.K., Vasirani M., and Aberer K.: "When bias matters: an economic assessment of demand response baselines for residential customers," *IEEE Trans. Smart Grid*, 2014, vol. 5, no. 4, pp. 1755–63.

[24] Jazaeri J., Alpcan T., Gordon R., Brandao M., Hoban T., and Seeling C.: "Baseline methodologies for small scale residential demand response," in *2016 IEEE Innovative Smart Grid Technologies-Asia (ISGT-Asia)*, 2016, pp. 747–52.

[25] Coughlin K., Piette M.A., Goldman C., and Kiliccote S.: "Estimating demand response load impacts: evaluation of baselineload models for non-residential buildings in California," Ernest Orlando Lawrence Berkeley NationalLaboratory, Berkeley, CA, US, 2008.

[26] Larsen E., Rosenørn K., and Jónasdóttir A.: "Baselines for evaluating demand response in the EcoGrid 2.0 project," 2019.

[27] Zhou X., Yu N., Yao W., and Johnson R.: "Forecast load impact from demand response resources," in *2016 IEEE Power and Energy Society General Meeting (PESGM)*, 2016, pp. 1–5.

[28] Mohajeryami S., Doostan M., and Schwarz P.: "The impact of Customer Baseline Load (CBL) calculation methods on Peak Time Rebate program offered to residential customers," *Electr. Power Syst. Res.*, 2016, vol. 137, pp. 59–65.

[29] Müller F.L. and Jansen B.: "Large-scale demonstration of precise demand response provided by residential heat pumps," *Appl. Energy*, 2019, vol. 239, pp. 836–45.

[30] Li K., Wang B., Wang Z., Wang F., Mi Z., and Zhen Z.: "A baseline load estimation approach for residential customer based on load pattern clustering," *Energy Procedia*, 2017, vol. 142, pp. 2042–49.

[31] Zhang Y., Chen W., Xu R., and Black J.: "A cluster-based method for calculating baselines for residential loads," *IEEE Trans. Smart Grid*, 2015, vol. 7, no. 5, pp. 2368–77.

[32] Wang X., Wang Y., Wang J., and Shi D.: "Residential customer baseline load estimation using stacked autoencoder with pseudo-load selection," *IEEE J. Sel. Areas Commun.*, 2019, vol. 38, no. 1, pp. 61–70.

[33] Li K., Wang F., Mi Z., Fotuhi-Firuzabad M., Duic N., and Wang T., "Capacity and output power estimation approach of individual behind-the-meter distributed photovoltaic system for demand response baseline estimation," *Appl. Energy*, 2019, vol. 253, p. 113595.

[34] Sun M., Wang Y., Teng F., Ye Y., Strbac G., and Kang C.: "Clustering-based residential baseline estimation: a probabilistic perspective," *IEEE Trans. Smart Grid*, 2019 vol. 10, no. 6, pp. 6014–28.

[35] Weng Y., Yu J., and Rajagopal R.: "Probabilistic baseline estimation based on load patterns for better residential customer rewards," *Int. J. Electr. Power Energy Syst.*, 2018, vol. 100, pp. 508–16,.

[36] Park S., Ryu S., Choi Y., Kim J., and Kim H.: "Data-driven baseline estimation of residential buildings for demand response," *Energies*, 2015, vol. 8, no. 9, pp. 10239–59.

[37] Lin Y., Mathieu J.L., Johnson J.X., Hiskens I.A., and Backhaus S.: "Explaining inefficiencies in commercial buildings providing power system ancillary services," *Energy Build.*, 2017, vol. 152, pp. 216–26.

[38] Morstyn T., Teytelboym A., and McCulloch M.D.: "Designing decentralized markets for distribution system flexibility," *IEEE Trans. Power Syst.*, 2018, vol. 34, no. 3, pp. 2128–39.

[39] Nguyen D.T., Negnevitsky M., and De Groot M.: "Pool-based demand response exchange—concept and modeling," *IEEE Trans. Power Syst.*, 2010, vol. 26, no. 3, pp. 1677–85.

[40] Ramos A., De Jonghe C., Gómez V., and Belmans R.: "Realizing the smart grid's potential: defining local markets for flexibility," *Util. Policy*, 2016, vol. 40, pp. 26–35.

[41] Olivella-Rosell P., *et al.*: "Optimization problem for meeting distribution system operator requests in local flexibility markets with distributed energy resources," *Appl. Energy*, 2018, vol. 210, pp. 881–95.

[42] Muthirayan D., Kalathil D., Poolla K., and Varaiya P.: "Mechanism design for demand response programs," *IEEE Trans. Smart Grid*, 2019, vol. 11, no. 1, pp. 61–73.

[43] Vuelvas J., Ruiz F., and Gruosso G., "Limiting gaming opportunities on incentive-based demand response programs," *Appl. Energy*, 2018, vol. 225, pp. 668–81.

[44] Kema D.N.V.: "Development of demand response mechanism: baseline consumption methodology-phase 1 results," 2013.

[45] Goldberg M.L. and Agnew G.K.: "Measurement and verification for demand response," *US Dep. Energy Wash. DC USA*, 2013.

[46] Goldberg M.L. and Agnew G.K.: "Protocol development for demand response calculation-findings and recommendations." *Prepared for the California Energy Commission by KEMA-Xenergy*. CEC, 2003.

[47] "Estimation Errors in Demand Response with Large Customers," *AEIC Load Research Committee*, 2009.

[48] "Evaluation of 2005 Statewide Large Nonresidential Day-Ahead and Reliability Demand Response Programs," Southern California Edison Company (SCE), 2006.

[49] Herter K., McAuliffe P., and Rosenfeld A.: "An exploratory analysis of California residential customer response to critical peak pricing of electricity," *Energy*, 2007, vol. 32, no. 1, pp. 25–34.

[50] Wolak F.A.: "Residential customer response to real-time pricing: The anaheim critical peak pricing experiment," 2007.

[51] Pratt R.G., *et al.*: "The smart grid: an estimation of the energy and CO2 benefits," Pacific Northwest National Lab.(PNNL), Richland, WA (United States), 2010.

[52] Matsukawa I.: "Household response to incentive payments for load shifting: A Japanese time-of-day electricity pricing experiment," *Energy J.*, 2000, vol. 21, no. 1.

[53] Raw G. and Ross D.: "Energy Demand Research Project," AECOM; 2011.

[54] Wang F., Li K., Liu C., Mi Z., Shafie-Khah M., and Catalão J.P.: "Synchronous pattern matching principle-based residential demand response baseline estimation: Mechanism analysis and approach description," *IEEE Trans. Smart Grid*, 2018, vol. 9, no. 6, pp. 6972–85.

[55] Park S., Ryu S., Choi Y., and Kim H.: "A framework for baseline load estimation in demand response: data mining approach," in *2014 IEEE International Conference on Smart Grid Communications (SmartGridComm)*, 2014, pp. 638–43.

[56] Ko W., Vettikalladi H., Song S.-H., and Choi H.-J.: "Implementation of a demand-side management solution for South Korea's demand response program," *Appl. Sci.*, 2020 vol. 10, no. 5, p. 1751.

[57] Drezga I. and Rahman S.: "Short-term load forecasting with local ANN predictors," *IEEE Trans. Power Syst.*, 1999, vol. 14, no. 3, pp. 844–50.

[58] Shahidehpour M., Yamin H., and Li Z.: *Market operations in electric power systems: forecasting, scheduling, and risk management*. John Wiley & Sons, 2003.

[59] Lake C.: "PJM Empirical Analysis of Demand Response Baseline Methods." 2011.

[60] "Baselining the ARENA-AEMO Demand Response RERT Trial," Oakley Greenwood, 2019.

Chapter 8
Transactive energy industry demand response management market

K.S. Swarup[1] *and T. Vidyamani*[1]

In a smart grid paradigm, the concepts of demand response (DR) and transactive energy (TE) are used to optimize the consumption and generation in the power networks. In this chapter, two models for DR are analyzed based on the well-known Cobb–Douglas utility function. Both models maximize their utility, subject to different constraints. A time-of-use price-based DR program is employed. Restructuring in the electricity sector, with an increase in renewable energy resources and distributed energy management technologies, offers the potential for significant improvement in the efficiency of power systems through the TE framework. In a TE framework, prosumers of all sizes can participate in the double auction electricity markets via automated home energy management systems. Heating, ventilation, and air conditioning (HVAC) and energy storage devices are the two important loads in residential buildings that account for a large proportion of building energy consumption. A two-way exchange of energy and information is possible with the current advent of communication systems and net metering. In this work, we consider the case of solar photovoltaics (PV), HVAC, and energy storage devices (electric vehicles and battery energy storage systems) of prosumers participating in the retail real-time double auction market. The problem is formulated as maximization of social welfare subject to power balance and network constraints. Simulation studies and results are presented for the modified IEEE 13 node distribution system.

8.1 Demand response

Dynamic pricing and demand response (DR) programs play a key role in the modern smart grid environment to manage peak loads. The most important constraint of the power system, i.e., supply-demand balance constraint, can be met either by optimizing the generation side or the demand side. DR programs focus on the consumption side of the network. DR is defined as "Changes in electric usage by end-use customers from their normal consumption patterns in response to changes in the price of electricity

[1]Department of Electrical Engineering, IIT Madras, Chennai, India

over time, or to incentive payments designed to induce lower electricity use at times of high wholesale market prices or when system reliability is jeopardized" [1]. DR strategies are generally classified into two types:

1. Incentive-based programs
2. Price-based programs

Price-based DR programs include,

1. Time of use (TOU)
   TOU is a static price schedule. Higher rates during peak consumption hours and lower rates during partial peak and off-peak hours
2. Critical peak pricing (CPP)
   Pre-specified higher price during a limited number of hours or days per year
3. Extreme day pricing (EDP)
   Higher price for the whole 24 h which is known only day-ahead
4. Real-time pricing (RTP)
   Market prices are forwarded to end-use customers

8.2 Transactive control

Concentrating only on the consumption side might not fully exploit the capabilities of future smart grids. Thus there is a need for a framework that focuses not only on the consumption side but also on the rate of generation in both grid and demand sides [2]. Traditionally the flow of energy was from wholesale power plants to transmission systems and then to the end-use consumers of the distribution system. The flow was uni-directional, but now with the increase in the percentage of variable and intermittent distributed energy resources (DERs), the way the power system operates has changed. There is a bidirectional power flow due to the penetration of DERs such as distributed generation (DG), Battery Energy Storage System (BESS), and electric vehicles (EVs) in the distribution system.

There can be three control strategies established between the utility and the customer [3].

1. Passive control
   No real time transfer of price signal.
2. Active control
   Real-time price signals are sent from utility to end-use appliances which then respond based on their comfort.
3. Transactive control
   This control involves two-way communication in which end-use loads can bid for their demand and price is determined based on the bids from buyers and sellers.

Load pattern and the peak demand of the residential building will be different in each scheme [4]. Transactive control operation with HVAC as a transactive agent is shown in Figures 8.1 and 8.2. Where Π is the electricity price, and T is the temperature of

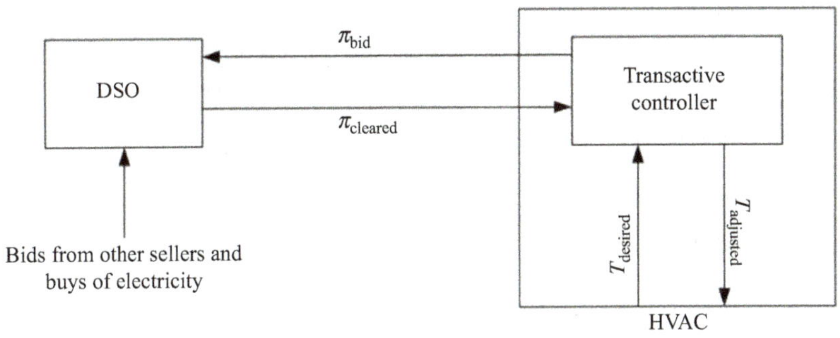

Figure 8.1 Transactive control

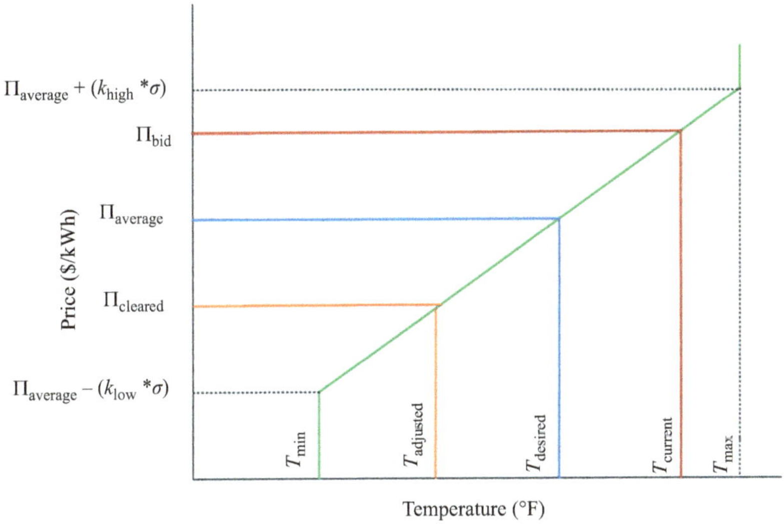

Figure 8.2 Thermostatic response for transactive control

HVAC. More details about the derivation of bidding price and adjusted temperature of HVAC from Figure 8.2 can be found in [5].

8.3 DR modeling and simulation results

Classical demand theory is a branch of microeconomics, and it is about the study of consumer demand in a market economy. Important tools in this theory are utility functions and two constraints, i.e., physical and budget constraints. In [6], authors have proposed Cobb–Douglas utility function based DR subjected to budget constraints. Here, two models for DR are analyzed based on the well-known Cobb–Douglas utility function. Based on the willingness to participate, consumers are classified into three types i.e., high flexible, medium flexible ,and low flexible. Results are discussed by considering different levels of participation in the program. The utility function is subjected to budget constraint in model A and total consumption constraint in model B.

8.3.1 Model A

In model A, the utility function is subjected only to budget constraint. Utility function and budget constraint of model A is given by,

$$\text{Maximize } U(q_1, q_2) = q_1^{\alpha} q_2^{\beta} \qquad \alpha > 0, \beta > 0, \alpha + \beta = 1 \tag{8.1}$$

subject to

$$I = p_1 q_1 + p_2 q_2 \tag{8.2}$$

Lagrange function is given by,

$$L = q_1^{\alpha} q_2^{\beta} + \lambda(I - q_1 p_1 - q_2 p_2) \tag{8.3}$$

$$\frac{\partial L}{\partial q_1} = \alpha q_1^{\alpha-1} q_2^{\beta} - \lambda p_1 = 0$$

$$\frac{\partial L}{\partial q_2} = \beta q_1^{\alpha} q_2^{\beta-1} - \lambda p_2 = 0$$

$$\frac{\partial L}{\partial \lambda} = I - q_1 p_1 - q_2 p_2 = 0$$

From the above equations,

$$q_1 = \frac{I\alpha}{p_1} \tag{8.4}$$

$$q_2 = \frac{I\beta}{p_2} \tag{8.5}$$

Demand functions q_1 and q_2 depends on total budget, prices of electricity and elasticity parameter.

Where,
q_1 is the peak period demand
q_2 is the off-peak period demand
p_1 is the peak time price
p_2 is the off-peak time price
Q is the total demand
α, β is the elasticity parameter
U(.) is the utility function
I is the budget/bill amount (₹/kWh).

8.3.2 Model-B

In model B, the utility function is subjected to total consumption limit constraint. The utility function and constraint of model B is given by,

$$\text{Maximize } U(q_1, q_2) = q_1^{\alpha} q_2^{\beta} \qquad \alpha > 0, \beta > 0, \alpha + \beta = 1 \tag{8.6}$$

subject to

$$Q = q_1 + q_2 \tag{8.7}$$

Lagrange function is given by,

$$L = q_1^{\alpha} q_2^{\beta} + \lambda(Q - q_1 - q_2) \tag{8.8}$$

$$\frac{\partial L}{\partial q_1} = \alpha q_1^{\alpha-1} q_2^{\beta} - \lambda = 0$$

$$\frac{\partial L}{\partial q_2} = \beta q_1^{\alpha} q_2^{\beta-1} - \lambda = 0$$

$$\frac{\partial L}{\partial \lambda} = Q - q_1 - q_2 = 0$$

From the above equations,

$$q_1 = \alpha Q \tag{8.9}$$

$$q_2 = \beta Q \tag{8.10}$$

Demand functions q_1 and q_2 depend on total demand and elasticity parameters.

8.3.3 Simulation results and discussion

Performance of both the models is analyzed using a residential load profile shown in Figure 8.3. Based on the load profile, the daily time horizon is divided into two time periods i.e., peak period and off-peak period and prices of the same are given in Table 8.1. The average price is 3.83 per kWh. Peak time and off-peak time demand can be adjusted by changing the values of elasticity parameters α and β. To participate in the DR programs, each user should have smart meters with the capability of entering their willingness to participate in the program with the elasticity parameter α. It is known that $\alpha + \beta = 1$, so $\beta = 1 - \alpha$.

8.3.3.1 Analysis of model A

Peak demand and off-peak demand without DR is 28.8 kWh and 17.55 kWh and the bill amount is ₹196.65. Table 8.2 shows the total demand for various values of the elasticity parameter ranging from 0.1 to 0.9. We have a budget constraint, so the billing amount will be the same irrespective of the elasticity parameter. Total energy consumption without demand response is 46.36 kWh. As seen from Table 8.2, when α is greater than 0.7, the total energy is less than 46.36 kWh which is the energy consumption without demand response. Here the assumption is that no user wants the reduction in

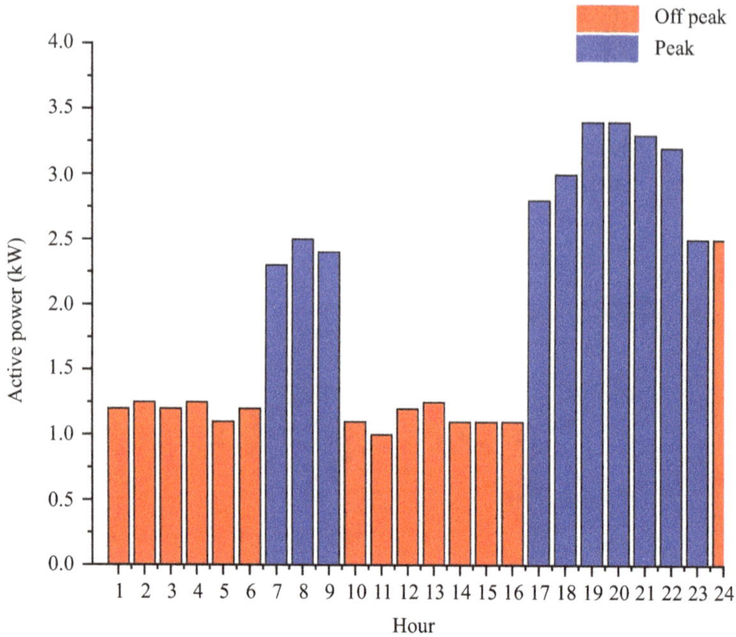

Figure 8.3 Load profile

Table 8.1 Price of electricity in different time periods

Demand level	q_1	q_2
Time period	6:00–9:00, 16:00–23:00	23:00–6:00 9:00–16:00
Price (₹ per kWh)	5	3

their energy consumption after participating in a demand response program. So the appropriate range for the elasticity parameter in model A is $0.1 < \alpha < 0.7$. Figure 8.4 shows that with increase in α, the amount of consumption increases in peak time. Figure 8.5 shows that with increase in α, the amount of consumption decreases in off-peak time. If the user selects the small value of elasticity parameter α, then the participation of that user in demand response program is more, thus he is more flexible. So his peak time consumption will be less and off-peak consumption will be more. With $\alpha = 0.1$, total energy consumption is 62.92 kWh which is 16.56 kWh more than the energy consumption without demand response program. Thus with the same bill amount, users can consume more energy if their willingness in participating demand response programs is more.

Table 8.2 Changes in the power consumption after implementation of model A

Elasticity parameter α	q_1 **(kWh)**	q_2 **(kWh)**	**Q (kWh)**
0.1	3.93	58.99	62.92
0.2	7.86	52.44	60.3
0.3	11.79	45.88	57.67
0.4	15.73	39.33	55.06
0.5	19.66	32.77	52.43
0.6	23.59	26.22	49.81
0.7	27.53	19.66	47.19
0.8	31.46	13.11	44.57
0.9	35.39	6.55	41.95

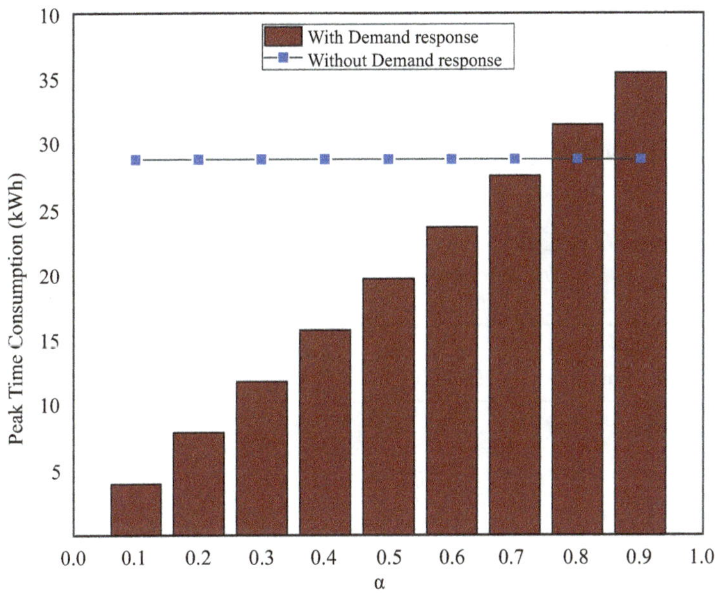

Figure 8.4 Model A—peak period consumption for different values of α

8.3.3.2 Analysis of model B

The main difference between model A and model B is the constraint it is subjected to. In this model, we have total consumption constraint. That is energy consumption before and after the implementation of demand response program will be the same. Table 8.3 shows the peak and off-peak time period power consumption of model B. As shown in Table 8.3, the bill amount has increased with an increase in the elasticity parameter α. Figure 8.6 shows the peak period consumption for different values of α.

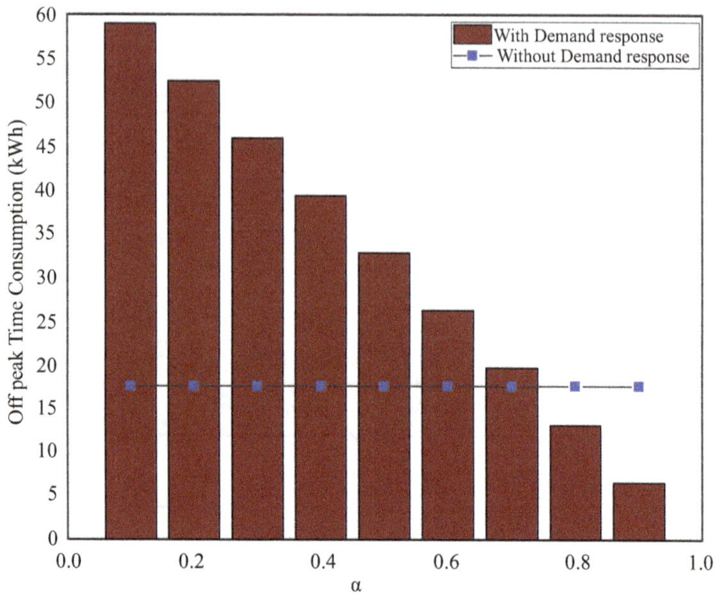

Figure 8.5 Model A—off peak period consumption for different values of α

Table 8.3 Changes in the power consumption after implementation of model B

Elasticity parameter α	q_1 (kWh)	q_2 (kWh)	Bill amount (₹)
0.1	4.63	41.71	148.29
0.2	9.27	37.08	157.59
0.3	13.90	32.44	166.82
0.4	18.54	27.81	176.13
0.5	23.17	23.17	185.36
0.6	27.81	18.54	194.67
0.7	32.44	13.90	203.94
0.8	37.08	9.27	213.21
0.9	41.71	4.63	222.48

Peak period consumption increases with an increase in the elasticity parameter. When α is greater than 0.6, q_1 is more than the peak period energy consumption without demand response. Figure 8.7 shows the off-peak period consumption for different values of α. Off-peak energy decreases with an increase in the elasticity parameter. With $\alpha > 0.6$, bill amount is also more than the usual one without demand response. So the appropriate range for the elasticity parameter in model B is $0.1 < \alpha < 0.6$. So it is clear that similar to model A, the user is more flexible if he chooses a small value

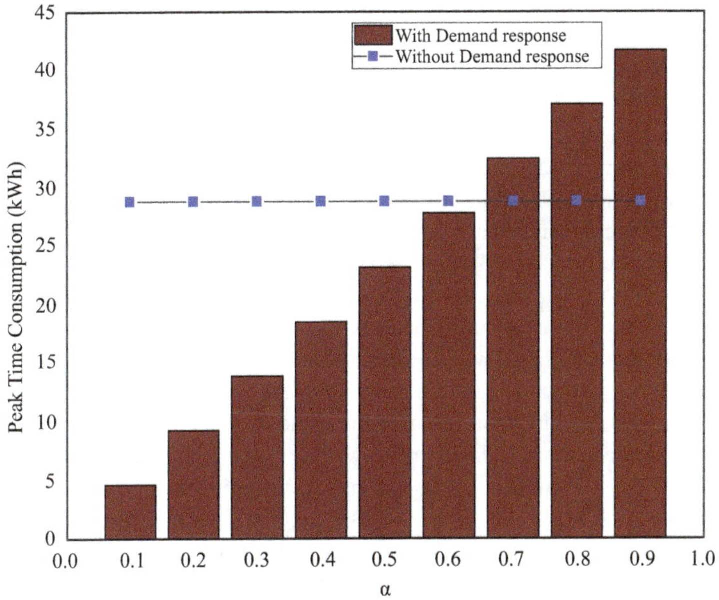

Figure 8.6 Model B—peak period consumption for different values of α

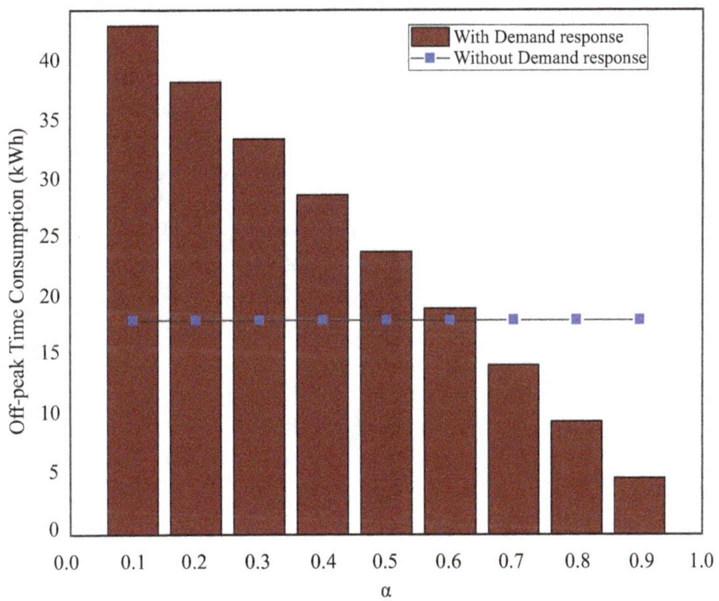

Figure 8.7 Model B—off-peak period consumption for different values of α

of elasticity factor. Thus users can consume the same amount of energy with less bill amount if their willingness in participating demand response programs are more.

8.3.3.3 Comparison of model A and model B

To participate in the demand response programs, consumers should have smart meters with inbuilt energy management system (EMS) software. Power demand equations and algorithms can be coded in the EMS systems. Demand response programs are user-friendly that consumers can join or leave the program at any time. In the digital display of EMS, the consumer has to enter elasticity factor α, which is the measure of their willingness to participate. In model A, the user will have the same bill amount before and after the implementation of demand response. If he is highly flexible, then most of his peak period loads will be shifted to off-peak period time. As an incentive for participation, he can use extra energy with the same bill money. In model B, though the energy consumption remains the same before and after implementation, a highly flexible consumer who is more willing in participating demand response programs will get a reduction in the bill by shifting his loads from peak period to off-peak period. It is assumed that EMS has the capability of changing from model A to model B when needed and vice versa in a suitable time interval. So the consumer can set his preference depends on whether he wants the same energy consumption with a reduction in the bill or more energy consumption with the same bill amount.

8.4 TE management

TE is the generalized form of price-based demand response. TE manages the rate of generation in both the grid and demand sides. GridWise Architectural Council (GWAC) defines TE as “A set of economic and control mechanisms that allows the dynamic balance of supply and demand across the entire electrical infrastructure using value as a key operational parameter” [7]. Decisions in TE Framework (TEF) are made based on the value, which is analogous to or literally economic transactions. Application of transactive control is well established in the wholesale market and in transmission level, but its application is largely missing in the distribution level and retail markets [8]. Transactive control accommodates two-way power flow and new generation using a decentralized supply model. It allows for faster transmission of information about supply, demand, and price across the entire infrastructure.

The TE framework helps the power system to evolve as a hybrid system of the wholesale energy market (WEM) and local energy markets (LEM) as shown in Figure 8.8. The wholesale energy market operator manages the network and power balance at the transmission level, and the local energy market operator/distribution system operator (DSO) is responsible for managing the DERs and power balance at the distribution level [9].

In a TE management system, transactive agents negotiate their actions with each other through double auction transactive markets. A double auction market mechanism computes the clearing price and quantity based on the bids and offers from buyers and sellers of energy. A transactive agent may be a residential house, industrial or

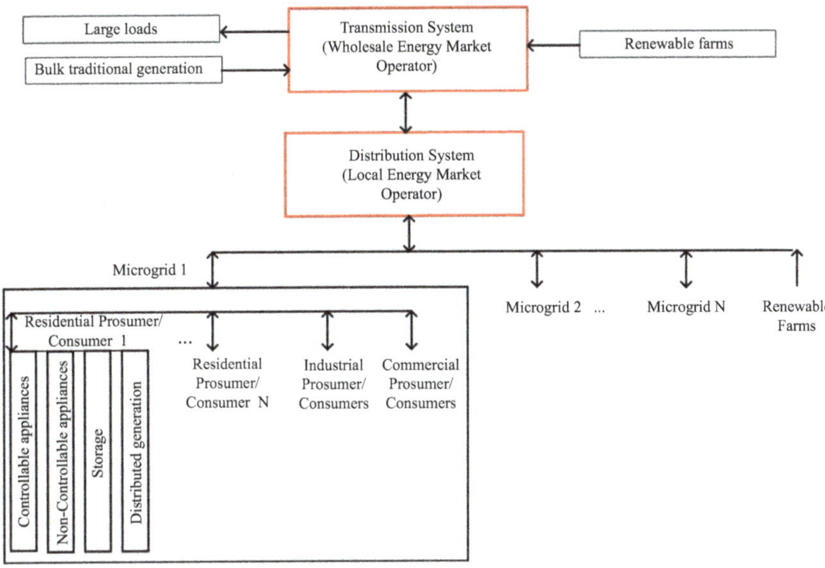

Figure 8.8 TE framework

commercial building, microgrid, demand response aggregator, microgrid aggregator, renewable farm, etc. As seen in Figure 8.8, the transactive control concept can be applied in all levels of the power system from residential to transmission level, and it facilitates the integration of the local energy market and wholesale energy market [2].

At the bottom level of the TE framework, we have residential houses, no matter their sizes, they can participate in the retail or local energy market and modify the schedule of its controllable appliances based on the cleared market price and user's local information. The TE concept can also be applied inside a single building where each zone can bid or compete among themselves [10]. Retail markets can be operated for only one microgrid (MG) as well as for the group of microgrids [11]. An energy management scheme for residential microgrids has been proposed in [12,13]. Transactive control helps in congestion and voltage management in the distribution system by considering the network operational constraints [14]. There are forward and spot transactions in transactive markets. Forward markets operate by relying on future delivery, and spot market is used for instant delivery.

GridWise Olympic Peninsula project [5] is the field demonstration of the transactive control concept in the distribution network with the market clearing interval as 5 min. Residential electric water heaters and Heating Ventilation Air Conditioning (HVAC) loads, commercial building HVAC loads, municipal water pump loads, and several distributed generators were coordinated to manage the congestion in the distribution feeder. The bid price of HVAC has been calculated based on the consumer's desired temperature set-point, comfort setting factor, and temperature limits. The temperature set-point will be adjusted based on the market-clearing price [15].

A reinforcement learning-based model-free bidding strategy for HVAC systems in double auction retail markets has been proposed in [16].

Authors in [17] formulated the bidding strategies of Electric Vehicles (EVs) to participate in the retail real-time double auction market. The bid price of EV has been calculated based on the consumer's comfort setting, arrival time, departure time, and the battery SOC level and limit values. Comfort setting factor (k) reflects the consumer's flexible nature whether he wants more comfort or a reduction in the bill. The bidding and offering strategies for EV participation in the real-time retail market have been proposed in [18]. The offering strategy has been developed by considering the wear price of the battery. In [19], the authors proposed the bi-level bidding model for residential prosumers in the day ahead TE market using the flexibility of BESSs. An energy storage model has been developed by incorporating the operational constraints and degradation of storage units when they undergo frequent charge-discharge cycles.

EV charging and HVAC systems account for a large proportion of energy consumption. Coordinated dispatch of EV and HVAC has been proposed using model predictive control in [20] and [21] to reduce the cost and to accommodate the uncertainties in the PV supply. TE framework can have centralized or decentralized energy trading [22]. In a centralized energy trading, all microgrid participants send their energy bids and offers to the LEM operator. LEM operator aggregates all the bids and offers to determine the market-clearing price while satisfying the network constraints. The cleared market price is sent back to the participants to initiate the transactions.

In this work, each prosumer is assumed to have PV, HVAC and Energy storage devices (EV, BESSs). A real-time market framework is developed that enables prosumers to adjust their willingness to pay/accept according to their comfort in a user-friendly manner. Prosumers can participate in the retail market via their HEMS. The DLMPs are calculated for the network with price-responsive devices by solving ACOPF equations.

8.5 Methodology

In a TE framework, the market participants submit their "price-quantity" bids/offers for electricity demand or supply in each time interval. Interaction of retail market with the wholesale market is incorporated in the developed model. The independent system operator in the wholesale market clears the market and sends the clearing prices of the day-ahead market known as locational marginal prices (LMPs) to all participants. The DSO, associated with retail market operations, calculates the distribution locational marginal prices (DLMPs) for the next day based on the received LMP and forecasted net demand. The calculated DLMP will be an uncertain parameter because of the uncertain nature of the net demand of distribution systems. In this work, the uncertain nature of DLMPs is defined by using the normal probability distribution function. DSO estimates the mean and standard deviation of the price and sends these signals to all the prosumer buildings as transactive incentive

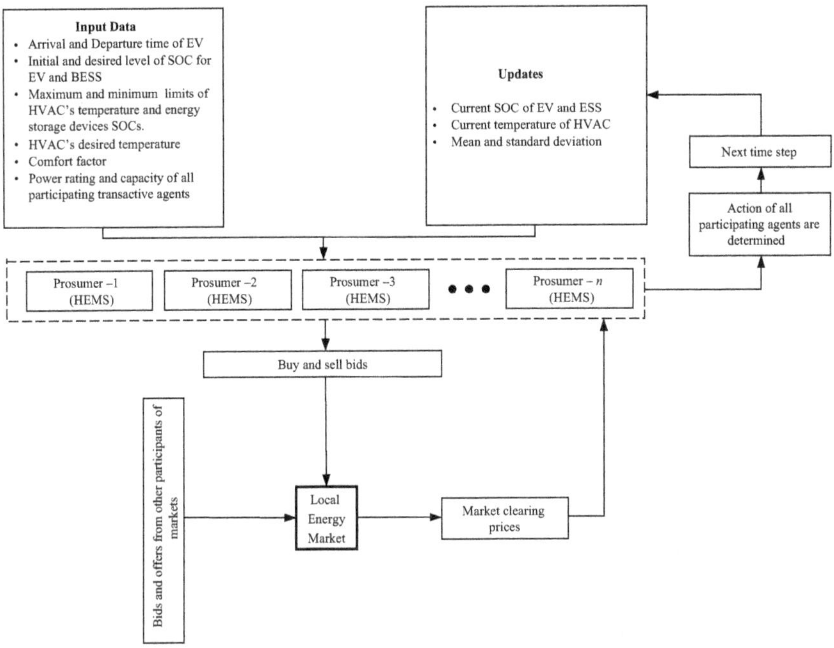

Figure 8.9 Flowchart of developed TE-based prosumer scheduling

signals (TIS). In each market time step, HEMS of prosumer buildings estimates the bid/offer prices of PV, energy storage devices, and HVAC based on the received prices, comfort settings, required and current SOC levels of energy storage devices, and temperature set-point limits of HVAC. Then, these estimated bids/offers are sent back to the DSO as transactive feedback signals. DSO clears the retail market by maximizing social welfare considering the network constraints. Based on the market-clearing prices, the real-time action of PV, HVAC, and energy storage devices will be determined.

For energy storage devices, if the local market-clearing price (MCP) is lower than the bid price, they will be charged. If the local market-clearing price is above the offer price, they will be discharged. Otherwise, they will be in idle mode until the next time step. For HVAC, in cooling mode of operation, if the MCP is greater than its bid price, the cooling set point will be moved higher, decreasing its consumption and if the MCP is less than its bid price, the cooling set point will be lowered further to take advantage of low market prices. PV sells the power only when it exceeds the self-consumption. If the local market-clearing price is greater than the PV's offer price, it sells power. Otherwise it will be stored in the BESS for future use. This section presents the bidding/offering strategies of energy storage devices, HVAC, and PV used in the proposed market clearing problem.

8.5.1 Bidding/offering strategy of energy storage devices (ESD)

Energy storage devices include EV batteries and BESSs. This work uses the model developed in [17] for the bidding strategy of EV and [18] for the offering strategy of ESD. Bid price model of ESD is given by,

$$\lambda_{ESD}^{buy} = \bar{\lambda} + (\sigma \times k \times \Delta ESD) \tag{8.11}$$

$$\Delta ESD = \frac{\text{Required time to fully charge}(t_{required})}{\text{Available time}(t_{available})} \tag{8.12}$$

$$t_{required} = \frac{(SOC_{desired} - SOC_{current}) \times E_{\max}}{P_{rated}} \tag{8.13}$$

$$t_{available} = t_{departure} - t_{current} \tag{8.14}$$

In (8.11), $\bar{\lambda}$ and σ are the mean and standard deviation of the price in the corresponding time step and will be sent by the retail market operator. k is the user's comfort control factor with the range of [0,1] and can be interpreted as the higher the value of k, the more comfort. $t_{required}$ can be calculated using (8.13), where $E_{\max}$ is the battery capacity in kWh and P_{rated} is the charging rate in kW. SOC is the State Of Charge of energy storage device. For EV battery, $t_{available}$ is the difference between the current time and the departure time. For BESS, $t_{available}$ is equal to 24 h as it will be available at home all the time. The offer price of ESD is estimated based on its discharge cost. The total discharging cost ESD is given in (8.15),

$$Cost_{dis} = Cost_{rech} + Cost_{wear} \tag{8.15}$$

First-term in (8.15) is the average cost of recharging during the following time steps, and the second term is the battery wear cost caused by the discharge. The detailed modeling of recharging cost and wear cost can be found in [18]. Offer price model of ESD is given by,

$$\lambda_{ESD}^{sell} = \frac{\bar{\lambda}}{\eta^{ch}\eta^{dis}} + 2 \times \frac{wp}{\eta^{dis}} \tag{8.16}$$

η^{ch} and η^{dis} are the charging and discharging efficiency of ESD. *wp* is the battery wear price.

The bidding and offering quantities of electricity by ESD are calculated by HEMS as given in (8.17) and (8.18) [18]:

$$P_{ESD}^{buy} = \min\left\{P_{rated}, \frac{\text{Residual capacity}}{\eta^{ch}\Delta T}\right\} \tag{8.17}$$

$$\text{Residual capacity} = (1 - SOC_{current}) \times E_{\max}$$

$$P_{ESD}^{sell} = \min\left\{P_{rated}, \frac{\eta^{dis}(SOC_{current} \times E_{\max} - E_{\min})}{\Delta T}\right\} \tag{8.18}$$

$E_{\min}$ is the minimum allowable SOC level.

8.5.2 Bidding strategy of HVAC

This work uses the model developed in [3,16] for the bidding strategy of HVAC. Bid price model of HVAC is given by,

$$\lambda_{HVAC}^{buy} = \bar{\lambda} + (\sigma \times k \times \Delta HVAC) \tag{8.19}$$

$$\Delta HVAC = \frac{T_{current} - T_{desired}}{T_{\text{maximum}} - T_{\text{minimum}}} \tag{8.20}$$

In this work, only the cooling mode of operation is explained. However, the working of heating mode is similar. In addition to the bid price model, HEMS should also calculate the bid quantity of HVAC. The actual demand of an HVAC system varies with many factors like the voltage, outdoor air temperature, house construction materials, and room temperature set point. So the exact demand model should consider the actual previous power demand of HVAC systems for the specified conditions and bid this quantity in the market. However, in this work, we consider the nameplate capacity of HVAC as demand quantity bid as detailed modeling of appliances is out of the scope of this work.

8.5.3 Offering strategy of PVs

$$\lambda_{PV}^{sell} = \bar{\lambda} + \sigma \tag{8.21}$$

$$P_{PV}^{sell} = \text{Max}(PV^{prod} - P_d, 0) \tag{8.22}$$

In (8.22), PV^{prod} is the predicted power generation from PV and P_d is the load demand. Like ESD and HVAC, we do not have any comfort factor for PV. So PV is bidding at the mean price scaled by standard deviation and PV sells power only when it exceeds the self-consumption as given in (8.22). It is assumed that the load demand and power generation from PV are predicted with acceptable accuracy before the bidding window.

8.6 Problem formulation

The considered test system is the radial distribution network with node 1 being connected to the substation. Power exchange with the upstream grid and network constraints are considered in the formulation. The objective function as formulated in (8.23) is the social welfare (SW) maximization problem:

$$SW = \sum_{n \in N\backslash\{1\}} \sum_{i \in I} \Pi_i(\lambda, P_i) - \lambda^{LMP} P_{grid}^{buy} + \lambda^{FIT} P_{grid}^{sell} \tag{8.23}$$

$$\begin{aligned}\Pi(\lambda, P) = {} & \lambda_{EV}^{buy} P_{EV}^{ch} + \lambda_{BESS}^{buy} P_{BESS}^{ch} + \lambda_{HVAC}^{buy} P_{HVAC} \\ & - \lambda_{EV}^{sell} P_{EV}^{dis} - \lambda_{BESS}^{sell} P_{BESS}^{dis} - \lambda_{PV}^{sell} P_{PV}\end{aligned}$$

$$0 \leq P_{n,i,EV}^{ch} \leq P_{n,i,EV}^{buy} \quad ; \forall i \in I, n \in N\backslash\{1\} \tag{8.24}$$

$$0 \leq P_{n,i,EV}^{dis} \leq P_{n,i,EV}^{sell} \quad ; \forall i \in I, n \in N\backslash\{1\} \tag{8.25}$$

$$0 \leq P_{n,i,BESS}^{ch} \leq P_{n,i,BESS}^{buy} \quad ; \forall i \in I, n \in N\backslash\{1\} \tag{8.26}$$

$$0 \leq P_{n,i,BESS}^{dis} \leq P_{n,i,BESS}^{sell} \quad ; \forall i \in I, n \in N\backslash\{1\} \tag{8.27}$$

$$0 \leq P_{n,i,PV} \leq \text{Max}(PV_{n,i}^{prod} - Pd_{n,i}) \quad ; \forall i \in I, n \in N\backslash\{1\} \tag{8.28}$$

where N defines the set of all nodes in the system and n is the index of nodes. L is the set of lines in the distribution network. (j, k) are the indices of buses, i is the index of prosumers, I is the set of prosumers at bus n. Ω_l^j is the set of all nodes which are connected to node j. λ^{LMP} and λ^{FIT} are the electricity price and Feed in tariff rate. From the social perspective, DSO desires to increase the sum of comfort obtained by each user and to decrease the cost. Equation (8.23) states that the scheduling will be optimal as the difference between total gross benefit of buyers and total cost of sellers i.e., surplus is maximized. Last two terms in (8.23) represent the cost and revenue of energy exchange with the main grid. The limits on prosumer's decision variables are given in (8.24)–(8.30)

- Energy exchanging constraints

$$0 \le P_{grid}^{buy} \le P^{grid,\max}.x \tag{8.29}$$

$$0 \le P_{grid}^{sell} \le P^{grid,\max}.\bar{x} \tag{8.30}$$

$$-Q^{grid,\max} \le Q^{grid} \le Q^{grid,\max} \tag{8.31}$$

$$x + \bar{x} \le 1 \tag{8.32}$$

Maximum allowable range of active power and reactive power exchange with the upstream grid is given by (8.29)–(8.31). Equation (8.32) implements the binary logic for the direction of flow of active power with the upstream grid.

- Power balance constraints

$$\sum_{i \in I} (P_{n,i,EV}^{dis} + P_{n,i,BESS}^{dis} - P_{n,i,EV}^{ch} - P_{n,i,BESS}^{ch} - P_{n,i,HVAC} + P_{n,i,PV})$$
$$-PD_{n,load} = \sum_{k \in \Omega_l^j} P_{jk}; \forall n \in N \backslash \{1\} \tag{8.33}$$

$$-QD_{n,load} = \sum_{k \in \Omega_l^j} Q_{jk} \quad ; \forall n \in N \backslash \{1\} \tag{8.34}$$

$$P_{in}^{grid} - P_{out}^{grid} = \sum_{k \in \Omega_l^j} P_{jk} \quad ; \forall n \in N\{1\} \tag{8.35}$$

$$Q^{grid} = \sum_{k \in \Omega_l^j} Q_{jk} \quad ; \forall n \in N\{1\} \tag{8.36}$$

Active and reactive power balances at the buses are formulated as (8.33)–(8.36). $PD_{i,load}$ and $QD_{i,load}$ are the unresponsive active and reactive power load at bus i.

- Electricity network constraints

$$P_{jk} = G_{jk} V_j^2 - G_{jk} V_j V_k \cos(\theta_j - \theta_k)$$
$$-B_{jk} V_j V_k \sin(\theta_j - \theta_k) \quad ; \forall \{j, k\} \in l \tag{8.37}$$

$$Q_{jk} = -B_{jk}V_j^2 + B_{jk}V_jV_k\cos(\theta_j - \theta_k)$$
$$-G_{jk}V_jV_k\sin(\theta_j - \theta_k) \quad ;\forall\{j,k\} \in l \tag{8.38}$$

$$-P_{jk}^{\max} \le P_{jk} \le P_{jk}^{\max} \quad ;\forall\{j,k\} \in l \tag{8.39}$$

$$-Q_{jk}^{\max} \le Q_{jk} \le Q_{jk}^{\max} \quad ;\forall\{j,k\} \in l \tag{8.40}$$

$$V^{\min} \le V_n \le V^{\max} \quad ;\forall\{n\} \in N \tag{8.41}$$

Equations (8.37) and (8.38) compute the real power and reactive power flow in the line. Equations (8.39)–(8.41) impose the limits on line flow and bus voltage magnitude. The market clearing prices are determined by solving the social welfare maximization problem. DSO sends these cleared prices to the HEMS in each prosumer house and then the real-time action of PV, energy storage devices, and HVAC will be determined by comparing the cleared prices and the submitted bid/offer prices. Energy storage devices will charge, if the retail market clearing price (MCP) is less than its bid price and if the MCP is greater than its offer price, it will discharge. Otherwise, it will be in idle mode until the next time slot.

The SOC of energy storage devices will be updated in each time slot as given in (8.42) below,

$$SOC_{new,ESD} = SOC_{old,ESD} + \frac{\eta_{ESD}^{ch} P_{ESD}^{ch} - \frac{1}{\eta_{ESD}^{dis}} P_{ESD}^{dis} \times \Delta T}{E_{\max}} \tag{8.42}$$

$$SOC_{\text{minimum}} \le SOC_{t,current} \le SOC_{\text{maximum}}$$

At every time slot, the SOC of energy storage devices should be within its maximum and minimum limits.

For HVAC, in cooling mode of operation, if the MCP is greater than its bid price, the cooling set point will be moved higher, decreasing its consumption and if the MCP is less than its bid price, the cooling set point will be lowered further to take advantage of low market prices. The temperature set point of HVAC will be updated in each time slot as given in (8.43) below,

$$T_{adjusted,HVAC} = T_{current,HVAC} + \frac{(\lambda^{clear} - \lambda_{HVAC}^{buy}) \times (T_{\text{maximum}} - T_{\text{minimum}})}{k \times \sigma} \tag{8.43}$$

$$T_{t,\text{minimum}} \le T_t \le T_{t,\text{maximum}}$$

8.7 Simulation results and discussions

For the simulation studies, a modified version of IEEE 13-node test distribution feeder [23] shown in Figure 8.10 is considered. For the sake of simplicity, it is assumed that the system under study is a balanced system, and just one transactive home from nodes 2, 3, 4, 6, 7, 9, 11, 12, and 13 are participating in the local energy market. Voltage regulators and capacitors from the original test feeder are not considered. The

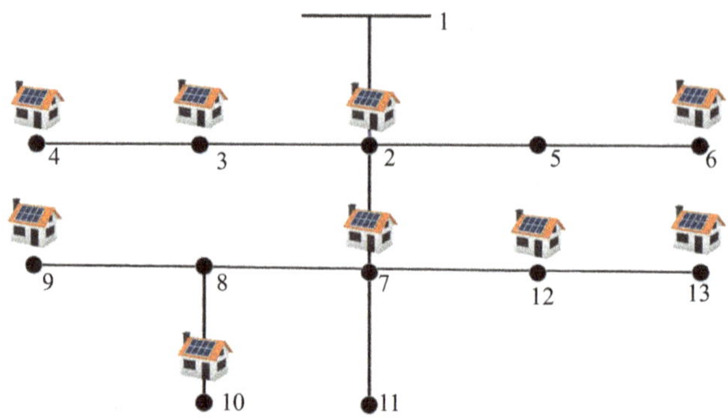

Figure 8.10 IEEE modified 13 node distribution feeder

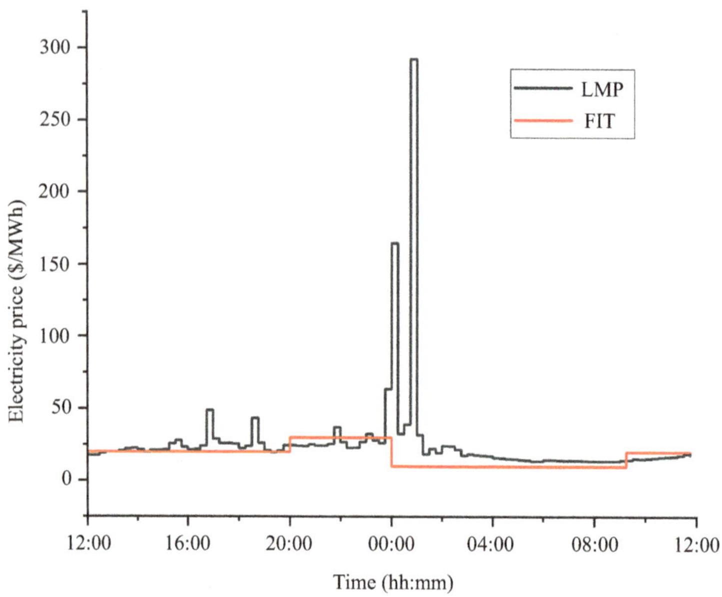

Figure 8.11 LMP and FIT rate of electricity price

developed social welfare maximization problem is solved using GAMS software. The electricity price (LMP) and feed-in-tariff (FIT) rate considered in this study are shown in Figure 8.11. All prosumers are assumed to be equipped with the same model of appliances. The comfort factor, initial SOC and desired SOC of energy storage devices, and the desired temperature of HVAC are assigned based on the prosumer's preference within the range defined in Table 8.4. Arrival and departure times of EVs are considered within the range defined in Table 8.4. Solar panel capacity and HVAC

Table 8.4 Prosumer parameters

Parameters	**Value**
Initial SOC of energy storage devices	0.3–0.6
Desired SOC of energy storage devices	0.7–1
SOC minimum and SOC maximum	0.1 & 1
Arrival time of electric vehicles	16:00–21:00 (hours)
Departure time of electric vehicles	06:00–8:00 (hours)
$E_{\max}$ and P_{rated} of EV	10 (kWh) and 4 (kW)
$E_{\max}$ and $P_{rated}$of BESS	32 (kWh) and 4 (kW)
η_{ch} and η_{dis}	0.95
Comfort factor	0.1–1
Wear rate	0.03 ($/kWh)
Rated power of HVAC	2.5 (kW)
Desired temperature of HVAC	(70–80) ° F
Minimum and maximum temperature limits	67° F and 84 ° F
Solar PV panel capacity	3 (kW)

rating of all the prosumer's household are assumed to be the same, and its value is given in Table 8.4. The values of other parameters used in the study like charging efficiency, discharging efficiency, battery wear rate, capacity, and power rating of energy storage devices are given in 8.4. The wear rate of energy storage devices is taken from [18]. The capacity and power ratings of energy storage devices are taken from [24]. The considered real-time market-clearing duration is 15 min. For all the time slots, the standard deviation of DLMP is assumed to be 0.1. The participation of prosumers in the local energy market will be started by submitting bids/offers to the DSO. The action of prosumer appliances will be determined by comparing the bid/offer prices with the market clearing prices by the HEMS. Figure 8.12 shows the local market clearing prices of all the time slots. In this work, we considered the available time of ESS as 24 h since, unlike electric vehicles, it will be available at home all the time and participating in the electricity market based on the prosumer's willingness. Figure 8.13 shows the SOC profile of BESSs, and Figure 8.14 shows the SOC profile of an electric vehicle during its available time at home. Energy storage devices have charged when the MCP is less than its bid price and discharged when MCP is greater than its offer price. Figure 8.15 shows the desired SOC level of prosumer EVs and SOC level at departure times. Based on prosumer's preferences, the SOC level at departure time has a slight variation with the desired level of SOC. The difference is around 5% which can be negligible. Figure 8.16 shows the temperature variation of HVAC at one of the prosumers who participated in the local energy market. The minimum and maximum temperature limits are 70° and 84°. As seen from Figure 8.16, whenever MCP is less than the bid price of HVAC, the cooling set point reduces than the current room temperature to take advantage of the low market prices. The current temperature is assumed to be the same for all time slots just to show the variation of adjusted temperature in accordance with the prices.

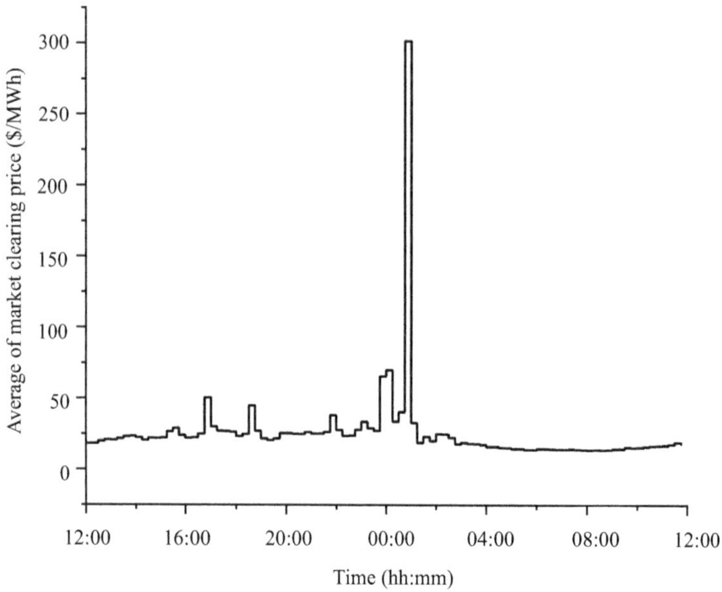

Figure 8.12 Average of market clearing prices

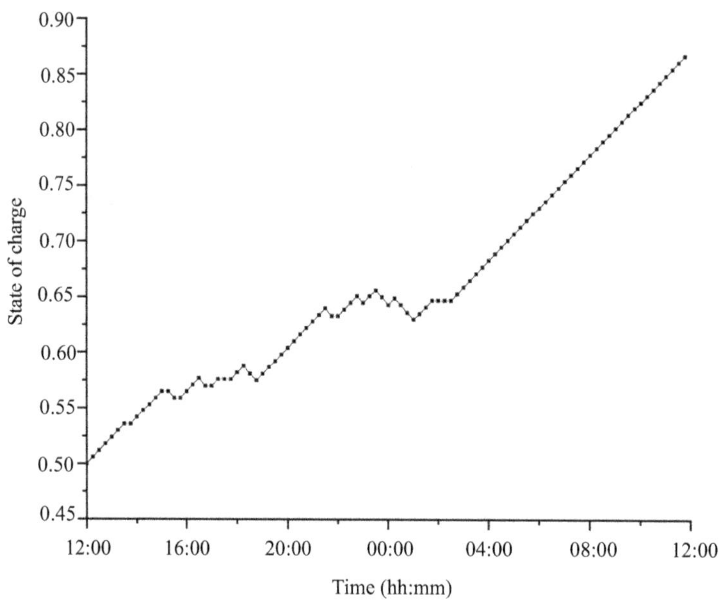

Figure 8.13 SOC profile of battery energy storage system

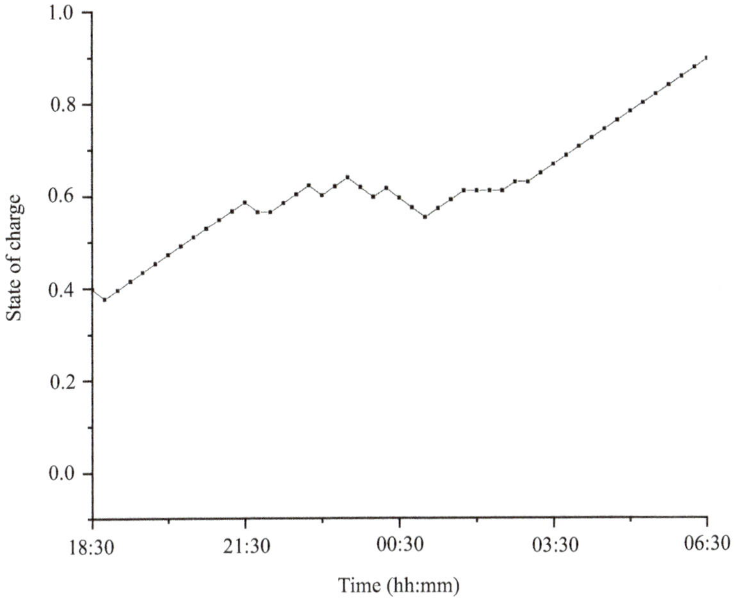

Figure 8.14 SOC profile of electric vehicle

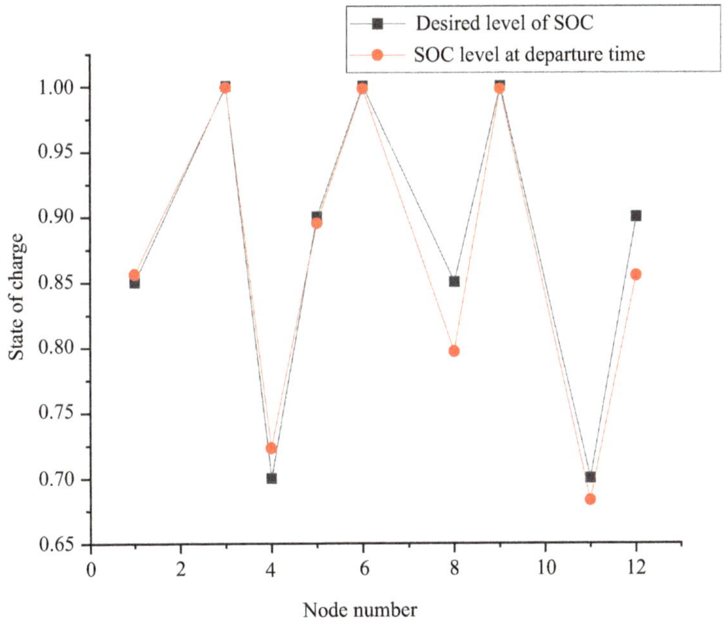

Figure 8.15 SOC level at departure time

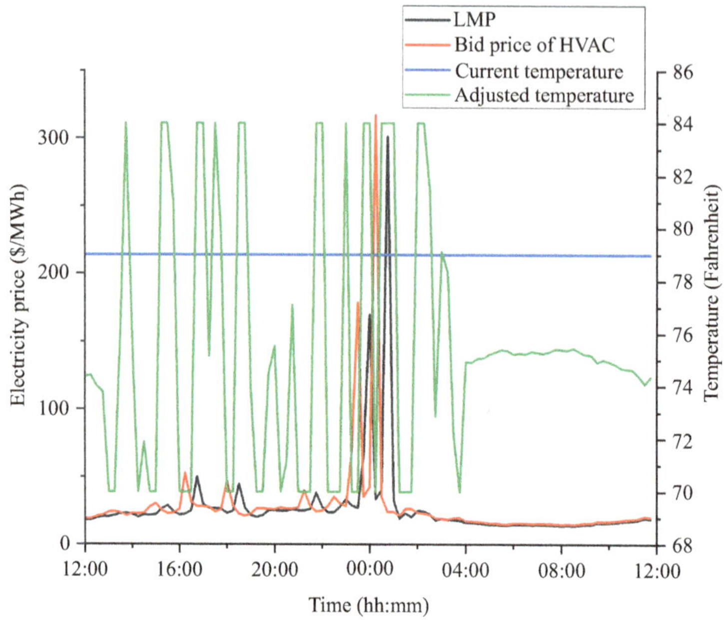

Figure 8.16 Operation of single HVAC transactive agent

8.8 Future works

In future studies, the linearization or convexification of ACOPF network constraints can be included in the model to ensure the global optimum. The developed model can be extended to include the other non-base loads of the prosumers. A decentralized optimization based market-clearing can be used to analyze the extension of the developed model.

8.9 Conclusion

This chapter analyzes two models for demand response programs based on the Time of Use price method. These models are compatible with all types of users with any level of flexibility. DR programs are applied only on the demand side to optimize the energy consumption in the network. Due to the shifting of the majority of loads from the peak period to the off-peak period in time-of-use price-based DR, the rebound effect may occur, which can be reduced by using real-time prices.

TE is the generalized form of price-based demand response which manages and controls the rate of generation in both grid side and demand sides. This chapter studies the scheduling of Prosumer's controllable appliance based on the TE concept. TE concept is two-way communication of data and energy. With more DERs at the distribution level, there is bidirectional power flow in the network. In this study, Prosumer is assumed to be equipped with HVAC, energy storage devices, and PV. The market is cleared based on the bids and offers from prosumers. Prosumers and the market operators can both benefit from the distribution market. The developed model is scalable, and it can be applied to the market with any number of participants.

References

[1] 2010 assessment of demand response and advanced metering staff report. Federal Energy Regulatory Commission; 2011. Washington, DC, USA.

[2] Abrishambaf O, Lezama F, Faria P, *et al.* Towards transactive energy systems: an analysis on current trends. Energy Strategy Reviews. 2019;26:100418. Available from: https://www.sciencedirect.com/science/article/pii/S2211467X19301105.

[3] Fuller JC, Schneider KP, Chassin D. Analysis of Residential Demand Response and double-auction markets. In: 2011 IEEE Power and Energy Society General Meeting; 2011. pp. 1–7.

[4] Vidyamani T, Swarup KS. Analysis of active and transactive demand response strategies for smart residential buildings. In: 2018 20th National Power Systems Conference (NPSC); 2018. pp. 1–5.

[5] Hammerstrom DJ, Ambrosio R, Carlon TA, *et al.* Pacific northwest gridwise testbed demonstration projects; Part I. Olympic Peninsula Project. Pacific Northwest National Lab (PNNL), Richland, WA, USA; 2008.

[6] Sharifi R, Fathi SH, Anvari-Moghaddam A, *et al.* An economic customer-oriented demand response model in electricity markets. In: 2018 IEEE International Conference on Industrial Technology (ICIT); 2018. pp. 1149–1153.

[7] Transactive System Part 1: Theoretical Underpinnings of Payoff Functions, Control Decisions, Information Privacy, and Solution Concepts; December 2017. US Department of Energy.

[8] Hu J, Yang G, Kok K, *et al.* Transactive control: a framework for operating power systems characterized by high penetration of distributed energy resources. Journal of Modern Power Systems and Clean Energy. 2017;5(3): 451–464.

[9] Rahimi FA, Ipakchi A. Transactive energy techniques: closing the gap between wholesale and retail markets. The Electricity Journal. 2012;25(8):29–35. Available from: https://www.sciencedirect.com/science/article/pii/S104061901200228X.

[10] Hao H, Corbin CD, Kalsi K, *et al.* Transactive control of commercial buildings for demand response. IEEE Transactions on Power Systems. 2017;32(1): 774–783.

[11] Daneshvar M, Mohammadi-Ivatloo B, Abapour M, *et al.* Energy exchange control in multiple microgrids with transactive energy management. Journal of Modern Power Systems and Clean Energy. 2020;8(4):719–726.

[12] Akter MN, Mahmud MA, Haque ME, *et al.* An optimal distributed energy management scheme for solving transactive energy sharing problems in residential microgrids. Applied Energy. 2020;270:115133. Available from: https://www.sciencedirect.com/science/article/pii/S0306261920306450.

[13] Marzband M, Fouladfar MH, Akorede MF, *et al.* Framework for smart transactive energy in home-microgrids considering coalition formation and demand side management. Sustainable Cities and Society. 2018;40:136–154. Available from: https://www.sciencedirect.com/science/article/pii/S2210670717307618.

[14] Hu J, Yang G, Bindner HW, *et al.* Application of network-constrained transactive control to electric vehicle charging for secure grid operation. IEEE Transactions on Sustainable Energy. 2017;8(2):505–515.

[15] Chassin DP, Fuller JC, Pratt RG, *et al.* Forward-looking transactive pricing schemes for use in a market-based resource allocation system. Google Patents; 2016. US Patent 9,245,297.

[16] Sun Y, Somani A, Carroll TE. Learning based bidding strategy for HVAC systems in double auction retail energy markets. In: 2015 American Control Conference (ACC); 2015. pp. 2912–2917.

[17] Behboodi S, Chassin DP, Crawford C, *et al.* Electric vehicle participation in transactive power systems using real-time retail prices. In: 2016 49th Hawaii International Conference on System Sciences (HICSS); 2016. pp. 2400–2407.

[18] Saber H, Ehsan M, Moeini-Aghtaie M, *et al.* Network-constrained transactive coordination for plug-in electric vehicles participation in real-time retail electricity markets. IEEE Transactions on Sustainable Energy. 2021;12(2):1439–1448.

[19] Nizami MSH, Hossain MJ, Amin BMR, *et al.* Transactive energy trading of residential prosumers using battery energy storage systems. In: 2019 IEEE Milan PowerTech; 2019. pp. 1–6.

[20] Zhao H, Xu Z, Wu J, *et al.* Optimal coordination of evs and hvac systems with uncertain renewable supply. In: 2019 IEEE 15th International Conference on Automation Science and Engineering (CASE); 2019. pp. 733–738.

[21] Wei T, Zhu Q, Maasoumy M. Co-scheduling of HVAC control, EV charging and battery usage for building energy efficiency. In: 2014 IEEE/ACM International Conference on Computer-Aided Design (ICCAD); 2014. pp. 191–196.

[22] Rayati M, Amirzadeh Goghari S, Nasiri Gheidari Z, *et al.* An optimal and decentralized transactive energy system for electrical grids with high penetration of renewable energy sources. International Journal of Electrical Power & Energy Systems. 2019;113:850–860. Available from: https://www.sciencedirect.com/science/article/pii/S0142061518336718.

[23] IEEE 13-node Distribution test feeder. [online]. Available from: https://site.ieee.org/pes-testfeeders/.

[24] Deilami S, Masoum AS, Moses PS, *et al.* Real-time coordination of plug-in electric vehicle charging in smart grids to minimize power losses and improve voltage profile. IEEE Transactions on Smart Grid. 2011;2(3):456–467.

Chapter 9

Industrial demand response opportunities with residential appliances in smart grids

Anam Malik[1] *and Jayashri Ravishankar*[1]

Supply-demand balance is imperative for the reliability of the power system. Inability to maintain this balance results in frequency deviation and system failure. The recent integration of renewable energy sources such as wind and solar have reduced inertia and variable output which leaves the power system at risk of disturbance while also reducing the controllability of generators. However, the latter-day demand response is coming across as an economical and effective way of adding to the reliability and security of power systems by managing the electricity demand of customers at times of severe power imbalance. This chapter carries out a detailed literature review of centralized and decentralized demand control approaches. As well as presents a novel demand control approach for providing frequency regulation by using domestic refrigerators as control loads. This chapter also carries out a detailed study of large-scale appliance level interval meter consumption data from Australia's largest network provider Ausgrid. Appliance level data is used in combination with household-level data to study the contribution of air-conditioners in summer peak demand. Clustering is performed on air-conditioner data to identify various air-conditioner load profile patterns. These patterns are then used with demand control strategies to study the possible load reductions from residential air-conditioner control across the Australian State of New South Wales.

Nomenclature

Symbols

*T*_therm	Thermostat temperature
*T*_comp	Compressor temperature
Total_Power	Total power being consumed by refrigerators in utility
Total_Power_CoHEM	Total power being consumed by refrigerators in a particular CoHEM
*P*_comp	Compressor power of refrigerators

[1] School of Electrical Engineering and Telecommunications, The University of New South Wales (UNSW), Sydney, Australia

f_ref	Nominal frequency of power system
f	Frequency of power system
M	Scaling parameter
load_o	Total number of ON refrigerator units at the previous iteration
df	Frequency error
load_c	New number of refrigerator units that need to be ON/OFF
Total_Units_OFF	Total number of OFF refrigerators in all three CoHEMs
New_load	New refrigerator load
Control_Total	Number of refrigerator units whose compressor cycle needs to be manipulated

9.1 Introduction

Proper functioning of the power system relies on maintaining continuous supply-demand balance. Supply-demand imbalance results in frequency deviation from the nominal value that may result in power system failure. Therefore, reducing power mismatch in real-time is critical for the best operation of power system. However, maintaining power balance always is not easy as many unforeseen circumstances such as distribution and transmission line outages, frequent changes in customer demand, generation outages etc., may occur [1].

Historically, automatic generation control (AGC) is used for maintaining supply-demand balance. AGC works on the principal that supply always follows demand. Output from multiple generators is manipulated such that supply-demand balance is maintained. However, the recent penetration of renewable energy sources in power systems has rendered AGC insufficient. Renewable energy sources such as wind and solar have reduced inertia and variable output which leaves the power system at risk to disturbance while also reducing the controllability of generators. Fifty per cent penetration of renewables in the power system reduces system inertia to half of its nominal value [2].

Literature in past has mentioned studies where AGC is able to maintain frequency within a narrow offset without renewables. However, with the penetration of 50% renewables in the power systems AGC alone failed to maintain frequency within desired limits [2].

To rely on AGC alone for frequency regulation in power systems with renewables, there will be a need to spend considerable additional cost on generators that offer responsive services at times of frequency deviation, resulting in huge economic burden and inefficiency of resources [3].

9.2 Demand peaks

The residential sector accounts for 30–40% of total energy consumption worldwide [4]. In the past few years, residential demand has been seen as a major contributor to electricity demand peaks recording as high as 45% in the UK [5] and more than 50% in South Australia [6] and New Zealand [7].

Electricity demand peaks occur when many customers coincidentally turn their heating or cooling devices on, on hot summer or cold winter days. These peaks are infrequent; however, the network businesses need to invest in additional generators and upgrade their network to meet such infrequent peak demand requirements [4]. A study of the load duration curve for South Australia reveals that demand equals or exceeds 60% of maximum demand for only 10% of the time [6].

Another study reveals that the zone substation of the Sydney region in NSW serves the top 10% of peak demand for only 0.5% of the times [8]. The recent increase in electricity prices in Australia has been attributed to the increase in network investment to meet peak demands. Such expenditures have resulted in a 50% increase in residential electricity bills in Australia [9].

9.3 Demand response

In the latter day, demand response (DR) is coming across as an economical way of complementing AGC. DR adds to the reliability and security of power systems by managing the electricity demand of customers at times of severe power imbalance. DR manipulates customer loads and maintains an even load profile while avoiding the need for network augmentation and huge investment in generators [10].

DR is believed to offer added benefits over AGC some of which are reduced response times, pollution free, economically viable, widely spread in distribution network, etc. [11].

It has been reported in a study [10] that with just 10% of consumers participating in the DR, great peak reductions of about 5.6% can be achieved. If 25% of consumers are active, the network losses reduce by 2.6%.

It is believed that just a 5% cut in electricity demand could have reduced electricity prices by one-half during the extreme electricity crisis seen by California in the years 2000–01. The reason being generation price considerably rises when generators run close to their maximum generating capacity. Therefore, curtailing customer demand during these severe electricity peaks can save generation costs resulting in reduced electricity bills for customers [1].

9.4 Thermostatically controlled loads (TCLS)

Thermostatically controlled customer appliances such as air-conditioners (aircons), fridges, freezers, and electric water heaters have large thermal inertia and show great demand response potential. Their large thermal inertia allows a delay in the appliance on and off operation for short time periods without affecting their function ability and customer comfort, therefore, making them suitable candidates for DR [12].

Considerable literature in past has confirmed the DR capability of TCLs. An experiment performed on 25 refrigerators confirmed their potential for secondary frequency control [13]. In another work [14], peak demand hours were saved in the memory of the fridge. During the peak hours, the fridge would turn off operation. However, the devised strategy failed to cater for demand peaks outside the hours set in fridge memory.

Some studies relied on centralized DR for TCLs. In a centralized DR approach a central, controller device communicates with all connected devices, signalling them to turn their operation ON or OFF. In [15], a centralized DR strategy was proposed where electric water heaters were used as control loads and were turned ON or OFF with the aim of eliminating frequency offset. Simulations were carried out both with and without wind generation. The simulation results verified the success of centralized DR approach with electric water heaters in eliminating frequency error. However, coincident signals sent out to all devices to turn ON or OFF at the same time resulted in the phenomena of synchronization, where the duty cycles of all devices sync up resulting in severe oscillations that can lead to power system failure. The centralized control approach reduces the chances of error but at the same time it has the disadvantage of increasing complexity of data processing tasks as well as communication cost on utility as a single controller device is responsible for communicating to all connected devices [16].

Much of the work considers decentralized DR strategies which rely on more than one centralized controller. The strategy involves deploying frequency measurement units in every household that continuously compare nominal frequency to real-time frequency. Any frequency deviation outside a narrow band of nominal frequency sends a signal to the connected household device to start or seize operation for a short time. Work in [17] suggested a stochastic decentralized approach for manipulating the operation of several TCLs for maintaining frequency close to nominal frequency. However, the study failed to control switching events of TCLs within a given time resulting in the possibility of TCLs to switch multiple times in a short time interval. Another decentralized demand control technique was suggested in [18]. Simulation results carried out on 1,000 HVAC units confirmed the performance of the proposed strategy in regulating frequency. Unlike a single controller in centralized control approach the presence of multiple frequency measurement units reduces the communication cost as well as data processing burden on the utility. However, the decentralized control approach has drawbacks like inaccurate measurements from frequency measurement units as well as the additional cost associated with installing these units in every household [16].

Few studies [19–22] introduced a hybrid control approach using aggregators as links between utility and customers. The aggregators negotiate incentives with customers for responding to signals by manipulating their device operations. In turn, the utility pays the aggregator for managing the supply–demand balance. However, the hybrid approach involved complex data processing and resulted in reduced profits for utilities as now they had to pay aggregators for executing frequency regulation for them. Table 9.1 summarizes the limitations of centralized, decentralized, and hybrid control techniques [23].

Considering the limitations highlighted in Table 9.1 a hybrid architecture is proposed in a case study that is expected to exploit the benefits of both centralized and decentralized demand control strategies without hampering utility profits [24].

Table 9.1 Limitations of centralized, decentralized and hybrid control techniques [23]

Centralized control approach	**Decentralized control approach**	**Hybrid control approach**
High cost associated with maintaining secure two-way communication network. Huge processing burden on centralized controller.	Local measurement of frequency signal with accuracy is difficult. Frequency measurement units are expensive to be installed in every house.	Involves three different tiers (utility, aggregator, and customers) resulting in complex data processing tasks that are difficult to evaluate.
Feasible only with limited number of control loads.		Reduced utility profits.

9.5 Case study 1: hybrid control approach for frequency regulation

Refrigerators are TCLs that have previously been used for DR. In [3], refrigerators were used as control loads for frequency regulation. Simulation results confirmed that refrigerators could provide ancillary services like spinning reserves.

9.5.1 Refrigerator modelling

Refrigerators are cooling devices that operate by maintaining compartment temperature (T_comp) within a nominal value of the thermostat temperature (*T*_therm) decided by the end-user. Figure 9.1 shows the operation of a refrigerator [24]. The refrigerator thermostat is set at 5°C and has an offset of ±2°C which implies that the refrigerator compartment temperature should stay between 3°C and 7°C. Maintaining compartment temperature within these limits is critical for the safety of food inside the fridge.

To utilize refrigerators as control loads for DR, additional refrigerator states have been added to the conventional ON and OFF states. Table 9.2 explains the condition for each refrigerator state as well as the availability of the refrigerators for load manipulation in each state.

The new added critical states prevent the sudden ON and OFF operation in refrigerators. Without critical states, if for example the compartment temperature hit 7°C and the compressor just turned ON, on receiving an OFF signal it will turn OFF and then immediately turn back ON to keep the temperature within limits. The critical states help avoid these sudden ON, OFF in refrigerators in turn adding to the robustness of the controller.

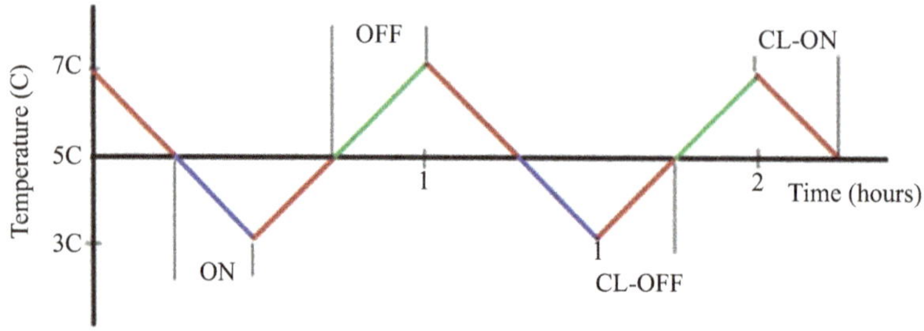

Figure 9.1 Compressor cycle and refrigerator states [24]

Table 9.2 Conditions for each refrigerator state and refrigerator availability for load manipulation

State	Condition for each state	Availability for load manipulation
State 0, OFF	T_comp > T_therm & compressor = OFF	Compressor is available to be switched ON.
State 1, ON	T_comp < T_therm & compressor = ON	Compressor is available to be switched OFF.
State 2, Critical Load OFF (CL-OFF)	T_comp < T_therm & compressor = OFF	NOT available for load manipulation.
State 3, Critical Load ON (CL-ON)	T_comp > T_therm & compressor = ON	NOT available for load manipulation.

9.5.2 DR controller description

DR at the customer level will require home energy management systems (HEMs) to be deployed in local households. HEMs are screens installed in households capable of measuring the energy consumption of customers in given time periods [25]. It will act as a link between utility and customers and will assist in load scheduling. Studies in [26,27] have verified the ability of HEMs in scheduling customer loads at times of low electricity price, therefore benefiting the customers. However, HEMs may cause volatility in demand and result in rebound peaks. Rebound peaks occur when many customers schedule loads at times of low electricity price resulting in a new peak demand at those times. This problem can be solved by using a cooperative home energy management system (CoHEM) that will communicate with the utility and will coordinate the operation of many HEMs with the aim of avoiding new peaks as well as down-turning utility revenues.

This study introduces a hybrid architecture where CoHEMs will be installed at distribution transformers and each CoHEM will be able to communicate only with

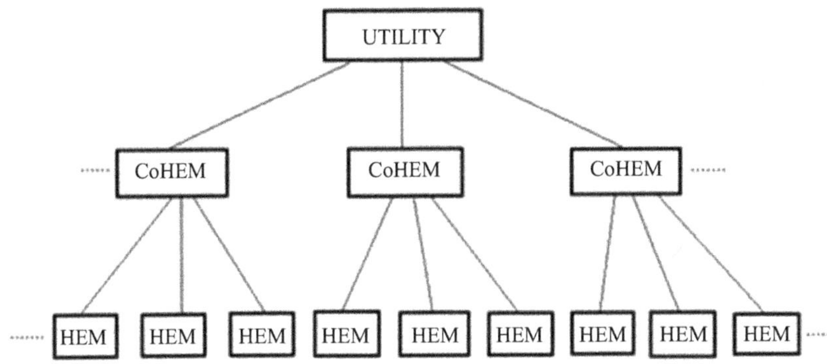

Figure 9.2 Recommended architecture with CoHEMs acting as link between utility and end-users [24]

HEMs connected to that distribution transformer [24]. Figure 9.2 shows the hybrid architecture of utility, CoHEMs, and HEMs.

Refrigerators are the control loads used for frequency regulation in this study. Refrigerators will continuously update HEM of their compartment temperature. HEM will use the refrigerator thermostat, offset, and compartment temperature to compute refrigerator state as given in Table 9.2. It will then forward the refrigerator state to CoHEM. On receiving refrigerator states from all HEMs connected to it, it will calculate the total power being consumed by refrigerators as shown in Figure 9.3. Each CoHEM will send their calculated total refrigerator load to the utility (Total_Power_CoHEM). On receiving the values from each CoHEM utility will calculate the total refrigerator load (Total_Power):

$$\begin{aligned}\text{Total\_Power} = {} & (\text{Total\_Power\_CoHEM1}) + (\text{Total\_Power\_CoHEM2}) \\ & + \underline{(\text{Total\_Power\_CoHEM3})}\end{aligned} \tag{9.1}$$

The utility continuously measures system frequency (f) and compares it against nominal frequency (f_ref). A deviation ($df = f - f$_ref) of ±0.05 Hz from nominal frequency is acceptable. Table 9.3 lists the various actions that need to be taken in case of frequency deviation.

In the event where system frequency is not within acceptable bounds, the utility will use Hill-Climbing Method to calculate the total number of refrigerators whose compressor operation needs to be manipulated to regulate frequency.

9.5.3 HillClimbing method

Hill-Climbing method uses a constant scaling factor 'M' that is used to curtail frequency error [15]. The number of refrigerator units required to be in ON state (load_c) is then calculated using the number of refrigerators at the previous iteration (load_o) and df. Load_c will be greater than load_o in case of df>0 suggesting that additional

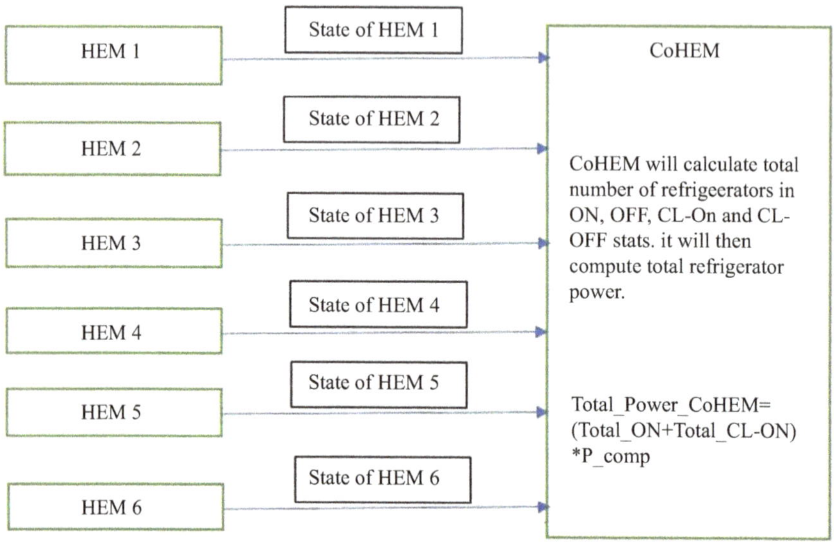

Figure 9.3 Steps required for calculating total refrigerator load at CoHEM

Table 9.3 Control action required based on frequency deviation value

Frequency deviation value	**Action required**
If abs|df| < 0.05	No action required.
If abs|df| > 0.05 and df < 0	Additional refrigerators need to be turned OFF to regulate frequency.
If abs|df| > 0.05 and df > 0	Additional refrigerators need to be turned ON to regulate frequency.

refrigerator units need to be turned ON. Load_c has to be greater than zero. However, it cannot be greater than the total number of refrigerators available to be turned ON:

$$\text{Total available units to be turned ON} = \text{load\_o} + \text{Total\_Units\_OFF} \qquad (9.2)$$

New_load can be calculated by multiplying load_c with refrigerator compressor power (P_comp). Also, Control_Total gives the total number of refrigerator units whose compressor operation needs to be manipulated. Figure 9.4 outlines all the steps involved in Hill Climbing method [24].

Once the Control_Total is computed, signals will be sent by the utility in a round-robin approach. The control signal will first be sent to CoHEM 1. In case where CoHEM 1 has a smaller number of refrigerators than that computed by the utility for manipulation, the control signal will be sent to CoHEM 2 and then to CoHEM 3 if required. Control signals will be sent out until the frequency is regulated within

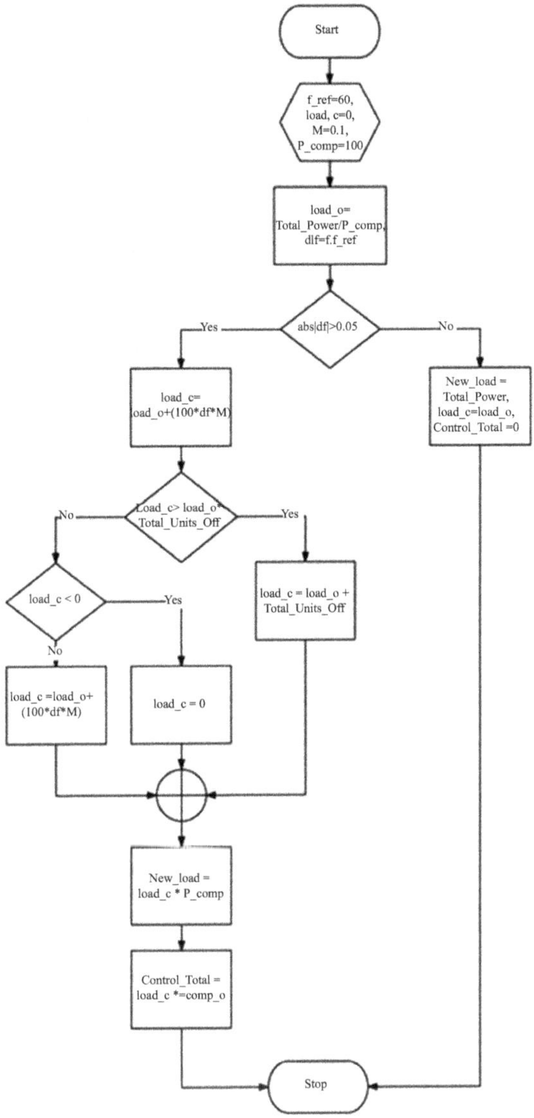

Figure 9.4 Hill Climbing method [24]

limits. Round robin approach ensures that the same CoHEM is not selected every time for frequency regulation.

9.5.4 System description

A power system with both wind and diesel generation is considered in Simulink for analysis. Twenty-four refrigerators each with P_comp of 100 W are considered.

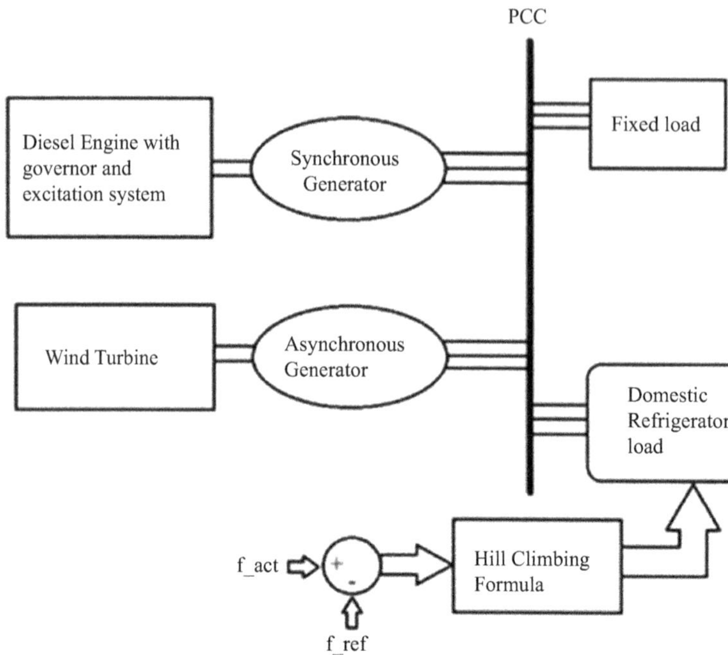

Figure 9.5 Configuration of proposed system [15]

The refrigerators are treated as control loads, whereas all other loads are fixed loads and do not alter their operation in case of frequency deviation outside nominal value. The system configuration is shown in Figure 9.5 and is a modified system from [15]. The latter only considered a conventional power system with the diesel engine only.

9.5.5 Simulation results

A fixed load of 6,800 W is considered for simulations. The control load is 2,400 W.

$$\text{Control load} = \text{Total number of refrigerators} * P\_\text{comp} \tag{9.3}$$

Control load = 24 * 100 = 2,400 W

Four different scenarios are considered to compare the performance of the system with and without DR. Two cases are considered where the wind speed is dropped resulting in reduced wind energy. In the other two cases, the fixed load is varied. Table 9.4 discusses the four different scenarios and compares frequency regulation results with only AGC and with both AGC and DR.

9.5.6 Discussion

Simulation results validate the superior performance of control cases over non-control cases in all four scenarios. It is evident from Table 9.4 that AGC alone is insufficient

Table 9.4 Frequency regulation performance comparison between the controller and non-controller cases under four different scenarios

	Scenarios without controller (AGC alone)	**Frequency regulation with the proposed controller (AGC + DR)**	**Frequency regulation manipulation**	**Control load**
Scenario 1 (Figure 9.6)	Wind speed reduced from 9.5 to 8 m/s resulting in wind power reduction of 1,300 W.	Frequency falls below 49.6 Hz and oscillates around that point. df > 0.05 Hz	Frequency restored in 1.5 s within a narrow dead band of 49.5–50.5 Hz.	Refrigerator load dropped from 1,100 W to 100 W by controller to regulate frequency.
Scenario 2 (Figure 9.7)	Wind speed reduced from 9 to 7.5 m/s resulting in wind power reduction of 1,500 W.	At 20 s: frequency dropped to 45.5 Hz. At 24 s: frequency dropped to 37 Hz.	At 20 s: frequency dropped to 48.7 Hz. At 24 s: frequency dropped to 47.5Hz.	Refrigerator load dropped from 1,500 W to 100 W by controller to regulate frequency.
Scenario 3 (Figure 9.8)	Load changed from 6,200 W to 6,800 W at 15 s for 1.5 s and reduced afterward to 6,500 W for the rest of the simulation. Wind power fixed at 1,700 W.	Without the controller frequency was regulated in 3 s.	With the controller frequency was regulated in a little over 1 s.	Refrigerator load dropped from 1,800 W to 300 W by the controller to regulate frequency.
Scenario 4 (Figure 9.9)	Load changed from 6,800 W to 7,700 W at 15 s. Wind power fixed at 2,500 W.	The frequency mismatch was 0.3 Hz. AGC alone was unable to regulate frequency.	The frequency mismatch was 0.3 Hz same as df in no controller case. Frequency regulated within acceptable bounds in 1.5 s.	Refrigerator load dropped from 1,600 W to 100 W by controller to regulate frequency.

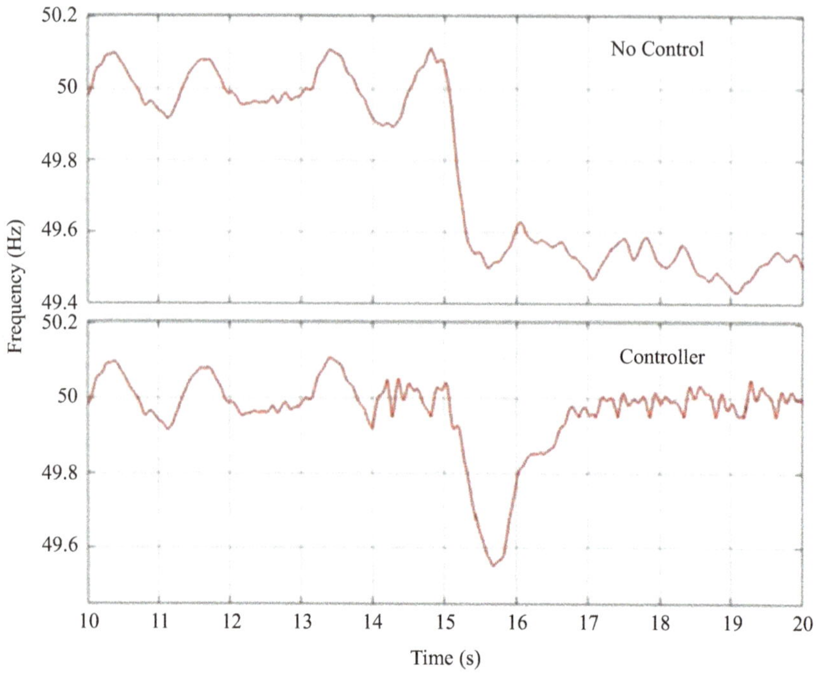

Figure 9.6 System frequency graph for no control and controller case under Scenario 1 [24]

in regulating frequency in a renewable-rich power system. However, AGC and DR together are better able to regulate frequency. In Scenario 2 both controller and non-controller cases were unable to maintain frequency within acceptable limits after a drop in wind energy. Even so the controller was able to restrict frequency drop to 47.5 Hz compared to the 37 Hz without the controller. In scenarios where both controller and non-controller cases were able to regulate frequency, the controller showed improved performance by regulating frequency in shorter time periods.

The proposed hybrid architecture is also believed to exploit the benefits of all three centralized, decentralized and hybrid controllers. Unlike centralized controller, the presence of multiple controllers (CoHEMs) reduces the computational burden on utility. Similarly, the proposed architecture offers advantages over decentralized controllers by using CoHEMs to coordinate the operation of HEMs avoiding rebound peaks. The suggested architecture is similar in concept to hybrid control systems where aggregators act as a link between utility and customers. However, with CoHEMs instead of aggregators the utility does not need to downturn its revenues by paying aggregators to carry out DR services on their behalf.

This case study highlights the effectiveness of TCL in regulating frequency. A clear understanding of key drivers of demand peak can help devise DR strategies for reducing power mismatch and hence the need for frequency regulation.

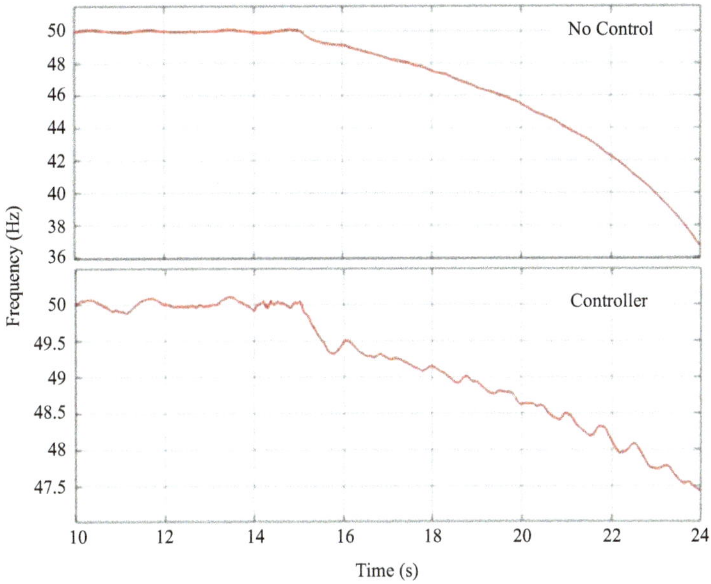

Figure 9.7 *System frequency graph for no control and controller case under Scenario 2 [24]*

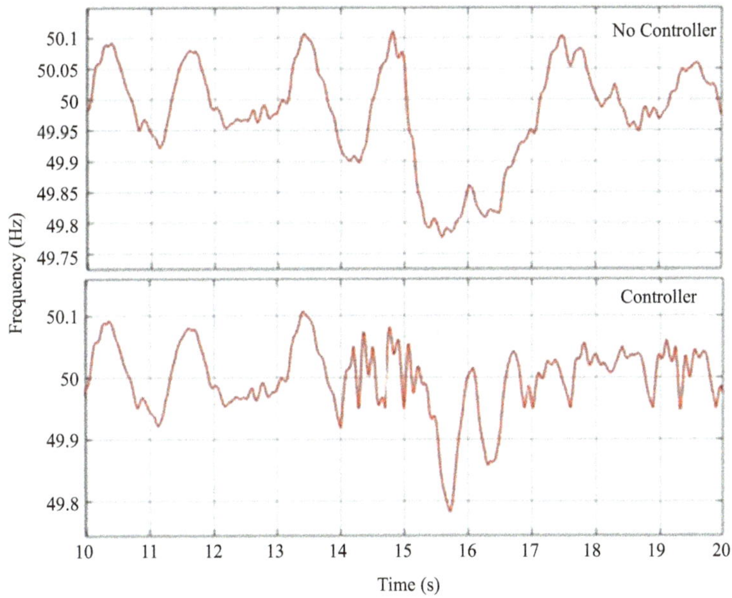

Figure 9.8 *System frequency graph for no control and controller case under Scenario 3 [24]*

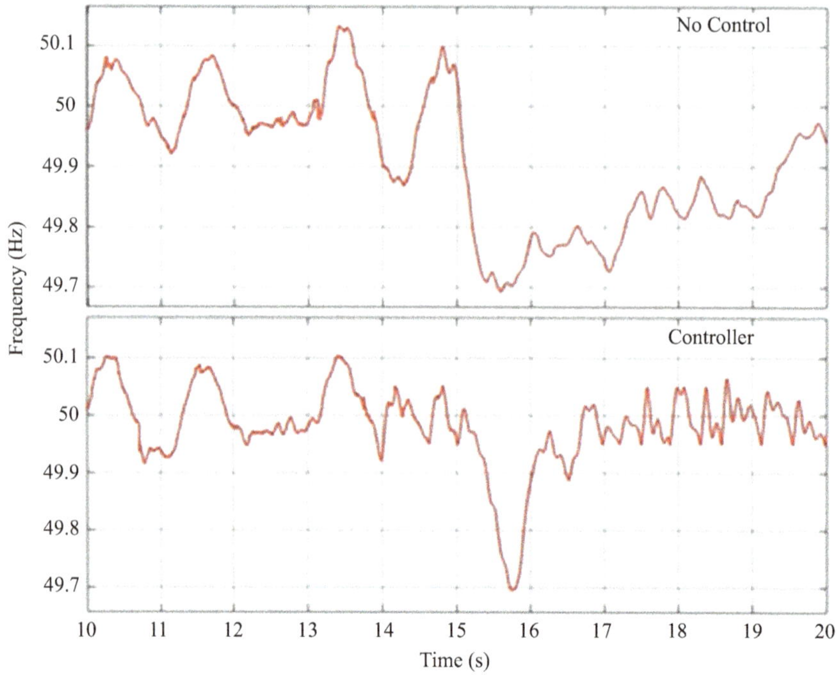

Figure 9.9 System frequency graph for no control and controller case under Scenario 4 [24]

9.6 Case study 2: appliance level data analysis of summer demand reduction potential from residential aircons

Aircons and electric water heaters have been identified as major contributors to peak demand in many jurisdictions on extreme weather days. Such extreme weather days in Australia pose risk of blackouts due to extremely high customer demand as well as high stress on the electricity network [28].

The true cost of aircons in Australia is identified in [29] where it is said that a $1,500 aircon can impose a cost of $7000 on electricity industry when adding to peak demand.

Many research studies have confirmed the DR capabilities of aircons and electric water heaters during extreme demand periods. However, much of the studies to date rely on interval metered household consumption data to estimate the contribution of aircons and electric water heaters to peak demand. Where aggregated data is successful in identifying demand peak times it fails to highlight the key appliances contributing to peak demand in those times. Appliance level interval metered data is much more valuable in studying the contribution of multiple household appliances to peak demand. However, very few of such data sets are available, and consumption is of course very context-specific.

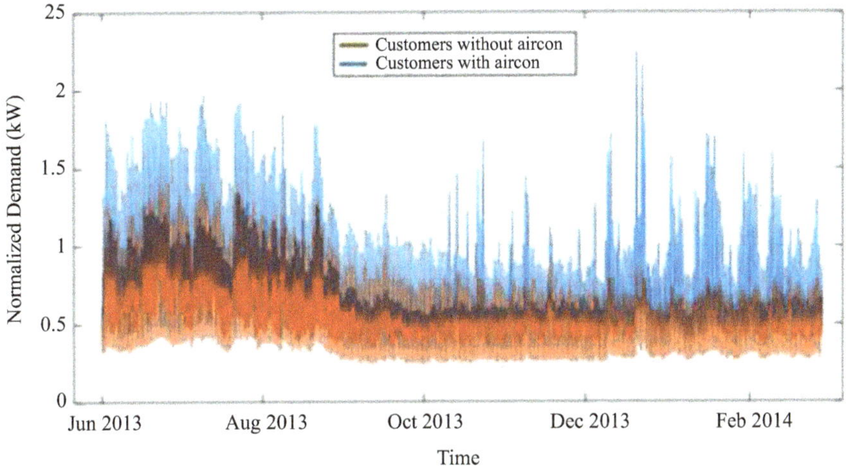

Figure 9.10 Normalized demand of customers in SGSC house level data with and without aircon [32]

In Australia, the Smart Grid Smart City (SGSC) project carried out by the Ausgrid distribution business has unlocked a great opportunity by providing half-hourly interval metered household data as well as appliance level data for a relatively larger sample of households in NSW, Australia. This dataset can be used for studying the actual contribution of various household appliances to peak demand.

SGSC data has previously been used in [30,31]. In [30] SGSC household interval metered data was used along with survey information to find the key peak demand contributors. Data analysis and modelling techniques suggested aircon ownership, number of household members, swimming pool and dryer ownership as being the major contributors to peak demand. Another work in [31] studied SGSC household interval metered data along with local weather forecast to identify aircon existence, energy consumption patterns and peak demand. However, household level data, survey results and weather forecasts are not very accurate in identifying particular appliance usage in peak demand as the appliance usage maybe correlated with other factors in the study. Therefore, the proper understanding of appliance level data is required.

This study is one of the first studies that used SGSC appliance level data to study aircon usage during summer peak demand times. The SGSC data in its original form was unworkable with problems like missing entries, unsynchronized data, and duplicate entries. Several steps were taken to treat the data. A data availability rate (a measure of how many of the total time periods being analysed were missing data) was calculated for the treated data. A smaller data availability rate suggests more missing data entries whereas a higher rate suggests high data quality. A data availability rate of 75% was chosen for the analysis. At this data availability rate, there were 127 unique customers and 64 aircons [32].

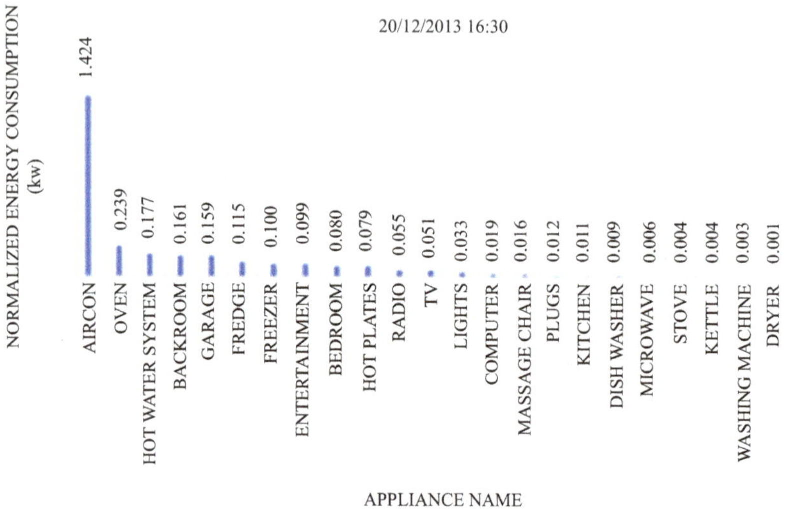

Figure 9.11 Normalized consumption (kW) of different appliances at the time of overall state summer peak, 20 December 2013 [32]

9.6.1 Summer peak demand analysis

Household consumption data was separated for customers with and without aircon ownership using SGSC household metered data along with survey results. Normalized demand of customers with and without aircons is shown in Figure 9.10 for a full year [32].

It is evident from Figure 9.10 that customers with aircon showed higher demand in summers. The demand for customers with aircon ownership peaked in December owing to extreme weather in Australia. The highest demand occurred on 20 December 2013 which was an extremely hot day. Average consumption of various household appliances was also studied over the peak period as shown in Figure 9.11 [32].

Aircon contribution is by far the highest during the state demand peak of December 2013 contributing over 80% more than other appliances.

To further build up analysis aircon consumption, appliance level and household level data are studied in Figure 9.12 [32]. The highest summer peak is highlighted and zoomed in to show the contribution of aircon load in appliance level and household peak demand. It is apparent that aircon load makes up most of the load in appliance level data and is a key contributor to household-level demand.

9.6.2 DR opportunities with aircons

A study of individual aircon loads averaged over three summer state peak days is shown in Figure 9.13 [32]. Aircon usage is distributed throughout with each aircon displaying varied usage patterns. For identifying the DR potential of aircons at peak times, it is important to first categorize aircons based on their usage patterns. K-means

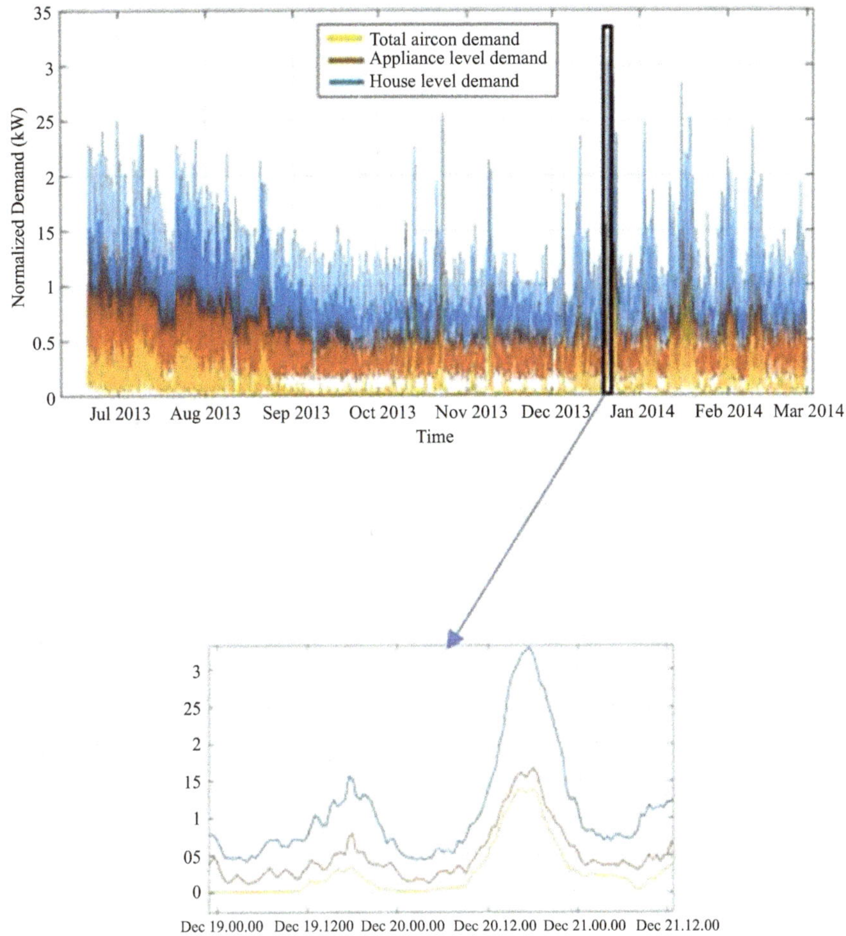

Figure 9.12 Normalized average contribution of monitored aircon load in total monitored appliance level demand, and overall (interval metered) household demand, including on the peak system demand day [32]

clustering is performed and aircons are categorized in six different clusters as shown in Figure 9.14 [32].

Aircons in clusters 1 and 4 do not show much load contribution at times of NSW state demand peak. Whereas aircons in clusters 2, 3, and 5 (58% of aircons) contribute significantly towards peak demand and show great DR potential. Aircons in cluster 6 are not operating at times of peak demand.

The combined operating power of aircons in clusters 2, 3 and 5 calculated over peak times over three summer peak days from the NSW state load for the year 2013–14 is 4.508 kW. A study conducted in [33] shows that increasing aircon thermostat by 1°C

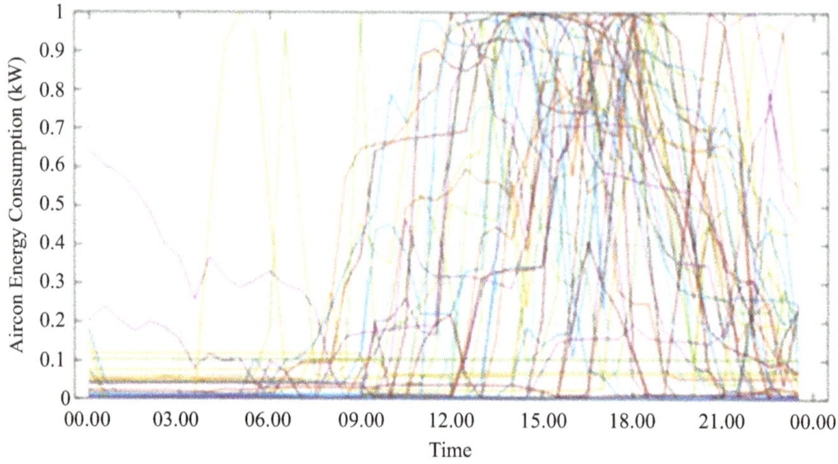

Figure 9.13 Individual household aircon consumption profile, averaged over the three highest summer state peak days [32]

during a DR event that lasts two hours can result in a 25% power reduction. Assuming similar responses, this case study will carry out the possible demand reductions by using 58% aircons for DR.

Increasing thermostat of 58% aircons by 1°C will result in a new operating power.

$$\text{New operating power} = \text{combined operating power} * 0.75 = 3.381\ \text{kW} \quad (9.4)$$

A survey conducted in (Ausgrid, 2015) on NSW aircon ownership rate suggests aircon ownership of roughly 1.66 million households in NSW. Upscaling share of clusters by number of households with aircon we estimate 967,000 aircons (58%) can be used for DR.

$$\text{Power reduction} = (\text{combined operating power} - \text{new operating power}) * 967,000 \quad (9.5)$$

Power reduction = (4.508-3.381) * 967,000 = 1.09 MW

Thermostat increase by just 1°C for residential aircons in NSW can help achieve peak demand reduction of over 1 MW.

9.7 Conclusion

This chapter has endeavoured to improve our understanding of the limitations of current frequency regulation strategies as well as peak demand studies. To overcome these limitations, several steps have been introduced to devise a novel frequency regulation strategy that does not suffer from previous problems. Also, a detailed study is carried out for appliance contribution to demand peaks and possible DR scenarios for peak reduction at the state level.

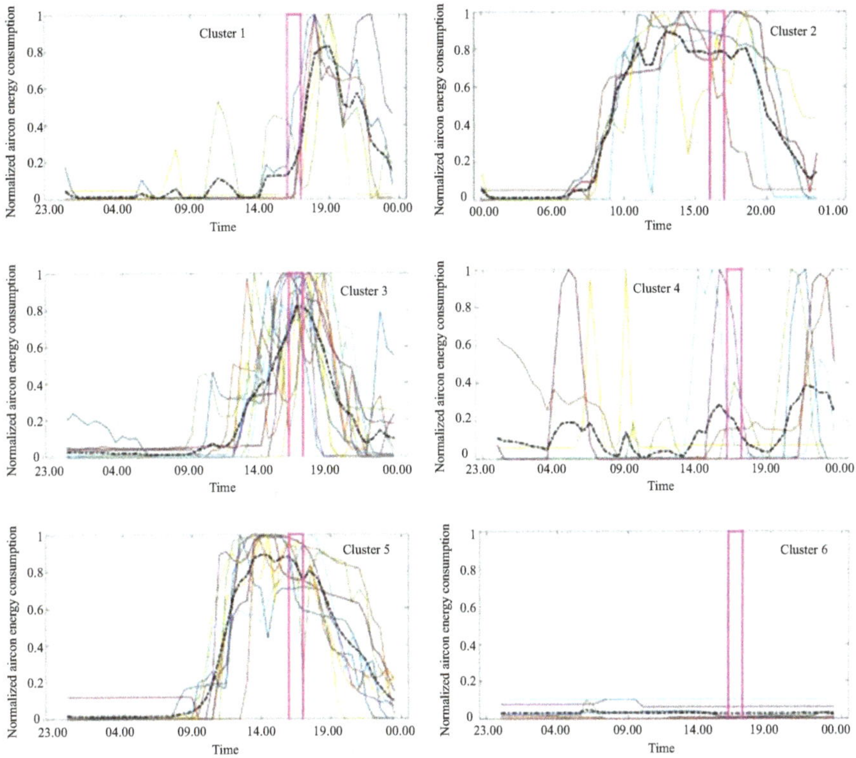

Figure 9.14 Clusters of aircon energy consumption profiles over the three highest summer state peak days. The black broken line shows the average normalized energy consumption of aircons in a cluster and the magenta bar shows the NSW state peak demand times [32].

The dataset undertaken for case study 2 is geographically limited. The aircon usage patterns as well as contribution to peak demand may vary in other NSW regions particularly where temperatures are more extreme. Also, the dataset from households across a larger geographical area would improve the quality of analysis and enhance understanding of aircons as key contributors to peak demand. Despite the limitations of this case study, and while the power reductions are just estimates the methods introduced are of greater potential relevance. It is hoped that this study will contribute to the work of utilities and DR service providers.

References

[1] Albadi M.H., El-Saadany E.F. 'A summary of demand response in electricity markets'. *Electric Power Systems Research*, 2008, vol. 78(11), pp. 1989–96.

[2] Liu T., Hill D.J., Zhang C. 'Non-disruptive load-side control for frequency regulation in power systems'. *IEEE Transactions on Smart Grid*, 2016, vol. 7, pp. 2142–53.

[3] Short J.A., Infield D.G., Freris L.L. 'Stabilization of grid frequency through dynamic demand control'. *IEEE Transactions on Power Systems*, 2007, vol. 22, pp. 1284–93.

[4] Haider H.T., See O.H., Elmenreich W. 'A review of residential demand response of smart grid'. *Renewable and Sustainable Energy Reviews*, 2016, vol. 59, pp. 166–78.

[5] Darby S.J., McKenna E. 'Social implications of residential demand response in cool temperate climates'. *Energy Policy*, 2012, vol. 49, pp. 759–69.

[6] 'SA-Department-for-Transport-Energy-and-Infrastruct.pdf', 2011.

[7] Electricity Commission. 'New Zealand electric energy-efficiency potential study'. Vol. 1. 2007

[8] Haghdadi N., Bruce A., Macgill I., Passey R. 'Impact of distributed photovoltaic systems on zone substation peak demand'. *IEEE Transactions on Sustainable Energy*, 2018, vol. 9(2), pp. 621–29.

[9] Australian Energy Regulator. 'State of the Energy Market'. 2017.

[10] Safdarian A., Fotuhi-Firuzabad M., Lehtonen M. 'Benefits of demand response on operation of distribution networks: a case study'. *IEEE Systems Journal*, 2016, vol. 10, pp. 189–97.

[11] Xu Z., Østergaard J., Togeby M. 'Demand as frequency controlled reserve'. *IEEE Transactions on Power Systems*, 2011, vol. 26(3), pp. 1062–71.

[12] Zheng Y., Hill D.J., Zhang C., Meng K. 'Non-interruptive thermostatically controlled load for primary frequency support'. *Power and Energy Society General Meeting (PESGM)*, 2016.

[13] Lakshmanan V., Marinelli M., Hu J., Bindner H.W. 'Provision of secondary frequency control via demand response activation on thermostatically controlled loads: solutions and experiences from Denmark'. *Applied Energy*, 2016, vol. 173, pp. 470–80.

[14] Benysek G., Bojarski J., Smolenski R., Jarnut M., Werminski S. 'Application of stochastic decentralized active demand response (DADR) system for load frequency control'. In *IEEE Transactions on Smart Grid*, 2018, vol. 9(2), pp. 1055–62.

[15] Pourmousavi S.A., Nehrir M.H. 'Real-time central demand response for primary frequency regulation in microgrids'. *IEEE Transactions on Smart Grid*, 2012, vol. 3, pp. 1988–96.

[16] Dehghanpour K., Afsharnia S. 'Electrical demand side contribution to frequency control in power systems: a review on technical aspects'. *Renewable and Sustainable Energy Reviews,* 2015, vol. 41, pp. 1267–76.

[17] Tindemans S.H., Trovato V., Strbac G. 'Decentralized control of thermostatic loads for flexible demand response'. *IEEE Transactions on Control Systems Technology*, 2015, vol. 23, pp. 1685–700.

[18] Rui X., Liu X., Meng J. 'Dynamic frequency regulation method based on thermostatically controlled appliances in the power system'. *Energy Procedia*, 2016, vol. 88, pp. 382–88.
[19] Papavasiliou A., Hindi H., Greene D. 'Market-based control mechanisms for electric power demand response'. In *Proceedings of the IEEE Conference on Decision and Control (CDC)*, 2010, pp. 1891–98.
[20] Hindi H., Greene D., Laventall C. 'Coordinating regulation and demand response in electric power grids using multirate model predictive control'. In *Proceedings of the IEEE PES Conference on Innovative Smart Grid Tech.*, 2011, pp. 1–8.
[21] Gkatzikis L., Koutsopoulos I., Salonidis T. 'The role of aggregators in smart grid demand response markets'. *IEEE Journal on Selected Areas in Communications*, 2013, vol. 31, pp. 1247–57.
[22] Kim H., Thottan M. 'A two-stage market model for microgrid power transactions via aggregators'. *Bell Labs Technical Journal*, 2011, vol. 16, pp. 101–07.
[23] Malik A., Ravishankar J. 'A review of demand response techniques in smart grids'. *IEEE Electrical Power and Energy Conference (EPEC)*, 2016, pp. 1–6.
[24] Malik A., Ravishankar J. 'A hybrid control approach for regulating frequency through demand response'. *Applied Energy*, 2018, pp. 1347–62.
[25] Siano P. 'Demand response and smart grids—a survey'. *Renewable and Sustainable Energy Reviews*, 2014, vol. 30, pp. 461–78.
[26] Chang T.H., Alizadeh M., Scaglione A. 'Coordinated home energy management for real-time power balancing'. In *Power and Energy Society General Meeting*, 2012, pp. 1–8.
[27] Chang T.H., Alizadeh M., Scaglione A. 'Real-time power balancing via decentralized coordinated home energy scheduling'. *IEEE Transactions on Smart Grid*, 2013, vol. 4(3), pp. 1490–504.
[28] Wilkenfeld G. 'A National Demand Response Strategy for Small Airconditioners'. 2004.
[29] 'Energy White Paper 2012'. 2012.
[30] Fan H., MacGill I.F., Sproul A.B. 'Statistical analysis of drivers of residential peak electricity demand'. *Energy Building*, 2017, vol. 141, pp. 205–17.
[31] Smith R., Meng K., Dong Z., Simpson R. 'Demand response: a strategy to address residential air-conditioning peak load in Australia'. *Journal of Modern Power Systems and Clean Energy*, 2013, vol. 1(3), pp. 223–30.
[32] Malik A., Haghdadi N., MacGill I., Ravishankar J. 'Appliance level data analysis of summer demand reduction potential from residential air conditioner control'. *Applied Energy*, 2019, pp. 776–85.
[33] Hu M., Xiao F. 'Investigation of the demand response potentials of residential air conditioners using grey-box room thermal model'. *Applied Energy*, 2017, vol. 105(2017), pp. 2759–65.

Chapter 10

Modelling and optimal scheduling of flexibility in energy-intensive industry

Roman Cantu[1], *Emilio José Palacios-García*[1], *and Geert Deconinck*[1]

10.1 Introduction

Current environmental trends such as the rapid penetration of renewable energy resources (RES) and decommissioning of controllable but polluting generators are putting stress on the reliable operation of electricity systems. This reduction of flexibility and the increase in volatility in the supply-side call for compensation from other sources in the grid. Although the development of energy storage systems (ESS) is creating an opportunity to relax the energy balance constraints in the grid, it is currently not sufficient to solve the constantly growing need for flexibility [1].

Many researchers identify demand-side flexibility as a feasible approach to tackle this problem [2]. Therefore, it has been under research for some years already. However, most studies focus on commercial and residential consumers only [3,4]. Even though these sectors are usually less constrained than the industrial sector, which boosts their capacity to shift or curtail their demand, industry accounts for approximately 42% of the worldwide electricity consumption [5]. Hence, the industry shows flexibility potential that could be aggregated with other consumers' to provide greater grid balancing capacities.

Multiple constraints bound the ability to alter production plans in industrial facilities from technological, economic, or practical perspectives. While power generation facilities comprise a single or only a few processes, industrial production usually compounds several interdependent processes [6,7]. Such inter-dependencies appear as shared input or output resources [8], sequential structures in the production line [9], mutually exclusivity, etc. Following this line, some actions in industrial plants provide flexibility that results in a feasible increment or decrement of local energy consumption, i.e., adjustments in production line order, machine re-assignments, worker shifts, rescheduling of processes, capacity investments, among others.

Furthermore, an industrial energy consumer does not concentrate its optimisation efforts on its energy consumption [1,10]. Energy has been historically perceived

[1]KU Leuven ESAT-ELECTA, EnergyVille, Leuven, Belgium

rather as a resource that is constantly available at a fixed price. These industrial consumers prefer to focus on profit maximisation, which depends more strongly on bottlenecks and idle times in their production processes. However, this focus is slowly changing due to the current trends in energy systems and the global objectives towards decarbonising human activities, simultaneously increasing the importance of energy use in production planning and scheduling.

Various authors have proposed approaches to model and schedule industrial demand-side flexibility. Their models are a basis for an inclusive modelling methodology which combines all the relevant characteristics and remains simple but applicable across energy-intensive industrial sectors.

This chapter exposes the need for an industrial flexibility model and its requirements by identifying characteristics from contemporary grid balancing approaches. Section 10.2 gives an introduction of the need for flexibility and provides a comparison between the flexibility of diverse sectors of players in the electricity system. Considering the differences between other flexible resources and industrial loads, Section 10.3 reviews characteristics and existing formulations that should be part of an industrial energy system model. Once identified, these characteristics give the foundations for a high-level modelling framework for an energy-aware production planning formulation, whose description is included in Section 10.4. Finally, a case study in Section 10.5 exemplifies the use of the framework in a production line, where the flexible operation of the system results in economic benefits by performing a two-step sequential scheduling. This is on the day-ahead and intraday markets (DAM and IDM, respectively).

10.2 Understanding flexibility across electricity consumer sectors

Historically, flexibility has been sourced from the supply-side, considering the demand profiles as variable signals that require to be predicted and matched by the total available generation while keeping some capacity to cover for forecasting errors. On the supply side, flexibility from conventional generators is evident since their technological constraints are enough to know the upwards and downwards limitations. Furthermore, power plants produce and sell electricity to consumers intending to maximise their profit. In contrast, system users on the demand-side only perceive electricity as a utility required to fulfil some specific tasks, usually without considering its price variation [10] and only focusing on their primary business objective.

Demand-side management (DSM) comprises a set of practices to plan, implement and monitor energy-consuming activities to shape the consumption pattern as desired [11]. These practices include but are not limited to energy efficiency policies, demand response (DR) programs and on-site backup generation strategies. Multiple authors identify industrial DSM as one of the most cost-effective practices to cope with the flexibility requirement of current electricity systems [2]. This observation results from the intensive power consumption in this sector, their already available monitoring and control systems, the currently low participation of this type of consumers in balancing services and the increasing economic incentives for ancillary services providers [9].

DR is the most relevant DSM practice to tackle potential imbalances in the grid. It encloses actions that influence the magnitude and timing of a load's power consumption to balance supply and demand in the electricity system. The trigger for these measures is incentives such as time-varying electricity prices (price-based) or direct load control from the system operators after capacity contracting (incentive-based) [12–14]. Hence, DR programmes often bring external incentives to exploit the energy flexibility in the consumers' premises.

Both price- and incentive-based DR can be implemented in all electricity markets, depending on the type of consumer. For large commercial and industrial consumers, the DR signals come from wholesale electricity markets such as spot markets, ancillary services markets and capacity markets [15,16]. In contrast, small consumers receive stimuli to participate via Real Time Pricing (RTP), Time of Use (TOU) or Critical Peak Pricing (CPP) mechanisms since they are often subject to retail markets through an electricity supplier.

Recent studies explore the application of DR programs to tackle the flexibility needs in the grid. However, most of these studies focus on the residential and commercial sectors and ignore the high potential of the industrial sector, which is studied by only a few works [3,17]. However, the electricity consumption in the industrial sector accounted for approximately 42% of the total world's consumption and 37% of the European's consumption (EU-28) in 2018 [5]. As a large consumer, this sector has an inherent responsibility to participate in stability and reliability measures for the electricity system.

In addition to the fewer constraints in commercial and residential loads, DR studies in such sectors have been enabled by features such as the high energy intensity of heating, ventilation, air conditioning, and refrigeration technologies and their associated thermal inertia. Additionally, controllable appliances provide a high load shifting potential since most of their operations are not tightly time-constrained (e.g., washing machines, dishwashers, drying machines, electric vehicles, etc.) [18].

Energy flexibility in industrial contexts is more complex to quantify and operate since several factors influence its potential, which can be internally and externally bounded by technological, practical, and economic limitations as depicted in Figure 10.1. Furthermore, some of the feasible commercial flexibility might not comply with the requirements of the power system, which shrinks the consumer potential to participate in balancing markets.

Industrial processes usually comprise sub-processes that transform raw materials into final products or take part in this transformation as an intermediate step. Moreover, many of these processes and sub-processes are interdependent, which further shrinks their combined feasible operational region [6,7]. These links between processes often appear as shared resources between multiple processes such in the case of successive processing stages where the output of a process is the input of the next one [9]. Thus, efficient and effective scheduling of industrial loads requires consideration of all these interactions.

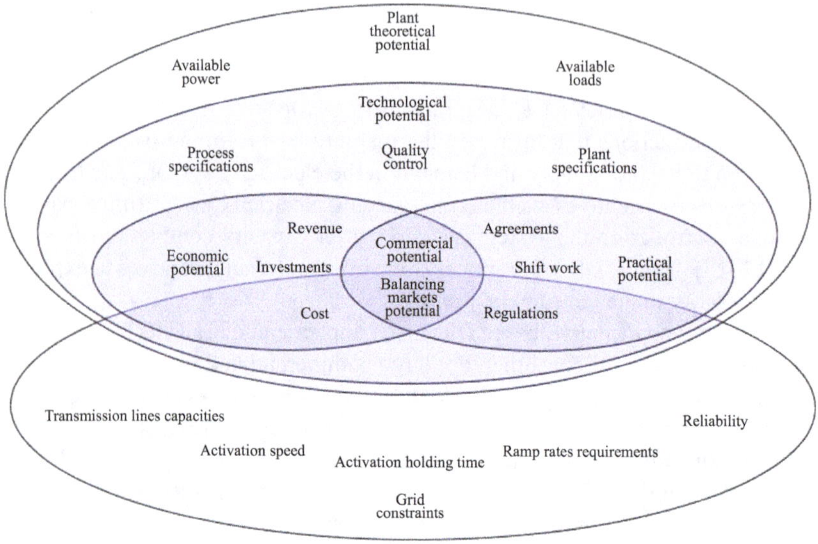

Figure 10.1 Industrial flexibility potential domains and influencing factors binding its feasible commercial capacity. Adapted and translated from [19].

10.3 Basis for an industrial flexibility model

Model is a term used for a mathematical structure that aims to reproduce the characteristics of a system. It should be accurate but simple enough to enable its use in practical applications [12]. Thanks to models, some dynamics, relationships, and behaviours of a system can be studied, explained, and optimised. The quality of the results from using a model will depend on the model's accuracy to reproduce real-world characteristics and its applicability on different systems. According to Maher and Williams [20], there are multiple types of models: mathematical programming, simulation, network planning, econometric, and time series models. We aim to build a mathematical programming model that enables operation and design optimisation of energy-flexible industries. In this regard, the size of the feasible solution space, the optimality of the results and the simplicity of the solution algorithm depend on the formulation selected [21].

A useful industrial flexibility model results from the analysis of characteristics describing: grid requirements, operation of supply-side flexibility (which has been extensively studied) and models describing the demand-side flexibility in other sectors (i.e., commercial and residential). This analysis is crucial to ensure compatibility between diverse sources and to foster the integration of industrial demand-side flexibility into balancing mechanisms. In this section, the characteristics reviewed target formulations of demand in electricity systems with an emphasis on those that allow for flexible consumption decisions. The first appearances of such characteristics in the text are emphasised with a bold font. At the end of this section, Table 10.2 describes these characteristics to clarify what these features represent in a system.

10.3.1 European grid balancing services

The definition of products' bids in balancing markets can be used as a starting point to build an industrial flexibility model since this is one of the mechanisms in which flexibility generates value from both perspectives: the grid system operator's and the industrial consumer's. Aspects related to the amount of power, reaction times, holding duration and transition times are part of the parameters that define the flexibility capacity according to European reserve markets, which is expected to enter in operation in the near future. Note that the reserves mechanisms described in this section tackle transactions between transmission system operators (TSOs) at a European level. These mechanisms aim to enhance the cooperation between TSOs in charge of different control blocks to improve the reliability of the coupled European grid. However, they still help understand the characteristics required in an electricity flexibility mechanism because traded products will comprise a portfolio of previously procured (by TSOs) downstream flexible loads and generators. Figure 10.2 depicts the relevant aspects in a generic graph of a flexibility source. Most of the parameters shown here have restrictions according to the reserve market as stated by the European Network of Transmission System Operators for Electricity (ENTSO-E). In this figure, the quantity (vertical) axis measures the power deviation offered by the balancing service provider (BSP). All the loads and generators in the system must be in a BSP portfolio. BSP are responsible for the imbalance caused by their portfolios after gate closure of the DAM and IDM. Thus, the TSO charges or pays a BSP according to its

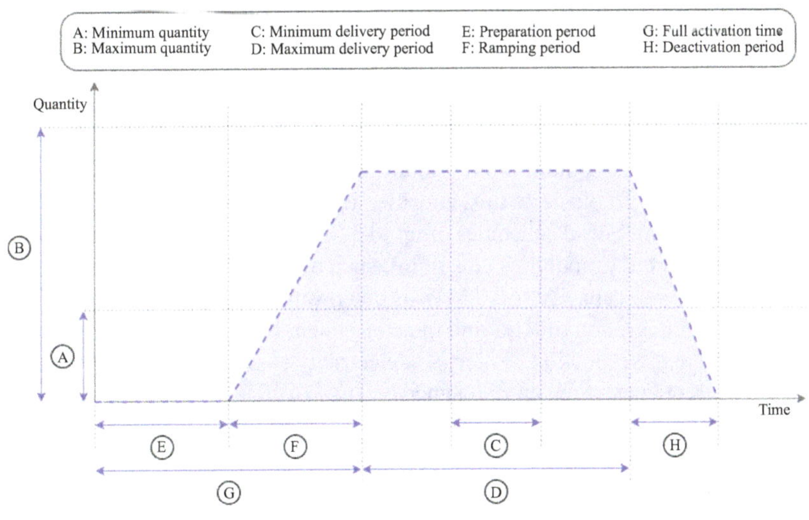

Figure 10.2 Visual representation of characteristics of energy products in reserves markets [22]

position after delivery. This position is a balance that considers the DAM and IDM commitments and the physical production/consumption measured after delivery.

Before going into the details of each characteristic, it is worthwhile introducing the various reserves markets here. All markets have multiple names, according to the country where they operate. First, the frequency containment reserve (FCR), also known as primary reserves or R1, is the first measure against imbalances in the grid. FCR's objective, as its name implies, is to contain any frequency changes via activation of fast flexibility sources to prevent any further drops or rises in the frequency. These sources are activated automatically in the synchronous area by controllers implemented in the FCR-contracted generators or loads in a decentralised manner. Second, there are the automatic frequency restoration reserves (aFRR), also known as secondary reserves or R2. Once the frequency is stable thanks to the FCR, the aFRR comes into play to drive the frequency back to its correct value (50 Hz in Europe) employing automatic controllers. Third, the manual frequency restoration reserves (mFRR), also known as tertiary reserves or R3, operate in combination with the aFRR with the same objective. Both aFRR and mFRR are activated locally (where the imbalance occurred) by the TSO within the load frequency control area (LFC area). There is a difference in full activation time (FAT) between these last two types of reserves. Thus, the aFRR is activated automatically first but, minutes afterwards, the TSO manually activates some reserves from the mFRR contributing to the imbalance correction within the LFC area and releasing some of the capacity used from the aFRR. Lastly, some TSOs use replacement reserves (RR) to release the capacity of both of the frequency restoration reserves, making them available again for future imbalances. At the end of this section, Table 10.1 summarises the constraints in the TSO–TSO reserves markets in Central Europe.

The first parameter to address is the reserved quantity which has **minimum and maximum limits**. These limits apply in FCR and aFRR, where the minimum is 1 MW for both and the maximum 25 MW (for each indivisible bid) and 9,999 MW, respectively. These minimum and maximum requirements in different markets, combined with the unavoidable consumption or generation limitations of assets, make this an important parameter to characterise flexibility. Additionally, all the balancing products in the market contain a bid granularity of 1 MW, making a portion of the flexibility unusable. To illustrate this, consider a generator with a free capacity (after committing their day-ahead schedule) of up to 5.5 MW upwards and 2.1 MW downwards. Due to the bid granularity constraint, such a generator can only offer 5 MW as upward regulation capacity and 2 MW as downward regulation capacity.

Due to the integration of RES and the consequent displacement of conventional generation, there is a reduction of inertia in electricity systems that are causing faster variations in frequency when an imbalance occurs [16]. This phenomenon translates into a fast response requirement from flexibility sources, which highlights the importance of **ramping rates** in the characterisation of these sources. Depending on the intrinsic reaction time of flexible resources, they can participate in different balancing reserve markets: FCR, aFRR and mFRR, which are required to achieve full activation times of maximum 30 sec, 5–15 min, and 12.5 min, respectively. The variation in the full activation times for aFRR depends on the country as depicted in Figure 10.3, but

Table 10.1 Characteristics, constraints, considerations and differences between TSO–TSO reserves markets in Central Europe. Sources: ENTSO-E Balancing Report 2020 [22] and ENTSO-E Market report 2021 [26].

Characteristic	FCR (FCR cooperation)	aFRR (PICASSO)	mFRR (MARI)	RR (TERRE)
Minimum quantity	1 MW	1 MW	–	–
Maximum quantity	25 MW [1]	9,999 MW	–	–
Bid granularity [4]	1 MW	1 MW	1 MW	1 MW
Minimum delivery period	15 min	–	5 min	*
Maximum delivery period	30 min	–	–	*
Full activation time (FAT)	30 s	5 min [3]	12.5 min [2]	30 min
Preparation period	–	–	–	0–30 min
Ramping period	–	–	–	0–30 min
Deactivation period	–	$\leq$ FAT	–	–
Validity periods	4 h	15 min	15 min	*
Auction frequency	6×daily	6×daily	6×daily	24×daily
Gate opening time (GOT)	–	D-1 @12:00	D-1 @12:00	T-70 min
Gate closure time (GCT)	–	T-25 min	T-25 min	T-55 min
Product bid horizon	6×4 h	6×4 h	6×4 h	1 h

[1] For indivisible bids.
[2] Homogenised after 24 July 2022.
[3] Homogenised after 18 December 2024. Currently, each TSO defines their FATs for aFRR.
[4] Activation request can be lower than the granularity.
* Only discrete delivery periods are available: 15, 30, 45 or 60 min.

shall be homogenised to 5 min by December 2024 to comply with the requirements of ENTSO-E [22].

Ramping rates stand out as features that characterise the flexibility from extensively studied sources such as those in the supply-side [24,25]. These features describe the maximum rate of change at which the flexible generator/load can change its power production/consumption. They result from the intrinsic inertia of processes, which physically limit the speed of change of their operation set-point. Furthermore, these changes can happen in both directions: an increase or decrease in the power production/consumption, each associated with a corresponding ramping rate. Hence, two different parameters are often used and identified as **ramp-up and ramp-down** rates, respectively. These two parameters can be visualised in Figure 10.2 as the slopes of the lines during periods F and H.

In that figure, it is also possible to observe how the duration of the reserves is another factor to consider in the bids. This is because network operators require the BSPs to maintain their balancing power activated in order to correct the frequency rate of change, its stable value, or to give time to other BSPs to take over, accordingly to the requirements of the specific type of reserve. For instance, BSPs contracted as FRR must operate once activated until units in the RR reach their full activation. If

Table 10.2 Summary of characteristics identified as a basis for a generic industrial energy planning model and their relation to our framework

Characteristic	Modelled by*	References	Description
(1) Max/Min power	Flow boundaries	[12,15,22, 24,27–34, 37]	(see Section 10.3.1)
(2) Ramping rates	Ramping rates	[12,15,17, 22,27–29, 32,33,37]	(see Section 10.3.1)
(3) Minimum up/down times	Minimum up-time and down-time	[8,12,15, 17,22,27, 29,31,32]	(see Section 10.3.1)
(4) Startup costs	Start-up detection	[12,15,27]	Costs incurred due to, e.g., startup energy requirements, set up costs, wear of machines, etc.
(5) Transmission constraints	–[1]	[27]	Safe limits of power transmission elements including the underlying power distribution dynamics across the grid.
(6) Min/max active duration	Minimum up-time[2]	[8,12,15, 17]	Minimum and maximum duration of flexibility activation. Complementary to (3).
(7) Max number of activations	–[3]	[12,15,17, 30]	Maximum number allowed of activation instances for the flexibility measure.
(8) Flexibility activation costs	–	[12,15]	Costs related to opportunity costs, penalties and fees as a consequence of short notice changes, changes in work shifts, etc.
(9) Energy level at deadline	–[4]	[12,15,17, 27,30,32]	Required level of energy in an ESS that must be acquired by the end of the time horizon.
(10) Energy capacity limits	Storage capacity limits	[8,12,15, 17,27,28, 30,31,33–35]	Feasible range of energy levels that an ESS might achieve during operation.
(11) (Dis)charge efficiencies	–[5]	[8,12,15, 17,28,31, 33]	Efficiencies associated with charging cycles of ESSs.

* Framework block or constraint that models each characteristic.
[1] Not considered within the scope of the planning and scheduling model, thus, not modelled.
[2] Only the minimum up-time is modelled in the framework, the maximum duration can be given by the user bid after characterisation of feasible flexibility.
[3] It is not included in the framework since we believe that this restriction might be a result of the availability of resources and interdependencies between processes.
[4] Easily implemented with a custom constraint for the storage variable at the last time-step.
[5] Can be modelled by a combination of blocks in the framework.

(Continues)

Table 10.2 Summary of characteristics identified as a basis for a generic industrial energy planning model and their relation to our framework. (Continued)

Characteristic	Modelled by*	References	Description
(12) Energy drain	–[5]	[12,17,28, 31,33]	Energy lost with time in an ESS due to its physical interactions with the environment or other reactions even when the ESS is not in use.
(13) Energy conversion efficiencies	Conversion	[17,28,31, 33,35,37]	Energy losses incurred during energy conversion between carriers.
(14) Energy balance	Flow balance	[8,28,31, 35,37]	Restriction related to the law of conservation of energy where the sum of energy inputs is exactly equal to the sum of outputs.
(15) Startup and shutdown profiles	–	[29,31]	Specific pattern of processes' electricity consumption during the start-up or shutdown transitions.
(16) Transport capacity limits	Flow boundaries	[28,31,37]	Limits in the rate of energy carriers that can be transported through the energy transmission systems.
(17) Flow dependencies	Pools	[34]	Functions that interrelate two or more flows of resources.
(18) Flow equations	Flow balance	[31,34]	Restrictions analogous to (14), including non-energy resources.
(19) Processes interdependencies	Pools	[12,17,34]	Constraints that relate the operation state of multiple processes. Complementary to (17).
(20) Earliest start time	–	[8,12,17, 37]	Earliest possible time-step at which a process can be activated.
(21) Multiple resources	Whole framework	[17,31]	Approach that considers material and energy resources as requirements to fulfil the production objectives of a plant.
(22) Uncontrollable processes	Conversion	[12,17,28, 31]	Identification of the processes whose operation cannot be modified.
(23) Operation modes	Discrete operation levels	[12,17,27, 32,34,35, 37]	Set of modes in which a process can operate, e.g., a mode for each different product, or a discrete number of multiple safe and stable operational modes.

* Framework block or constraint that models each characteristic.

the holding time of this FRR is not long enough, a further imbalance would appear in the system, calling for FCRs to operate and start the whole process again. A reserve product duration is a feature of the bids offered by BSPs in the reserves markets. Furthermore, a minimum and a maximum delivery periods length enclose the possible

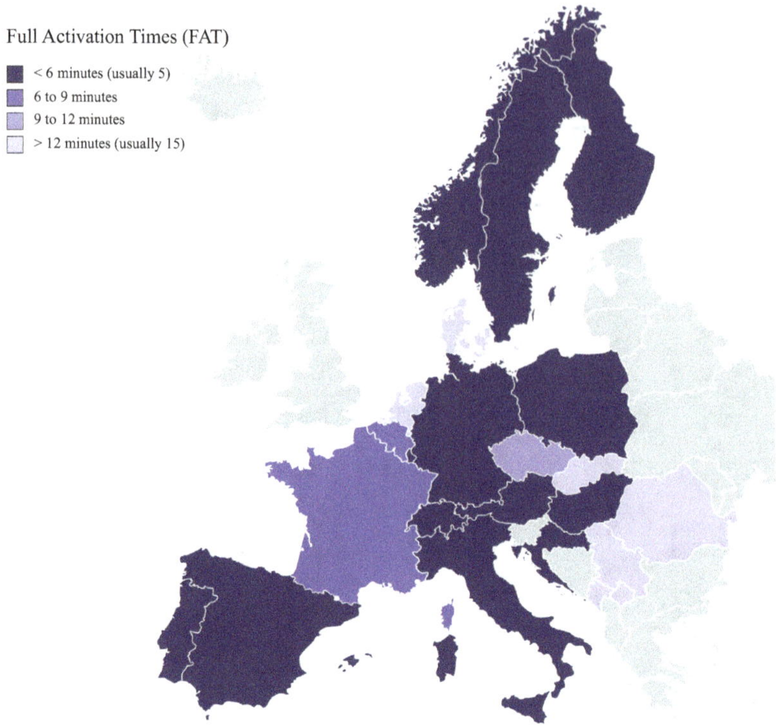

Figure 10.3 Full activation times of automatic frequency restoration reserves (aFRR) in different countries in Europe [23]

duration of reserve bids as depicted in Figure 10.2. These boundaries often translate to flexible loads models as **minimum-up times**.

10.3.2 Models in contemporary research

This section describes models that other authors have used to represent demand-side flexibility mathematically in the literature. They cover models with different aims, ranging from optimal electricity system operation to optimal schedule of demand-side loads from the consumer's perspective.

10.3.2.1 Models focusing on system operation

To start, we introduce multiple works that model flexible assets in both the supply and demand sides from a system perspective. They show different features that a model must have to enable the optimisation of the electricity system operation. Reza *et al.* [27] introduced a multi-objective security-constrained unit commitment (SCUC) problem formulated in the form of a mixed-integer linear program (MILP) considering thermal and hydro generators. On the one hand, this formulation is related to industrial demand-side flexibility since the hydro-generator is an energy-constrained asset, which resembles features in production plants such as products and energy storage techniques. On the other hand, the thermal generators show formulations related to the characteristics required in reserves markets such as limits in power, ramp rates and minimum up-times. Simultaneously, it shows other technical features of machines, e.g., the **minimum down-times, start-up costs** and the **transmission constraints** in the network that are essential in unit commitment (UC) problems (in this case, formulated according to the Direct Current (DC) power flow model).

Similarly, Parvania *et al.* [15] modelled a SCUC problem to integrate DR resources in wholesale electricity markets. For this purpose, the flexible loads are modelled according to four common flexibility products: load curtailment, load shifting, on-site generation and utilisation of ESSs. The characteristics of each type of product are assumed to be known and non-time varying (e.g., load curtailment power quantity is constant and known throughout the optimisation horizon), which might not be accurate in industrial sites. Each product type is associated with a set of constraints that bound the feasible solution set. The load curtailment product includes constraints on **minimum and maximum duration of load reduction, maximum number of curtailments and the activation cost**. Load shifting includes the quantity of load reduction, its associated cost, the minimum and maximum duration of load reduction and **an unchanged final energy consumption**. The on-site generation units contain constraints on minimum and maximum generation power limits, ramp-up and -down limits and minimum on and off times. Finally, the ESSs are modelled equally but including **energy capacity limits, (dis)charge efficiencies** and a limit in the minimum number of daily charging/discharging cycles.

From a different perspective, Heussen *et al.* [28] introduced their power nodes framework, which considers conversion and storage of multiple energy types. The main objective of this work is to represent the characteristics and limitations of energy after/before conversion from/into electricity in a unified framework to facilitate a more effective operation strategy in the electric network. With this, power grid optimisation problems can consider the storage and conversion of non-electrical energy. This model is versatile since the same formulation can describe many types of grid users, depending on the constraints and parameters chosen by the modeller. A state-dependent **energy drain** in the storage system and **energy conversion efficiencies** are the new constraints and parameters identified in this formulation in comparison with the ones previously discussed.

Lastly, another MILP model was used by Wang *et al.* [29] inspired by operational models with system balance constraints and the main operational features of assets connected to the grid. In this model, generators are described by a combination of maximum and minimum power limits, ramp-up and ramp-down limits, and minimum online and offline times. The authors used this model to find operational bottlenecks under multiple system conditions and the optimal cost-effective energy storage technology investment to correct them.

10.3.2.2 Flexible residential and commercial demand models

Lessons learned in other demand sectors can serve as a source of inspiration as well. On the one hand, these models can show the required characteristics of loads to be operated either as an aggregated pool of flexible distributed resources or in retail electricity markets. On the other hand, these sectors are certainly more explored as DR participants, resulting in more mature models.

A virtual power plant (VPP) is an aggregator that creates a portfolio of electricity producers and consumers and jointly operates them in the spot or balancing markets to generate profit. Consequently, these companies must decide what sources are flexible and useful, and select the best for their portfolio. Once they created a portfolio, these companies must decide on an operation plan that optimises the profit of these resources. With this purpose, Petersen *et al.* [25] identified all the possible characteristics to assess the quality of a flexibility source. The elements they listed are related to the source's level and duration of the power deviation and its reaction time. More specifically, they are power capacity, duration of flexibility measure, activation time, base load requirements, deadlines, ramp rates, storage capacities, length of storage period and up and down-time limits. In a different study, Petersen *et al.* [30] selected four constraints to characterise each element in a VPP portfolio: power capacity, energy capacity, **energy level at deadline** and minimum up-time. According to the set of constraints that applies to each element, they classified them as (1) buckets: power and energy-constrained, (2) batteries: buckets with energy level at a deadline and (3) bakeries: batteries with a minimum up-time. As expected, for elements of the same size, a bucket provides greater-quality flexibility than a battery whose flexibility quality is greater than a bakery since more constraints are added. Adding constraints to a problem has one of two effects in the feasible solution space, it is either not modified or shrinks.

To simplify and generalise the modelling task, Fink *et al.* [31] defined a framework that contains a set of building blocks that can be used and tuned to resemble real systems. This framework allows them to consider devices that consume, produce, store, transport or convert energy, which enables the representation of multi-energy systems. Thanks to the general approach of defining the blocks, a great variety of systems can be represented using this method without the need for adaptations according to, for instance, the type of industry, the specific dynamics of certain processes or other behaviours that physics-based models would contain. The set of building blocks contain pools, non-controllable devices, converters, pipes, buffers and time-shiftable devices. Each of these blocks contains a set of equations that will be part of the objective function and constraints of the optimisation problem. Pools are crucial blocks in this framework since they maintain the **energy balance** for all energy types considered in the system by restricting the consumption to equal the production of each energy type at every time interval. All the devices are connected to pools according to the energy type that they consume or produce. Converters are on/off devices characterised by a minimum run-time and **specific start-up and shutdown profiles**. Pipes are elements that transport energy from and/or to a pool and feature a **transport capacity** and energy losses. Buffers are energy storage devices such as batteries or thermal storage systems that have storage capacity limits, energy losses during charging, storing and discharging, and charging and discharging rate limits. Time-shiftable devices have a predefined energy consumption profile whose starting time can be decided. Since the converters and time-shiftable devices have specific energy consumption profiles, a more controllable block could be added to model devices with multiple feasible operation points.

10.3.2.3 Energy-aware industrial consumers

The last source of inspiration in this section is models and formulations utilised directly in industrial energy management contexts. Following this reasoning, it is worthwhile to introduce the *industRE* project, which took part in the European Union's Horizon 2020 research program. The objective of this project was to identify and assess the flexibility potential of industrial electricity consumption to lower the barriers to the integration of RES into the electricity grid. One result from this research was a Demand Response Audit methodology that contains four steps: identification, quantification, valorisation and exploitation [32]. This methodology helps to map the flexibility of an industrial consumer by analysing its processes structures and quantifying their potential. In this context, *flexgraphs* were developed as a technique to quantify this potential. They are graphs representing the flexibility potential as a set of trajectories that the electricity consumption of the industrial site can follow. Thus, it helps to visualise the power, energy, ramping and modulating constraints of each process. In addition, all the individual flexgraphs can be easily aggregated into an overall flexgraph by adding all the feasible consumption patterns of each process. These graphs are well in line with the metrics identified by Ulbig and Andersson [33] as a *flexibility trinity*: ramp-rates, power and energy. According to these authors, the capabilities of flexible electrical loads or generators can be visualised as a two- or three-dimensional representation of these three parameters. Although this methodology shows great visualisations, it is constrained to elements in which each one of the axis limits is independent. During aggregation, this is even more critical when each of the aggregated elements is interdependent. For instance, in a system with two flexible sequential production processes, the flexibility of one of them is constrained by the operation of the other. Both can operate flexibly as long as there is enough buffer capacity or material availability for their inputs/outputs. However, if the first process stops operating for a long time, the second process will deplete its input stock. Thus, the operation of one process constrains the flexibility of the other. Although Ulbig and Andersson did not focus on industrial demand-side flexibility, this helps understand some flaws that might appear by approaching industrial consumption from the electricity flexgraph perspective.

Karlsson developed a method for analysis of INDustrial energy systems (MIND) as a computational decision tool for energy management tasks [34]. This method revolves around formulating a MILP optimisation problem that applies to several industrial sectors in four steps: identification of the system, construction of the model, solution of the problem and export of results. The formulation considers the energy system as a network of nodes and edges where each node represents a process, machine or whole industry while the edges show the flow of resources between different nodes. These resources can be energy such as electricity, heat or fuels, or materials like raw materials, intermediate, by- or final products. Furthermore, the functions that represent the interactions in the system are malleable and can be defined according to the identified nodes and edges. The abstract model includes a multi-objective linear function that can be adapted to optimise the system according to the user requirements. This abstraction in the model is what allows to apply the methodology to fit multiple industrial cases for a variety of objectives. Similarly, the predetermined constraints are adaptable to specific systems and contain formulations for connecting resource flows, their boundaries, resource storage dynamics, processes dynamics and logistical restrictions; but customised functions are also allowed. Among the predetermined constraints, it is possible to find **flow dependencies**, **flow equations**, flow relations, flow boundaries, final flow deadlines, storage equations, storage boundaries, batch processing constraints, and **mutually exclusive processes constraints**. However, there is no formulation for the interactions of continuous processes, causing a gap in the model since some industries operate non-batch production processes (e.g., continuous steel casting, steam production, etc.).

A more concrete model was developed by Waldemarsson *et al.* [35], specifically designed for an energy-aware supply chain of a pulp production company. In this case, the problem was again formulated as a MILP, considering the transportation of energy and material resources, and optimises decisions on material supply, production jobs allocation, distribution of resources, and energy imports and exports along the supply chain. Although the problem matches the behaviour of a specific pulp production company, according to the authors, it is extendable to other pulp companies. However, a disadvantage of such a concrete model is that most of the structure of the problem would require reformulation to match a new system. The constraints and variables in this problem include interactions related to the flow of materials, raw materials inventory, production, storage and sales of pulp, energy carriers and by-product energy-rich materials, and water consumption. Note that exports of by- and intermediate products, and utilisation of energy carriers/products, are considered in the problem as additional valorisation strategies.

Similarly, there is another concrete model developed by Zeng *et al.* to reduce energy consumption and emissions of air pollutant agents in an iron and steel plant [36,37]. They include features such as fuel selection to cover the heat demand, gas storage level control, minimum heat requirement, operation of Combined Heat and Power (CHP) plants, and ramp rate limits. Their mathematical formulation is based on the physics of the processes considered in the plant and includes constraints that model energy balances and operational boundaries for all heat or heat and power generation units. Although this formulation follows a different approach, it shows how some relevant characteristics in an industrial energy system operation can be mapped to a combination of integer and linear constraints, resulting in a MILP formulation. In this case, the optimisation problem does not consider the production processes. Instead, only auxiliary processes for energy production restrict the feasible operational states of the system, which implies that some potential flexibility offered by the process is not exploited because the production schedule (represented by the heat and power demand) is fixed.

The case presented by Pei *et al.* [8] described an interesting problem where the schedule of a battery manufacturing production line is modelled and optimised. Although most of the features included in the problem have been discussed already, the ageing process that the batteries must undertake before being ready to export creates a special problem that might not be present in other industries. The ageing process is a charging and discharging treatment that the newly built batteries receive to improve the overall stability of the battery pack. Hence, the operation of this process in a DR program would resemble an ESS that needs a charge and discharge cycle before a deadline, but that can be operated flexibly. This is true to a certain point since the power and energy required for each charging cycle are constrained to a certain value, which means that only the start time can be decided for each job performed by the ageing machine.

To create a model that resembles the energy system of an industrial plant, it is necessary to understand the type of processes and behaviours in the objective systems. In general, the processes can be clustered according to the variety of outputs that they can produce and their production volume. Pelzer *et al.* [24] identified five classes of industrial processes, which ordered from more variety-oriented to more volume-oriented are: customer-oriented processes, job shop processes, batch processes, repetitive processes and continuous processes. In addition to this contribution, they point out that industrial sites are often multi-energy systems whose flexibility is extended by coupling the on-site need of multiple energy carriers and considering their interrelations.

As it can be noted, each author represented industrial energy systems according to the approach that they reckoned best, usually in the form of a MILP. This procedure requires a high engineering effort to learn and apply multiple approaches, which might raise the barriers against DSM strategies adoption in the industry. In this regard, Barth *et al.* [17] tackled this issue by reviewing current demand-side modelling approaches and developing a unified framework to optimise the operation of demand-side resources across the residential, commercial and industrial sectors. There are 14 quantifiable features that must be included in a demand-side flexibility model: (1) time frame, (2) interruptible processes, (3) storage, (4) **interdependencies between processes**, (5) **earliest start time**, (6) deadlines, (7) production outputs, (8) **multiple resources**, (9) **uncontrollable processes**, (10) **operation modes**, (11) drains/losses, (12) on- or off-times limitations, (13) number of activations, and (14) ramping rates.

Finally, another attempt to homogenise the task of modelling power consumption in industrial loads was presented by Schott *et al.* [12]. This approach presumes to be general enough to describe generically all power demand sectors (i.e., industrial, residential and commercial) and more modern players such as electric vehicles. In this context, three building blocks were identified as the pillars of flexibility models: flexible loads, dependencies and storages. Each element contains a set of key figures that parametrise the system and delimit the feasible operational limits. Flexible loads are the main elements of a flexible system since they are devices whose operation point shows freedom to execute different actions. Dependencies model the interrelations between two or more flexible loads and can describe sequential or mutually exclusive behaviours. Storages are the buffers that allow flexible loads to be operated flexibly by shifting resources availability in time. Although this is an extensive framework, the multi-resource approach seems to be missing. We define a **multi-resource approach** as one where multiple energy carriers and material flows throughout the processes are considered in the formulation. This extends multi-energy systems to consider materials and intermediate products as elements that interact with and transform into each other.

10.4 Modelling framework formulation

A modelling framework is proposed by the authors of this chapter. This framework was developed by taking inspiration from the reviewed models presented in Section 10.3, intending to fill the gaps identified in these models. It is based on a multi-resource perspective that contemplates all the materials and energy resources and their flows, storage, and transformations throughout the system. Note that this is an energy-aware high-level production planning and scheduling model that intends to unify the modelling of demand-side flexibility across industrial energy users. Consequently, this is not a physics-based model that replicates the behaviour of specific processes. Instead, it assumes a quasi-static generic behaviour on all processes.

10.4.1 Definitions

Before diving into the formulation of the framework, this section states some auxiliary definitions to facilitate the comprehension of the model.

10.4.1.1 Flexibility

Energy flexibility can be ambiguous due to the different perspectives from which it can be framed. Thus, from an individual (energy consumer or producer) operation perspective, we define flexibility as the capability to deviate from a plan [30], on short notice, as a reaction to new information. This definition consists of three parts: (1) the capability to deviate, (2) the plan, and (3) the new information. First, the capability to deviate is the set of feasible actions that would result in a technically correct operation of the system, regardless of the impact this might have on the objective functions or KPIs of the company. Second, the plan is the a-priori set of actions that have been decided, usually as a result of solving an optimisation problem. Lastly, the acquisition of new information is the update of known parameters in the problem due to contingencies, forecast inaccuracies, and changes in external incentives that would result in a sub-optimal outcome if the plan remains unchanged.

10.4.1.2 Resources

In this framework, everything that can be stored, exchanged, supplied, consumed, and transformed into something else is denominated resource. Energy carriers, materials, and labour fall under this category and might be present in the model of an industrial system in the form of electricity, heat, fuels, raw materials, intermediate and final products, among others. Each resource is subject to a balance equality constraint that enforces the contemplation of on-site availability of resources.

10.4.1.3 Flow

Resources can flow to and from each of the components in the system. Flows are defined as these transfers of resources between components and are closely related to the conversion control variables in the optimisation problem formulation. Flow, storage and conversion control variables comprise the decision variables of the problem. Following our definition of flexibility, flows F are divided into two parts: the planned term F^* and the flexibility term ΔF. Although this does not affect the optimum solution of the problem, the separation of this variable helps to easily identify the solutions before and after new information is acquired and to quantify the flexibility activated. In practice, once the optimisation problem is formulated, it is solved by fixing the flexibility terms to zero. This solution is the plan, which is the first part of the definition of flexibility. Once the optimal plan is known, the acquisition of new information will cause an update in some parameters of the problem. This update can cause the plan to be sub-optimal or, in some cases, even infeasible. Then, the schedule is corrected by fixing the planned terms (F^*, S^* and x^*), unfixing their flexibility terms (ΔF, ΔS and Δx) and optimising the problem over the free decision variables.

$$F_{ft} = F^*_{ft} + \Delta F_{ft} \tag{10.1}$$

10.4.1.4 Storage

Some of the components in the framework include storage variables S that quantify the resources that they contain. Similar to the flow variable, this one is also divided into pre- and post-flexibility activation parts S^* and ΔS, respectively.

$$S_{st} = S^*_{st} + \Delta S_{st} \tag{10.2}$$

10.4.1.5 Conversion control variable

Lastly, the conversion control variables x are present in the *converter* blocks of the framework. In the most basic case, the value of these variables represents the per-unit capacity used in a transformation process, where a set of input resources transform into a set of output resources. Unsurprisingly, the conversion control variables x are also divided into two parts, similarly to flows and storages.

$$x_{ut} = x^*_{ut} + \Delta x_{ut} \tag{10.3}$$

10.4.2 Modelling blocks

This framework predetermines a set of building blocks to facilitate the modelling of industrial energy systems. Each block is parametrisable and contains a group of abstract constraints that use the block parameters to add concrete constraints to the system's optimisation problem. The accuracy of the model still relies on the ability of the user to resemble the target system. However, the design of each block contains an intuitive structure, which intends to reduce the possibility of mistakes or missing characteristics. The underlying constraints composing each block and their respective required parameters are described in this section.

10.4.2.1 Component: the parent class

Although components are not model blocks themselves, they are used to conglomerate the recurrent parameters and constraints between all blocks to avoid redundant explanations. The three blocks that inherit all the component characteristics are the converters, storages and networks. Every system is modelled over a uniform discrete time horizon that delimits the information contained in the problem. Thus, each variable in this problem spans over the whole horizon, showing its evolution in time. Additionally, this horizon is represented as the set of time-steps $\mathbb{T}$ and contains an index t for each instant in the problem. Table 10.3 gives an overview of all the parameters required for every component during its instantiation.

Flow boundaries

One of the simplest constraints is the flow boundaries, which model a process's maximum and minimum flow limitations. These limits can model both the technical limits of a process, the transportation capacities between processes or safe operative boundaries. Furthermore, a single process might contain multiple flow boundaries if it is associated with multiple flows as is the case for *Converters*. Additionally, some machines or processes might have feasible ranges that do not include the off-state in between the maximum and minimum (e.g., if $\underline{F}_f > 0$ and $\overline{F}_f > 0$). In these cases, the *Flow boundaries* constraint is replaced by the *On/off constraint*.

$$\underline{F}_f \leq F_{ft} \leq \overline{F}_f \qquad \forall f, t \in \mathbb{F}_c \times \mathbb{T} \tag{10.4}$$

Ramping rates

As it was discussed in Section 10.3, ramping rates are important characteristics to consider during the operation of flexible resources. This is especially true for resources that provide balancing reserves services since the type of reserve they can provide

Table 10.3 List of parameters associated to all components

Characteristic	**Representation in equations**	**Description**
Maximum flows	$\overline{F}$	Upper bound for each resource flow (in absolute value).
Minimum flows	$\underline{F}$	Lower bound for each resource flow (in absolute value).
Time-step length	Δt	Amount of time between contiguous time instances in optimisation horizon.
Set of indices	$\mathbb{F}_c, \mathbb{S}_c, \mathbb{U}_c, \mathbb{P}_c, \mathbf{d}$	Lists of the Flows ($\mathbb{F}_c$), Storages ($\mathbb{S}_c$), control variables ($\mathbb{U}_c$), Pools to which the component is connected ($\mathbb{P}_c$) and a device index (**d**) corresponding to the component.
Ramp-up and Ramp-down	$\overline{r}, \underline{r}$	Maximum upwards and downwards rate of change in each of the resource flows of the converter.
Number of elements in sets	n_f, n_s, n_u, n_t	Number of indices contained in sets $\mathbb{F}_c, \mathbb{S}_c, \mathbb{U}_c$, and $\mathbb{P}_c$, respectively.

depends on the reaction time capabilities of the flexible resource. Additionally, most industrial processes have an intrinsic stabilisation time during their transitions, usually associated with mass or thermal inertia

$$\underline{r}_f \leq F_{ft} - F_{f(t-\Delta t)} \leq \overline{r}_f \qquad \forall (f, t \in \mathbb{F}_c \times \mathbb{T} | t > t_1) \tag{10.5}$$

10.4.2.2 Converter

Converters are the main blocks in the framework since they model the transformation between resources. This block can resemble a processing step in a production line, a device that converts between energy types such as water boilers, etc. Its conversion equation is a multi-dimensional linear equation that relates the output flows to the input flows given a certain conversion rate, selected by the modeller. The parameters used for the converter blocks are summarised in Table 10.4.

Conversion

The transformation between resources along the production line is modelled by the *conversion* constraint in all *converters*. It is a linear equality constraint that relates resources to each other, making the problem a multi-resource approach. Briefly, the rate of conversion is ruled by two parameters defined in the converter: the conversion matrix **M** and the conversion vector **b**. The former gives the first-grade term of the flow equation and contains n_f rows (linked to each flow) and n_u columns (linked to each conversion control variable). The latter provides a constant conversion term that does not depend on the conversion control variables, which makes it useful to model non-controllable converters, or processes with a fixed operation constraint and little flexibility around this point.

$$\begin{bmatrix} F_{f_1 t} \\ \vdots \\ F_{f_{n_f} t} \end{bmatrix} = \mathbf{M} \begin{bmatrix} x_{u_1 t} \\ \vdots \\ x_{u_{n_u} t} \end{bmatrix} + \mathbf{b} \qquad \forall t \in \mathbb{T} \tag{10.6}$$

Conversion control variable limits

According to the definition of *conversion control variables* in the modelling framework, these variables give the per-unit level of operation of a *converter*. Thus, they

Table 10.4 List of parameters associated to converters

Characteristic	Representation in equations	Description
Conversion matrix	**M**	First-grade coefficient in the linear conversion equation.
Conversion vector	**b**	Constant term in the linear conversion equation.
Minimum up-time and down-time	t_u, t_d	Minimum number of time-steps that the component must remain in operation after start-up and remain off after shutdown, respectively.

are bound to take values between 0 and 1 or between −1 and 1 depending on whether the process is reversible over this variable.

$$\{-1 \text{ or } 0\} \leq x_{ut} \leq 1 \qquad \forall u, t \in \mathbb{U}_c \times \mathbb{T} \tag{10.7}$$

Discrete operation levels (optional)

Some industrial processes have a discrete action-space rather than a continuous one. Thus, its operation point can vary with a certain granularity. Another set of processes can operate in different modes, to produce a variety of outputs, to provide an idle state that allows a warm start, etc. Therefore, the discrete operation levels constraint is an optional feature that can transform a continuous action-space converter into a discrete one in a similar fashion to a step-wise linear constraint. In this case, a conversion control variable x_{ut} and its corresponding conversion matrix column are selected for each operation mode. The conversion control variable is then constrained to be a binary variable and an additional constraint prevents simultaneous activation of more than one of them.

$$x_{ut} \in \{0, 1\} \qquad \forall u, t \in \mathbb{U}_c \times \mathbb{T} \tag{10.8}$$

$$y_{\mathbf{d}t} = \sum_{u \in \mathbb{U}_c} x_{ut} \qquad \forall t \in \mathbb{T} \tag{10.9}$$

Binary variables

Three binary variables are used as indicators in the optimisation problem. The first one ($y_{\mathbf{d}t}$) gives the operating status of a process at every interval in the time horizon, where the possible status is on or off. The second ($z_{\mathbf{d}t}^{on}$) and third ($z_{\mathbf{d}t}^{off}$) variables are similar to each other and they are equal to one at the time-steps where the process was started after an idle period or turned off after a running period, respectively. These three variables are required to model the following features: *minimum up-time and down-time* constraints, the *on/off constraint* and the start-up costs.

$$y_{\mathbf{d}t} \in \{0, 1\} \qquad \forall t \in \mathbb{T} \tag{10.10}$$

$$z_{\mathbf{d}t}^{on} \in \{0, 1\} \qquad \forall t \in \mathbb{T} \tag{10.11}$$

$$z_{\mathbf{d}t}^{off} \in \{0, 1\} \qquad \forall t \in \mathbb{T} \tag{10.12}$$

Minimum up-time and down-time

Due to maintenance requirements, safe operational limits, setup dead time, etc., there can be processes that have a time-constrained operation or idle times. These times are modelled by the minimum up- and down-time constraints, making use of the indicator variables $y_{\mathbf{d}t}$. Since the optimisation horizon is finite, the problem has an inherent reduction of optimality near the start or the end of the problems caused by this constraint. This is because the constraint requires information from multiple steps, causing overestimated constraints in the vicinity of the start and end of the horizon. However, this limitation can be overcome by extending the time horizon, at the expense of a more computationally intensive solution.

$$(t_u - 1) \cdot (y_{\mathbf{d}t} - y_{\mathbf{d}(t-\Delta t)}) \leq \sum_{k=t+1}^{min(t+t_u,\, t_{n_t})} y_{\mathbf{d}k} \qquad \forall (t \in \mathbb{T} | t > t_1) \tag{10.13}$$

$$(t_d - 1) \cdot (1 + y_{\mathbf{d}t} - y_{\mathbf{d}(t-\Delta t)}) \geq \sum_{k=t+1}^{min(t+t_d,\, t_{n_t})} \qquad \forall (t \in \mathbb{T} | t > t_1) \tag{10.14}$$

Start-up detection and shutdown detection

This set of equations provides the logic to ensure that the start-up and shutdown indicators take the correct values during the solution of the problem.

$$z^{on}_{\mathbf{d}t} \geq y_{\mathbf{d}t} - y_{\mathbf{d}(t-\Delta t)} \qquad \forall (t \in \mathbb{T} | t > t_1) \tag{10.15}$$

$$z^{on}_{\mathbf{d}t} \leq y_{\mathbf{d}t} \qquad \forall t \in \mathbb{T} \tag{10.16}$$

$$z^{on}_{\mathbf{d}t} \leq 1 - y_{\mathbf{d}(t-\Delta t)} \qquad \forall (t \in \mathbb{T} | t > t_1) \tag{10.17}$$

$$z^{off}_{\mathbf{d}t} \geq y_{\mathbf{d}(t-\Delta t)} - y_{\mathbf{d}t} \qquad \forall (t \in \mathbb{T} | t > t_1) \tag{10.18}$$

$$z^{off}_{\mathbf{d}t} \leq y_{\mathbf{d}(t-\Delta t)} \qquad \forall (t \in \mathbb{T} | t > t_1) \tag{10.19}$$

$$z^{off}_{\mathbf{d}t} \leq 1 - y_{\mathbf{d}t} \qquad \forall t \in \mathbb{T} \tag{10.20}$$

On/off constraint

As mentioned previously, the *on/off constraint* is used instead of the *flow boundaries* to include the off-state in the feasible operation space of converters even if the range between their minimum and maximum flows does not include the zero. Moreover, this constraint enforces the flow values to zero when the converter must be off according to the binary indicator variable while giving the maximum and minimum on-state boundaries.

$$y_{\mathbf{d}t} \cdot \underline{F}_f \leq F_{ft} \leq y_{\mathbf{d}t} \cdot \overline{F}_f \qquad \forall f, t \in \mathbb{F}_c \times \mathbb{T} \tag{10.21}$$

10.4.2.3 Storage

Storage components can model ESSs or material/product stock. They are limited by the maximum and minimum flow and storage capacities, but their flow variable is free to take the value that optimises the objective function as long as the resource balance is maintained (see Section 10.4.2.5). A list of the parameters of storage blocks is given in Table 10.5.

Storage transition

Depending on the flow at instant t, the storage variable of the *storage* and *Network* components will increase or decrease according to the *storage transition* equality. At the first time-step, this equation uses the value of the initial storage level S_0 instead of $S_{s_i(t-\Delta t)}$. Note that in these constraints a special set of indices $\mathbb{I}_c$ relates each flow to

Table 10.5 List of parameters associated to storages

Characteristic	**Representation in equations**	**Description**
Maximum storage capacity	$\overline{S}$	Maximum storage capacity in the storage element.
Minimum storage capacity	$\underline{S}$	Constant term in the linear conversion equation.
Value of the initial storage level	S_0	Maximum upwards rate of change in each of the resource flows of the converter.

Table 10.6 List of parameters associated to networks

Characteristic	**Representation in equations**	**Description**
Exchange prices	π	Time-series values of the prices of flows from the network.

each storage in the component according to their associated resource. This is important since flows and storages of non-matching resources should not be mixed.

$$S_{s_it} = S_{s_i(t-\Delta t)} + F_{f_it}\Delta t \qquad \forall((f_i, s_i), t) \in (\mathbb{F}_c \times \mathbb{S}_c | i \in \mathbb{I}_c) \times \mathbb{T} \tag{10.22}$$

Storage capacity limits

Since the storage capacities are limited, this constraint gives the maximum and minimum values that the storage variables can take.

$$\underline{S}_s \leq S_{st} \leq \overline{S}_s \qquad \forall s, t \in \mathbb{S}_c \times \mathbb{T} \tag{10.23}$$

10.4.2.4 Network

In our modelling framework, a *network* is a specific type of storage component with an added transaction price (see Table 10.6). Thus, every flow from or to these components has an associated cost or revenue. All the constraints and parameters required are the same as for storage components, with the addition of a time-series array containing the import/export prices at each instant t. Mainly, this component is created: on the one hand, to keep track of imports and exports and their prices, allowing their easy inclusion in the objective function; on the other hand, to differentiate external and internal components of the plant and make the model more self-explanatory.

10.4.2.5 Pool

The last block in the framework is the *pool* (its parameters are given in Table 10.7), which is in charge of maintaining the balance between inputs and outputs of each resource at every time step. During modelling, the first step is to identify the pools

Table 10.7 List of parameters associated to pools in the system

Characteristic	Representation in equations	Description
Flows linked to pool p	$\mathbb{F}_p$	Set of indices containing all the flows f for a specific pool p.
Pools	$\mathbb{P}$	Set of all pools in the system.

in the system, since there can only be as many resources as pools. Sometimes, it might be useful to define two different pools for a single resource. That is the case when two sets of processes require the same resource but do not have direct access to each other. For example, suppose that process A and process B require material X as input, but process A is in location L1 and process B in location L2, from a practical perspective, defining a single pool for resource X is unrealistic since transportation from location L1 to L2 might not be possible, feasible, or might require additional use of resources. Instead, two pools can be defined, one for resource X_L1 and another one for resource X_L2.

Flow balance

This constraint is the only one implemented in this block and applies to all resources and all instances of time. The special set of indices $\mathbb{F}_p$ relate the flows from all components that correspond to the same resource pool.

$$\sum_{f \in \mathbb{F}_p} F_{ft} = 0 \qquad \forall\, p, t \in \mathbb{P} \times \mathbb{T} \tag{10.24}$$

10.5 Case study

As a proof of concept, the modelling and subsequent schedule optimisation of a steel powder production plant exemplify the use of the framework. This industrial case is based on the one presented by Yu *et al.* [7], introducing some slight adaptations and assumptions to fill information gaps in the system. The system consists of a single production line with sequential processes and buffer capacities in between, where the intermediate products are stored. The production line is depicted in Figure 10.4 and follows the next sequence: spraying atomisation, dehydration, drying, crushing, classification, magnetic separation, finish-reduction, a second crushing and classification steps and blending. All production processes show some degree of freedom thanks to their multiple operation modes, except the finish-reduction furnace that is a non-controllable load constantly in operation. While some of them are only capable of switching between on and off states (spraying atomisation, dehydration, drying and magnetic separation), the other processes have three possible operation modes (crushing, classification and blending). In addition, the auxiliary processes have

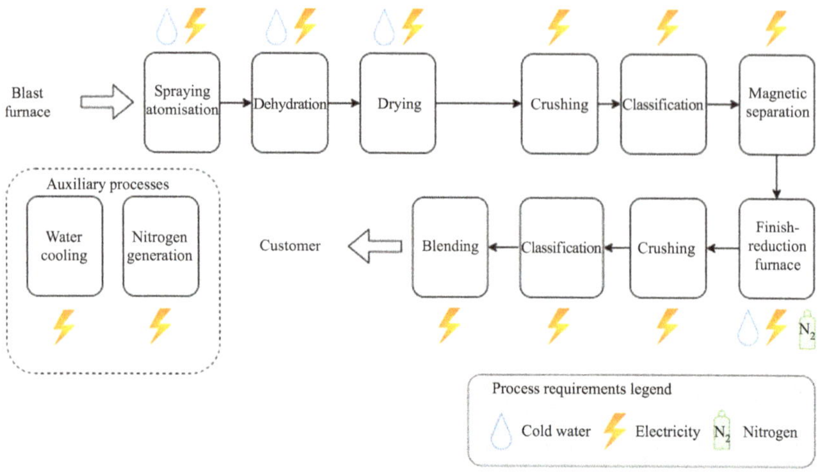

(a) *Sequence of the production line processes and their required utilities.*

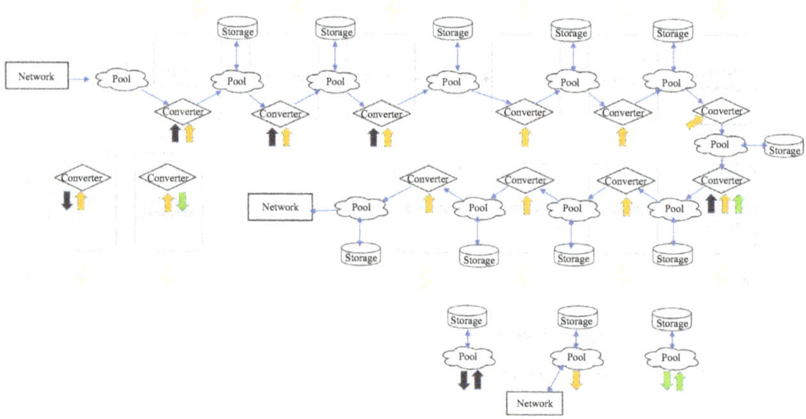

(b) *Modelling of the case study using the framework blocks. The coloured arrows indicate utilities (cold water, electricity and nitrogen) connections.*

Figure 10.4 Overview and modelling of a steel powder production line

a continuous operation space, ranging from the off-state to the nominal operating point. Electricity can be sourced externally from the power grid, but the cold water and nitrogen demands are self-supplied. All these utilities can also be stored locally for later use thanks to the presence of a battery energy storage system (BESS) and tanks for cold water and nitrogen.

The main objective of this problem is to maximise the profit of the production line. On the one hand, the revenues are given only by the export of electricity and final product. The final product is not bound to a specific demand and is taken as a free decision variable. On the other hand, the costs come exclusively from the imports of electricity. Although this problem should consider the material imports, it does not due to the lack of available information and the fact that the production line starts at the output of an own blast furnace. As a result, the molten iron coming from the blast furnace is assumed to be imported at zero cost. Electricity imports and exports are traded in the wholesale market, under the assumption of large and competitive markets whose prices do not depend on the decisions made by our production line. For this reason, past prices in the DAM and IDM in Belgium are used in this example. Most of the processes in the production line have two or three operation modes with different throughput and utility consumption rates. The finish-reduction furnace is the only non-controllable production process. Additionally, the auxiliary processes have continuous action-spaces, including the off state.

10.5.1 Model

Our model comprises 14 Pools, 12 Converters, 13 Storages and 3 Networks. First, the 14 resources in the Pools are the molten iron produced by the blast furnace, all the intermediate and the final products and electricity, nitrogen and cold water. Second, each of the Converters is associated with one of the processes in the production line or the auxiliary processes (see Figure 10.4). Third, the 13 Storages model the buffer capacity of each resource, except for the output of the blast furnace. Lastly, the networks model the electricity grid, the customer that imports the final product and the blast furnace.

To showcase the economic potential of flexibility in this system, it is exploited according to the definition given in Section 10.4.1.1. First, the capability to deviate is described by the feasible solution space of the optimisation problem. This space spans all possible combinations in the problem variables that would result in compliance with all constraints. Second, the plan is found by solving the problem in advance with available information. We assume that only the DAM electricity prices are available in advance. Thus, the plan only takes into account these prices, resulting in an optimal day-ahead schedule (optimum J_{plan}). Third, the IDM electricity prices for the whole day are acquired as new information and the system flexibility is used to correct deviations from the optimal schedule (optimum J_{flex}).

$$J_{plan} = \sum_{t\in\mathbb{T}} \left[\sum_{f\in\mathbb{F}_{cust}} F^*_{ft} \cdot \pi_{prod} - \sum_{f\in\mathbb{F}_{grid}} F^*_{ft} \cdot \pi_{DAM}(t) \right] \cdot \Delta t \tag{10.25}$$

$$J_{flex} = J_{plan} + \sum_{t\in\mathbb{T}} \left[\sum_{f\in\mathbb{F}_{cust}} \Delta F_{ft} \cdot \pi_{prod} - \sum_{f\in\mathbb{F}_{grid}} \Delta F_{ft} \cdot \pi_{IDM}(t) \right] \cdot \Delta t \tag{10.26}$$

The first objective function (J_{plan}) is maximised over the plan variables (F^*_{ft}, S^*_{st} and x^*_{ut}) while keeping the flexibility variables fixed at zero. On the contrary, the second (J_{flex}) is solved over the flexibility variables (ΔF_{ft}, ΔS_{st} and Δx_{ut}) keeping the planned variables fixed at the previous solution. In this formulation, $\mathbb{F}_{cust}$ is the set of flows connected to the customer, π_{prod} is the export price of the final product, $\mathbb{F}_{grid}$ is the set of flows to/from the grid, $\pi_{DAM}(t)$ is the time-varying DAM price and $\pi_{IDM}(t)$ is the time-varying IDM price (see Figure 10.5). Although the customer and

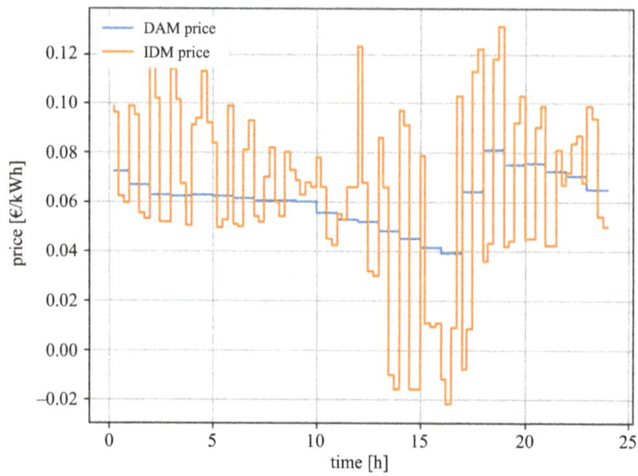

Figure 10.5 Comparison between DAM and IDM prices

the electricity grid are connected to only one flow variable each, the formulation of the objective functions is given for a general case that considers the possibility of additional flow variables. The most general profit formulation is the sum over all network flows and all time-steps of each flow multiplied by their respective (time-varying or constant) price and the time-step length. Nonetheless, a custom objective function can be designed according to the goals of the user.

10.5.2 Results

Shifting the use of electricity to better match the sequential markets while using the buffering capacity in the plant, allowed the system to reach a much better profit. These results show that although constrained, industrial energy flexibility can produce benefits for the system.

Some of the results are illustrated in Figure 10.6, which includes four charts. These are the evolution over the time horizon of the storage and flow variables (S_{st} and F_{ft}) related to electricity in the grid *network* and to the final product in the customer *network*. Starting with the electricity grid charts, Figure 10.6(a) represents the accumulated energy injected into the network. All these values are negative since there is a net consumption rather than production in the industrial plant. Figure 10.6(b) shows the power withdrawn from the electricity grid, thus, the industrial consumption. Contrastingly, Figure 10.6(c) shows the accumulated final products imported by the customer. Finally, chart (d) displays the customer's rate of product exports, which is negative since the customer always imports. In addition, the storage graphs (a) and (c) show the upwards and downwards flexibility used in different coloured areas. These areas are the change between the original plan and the updated schedule after receiving information on the IDM and exploiting the flexibility capacity. Similarly,

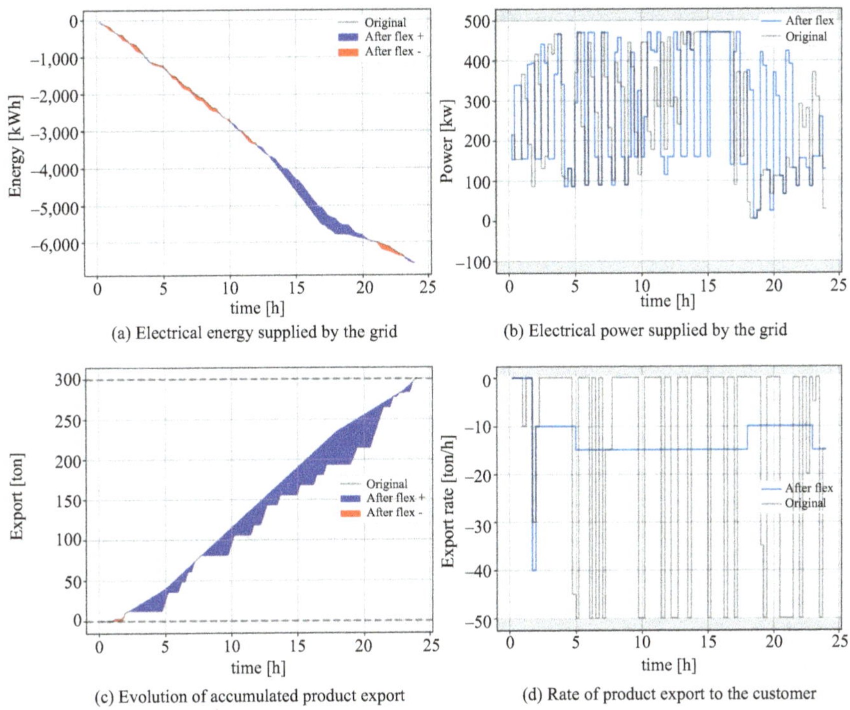

(a) Electrical energy supplied by the grid

(b) Electrical power supplied by the grid

(c) Evolution of accumulated product export

(d) Rate of product export to the customer

Figure 10.6 Schedule results after utilisation of flexibility to benefit from differences between DAM and IDM prices

the flow graphs (b) and (d) show the original plan and after-flexibility plan in different coloured lines. Note that the storage capacity limits in the storage variables and the flow boundaries in the flow variables are delimited by a dashed grey line and a pair of grey surfaces, respectively.

Upwards and downwards flexibility can be combined when participating in sequential markets to correct the production schedule according to the expected market prices. Figure 10.6(a) illustrates this argument, where decisions of reducing and increasing consumption at key time intervals resulted in the optimum operation of flexibility. These alterations to the original plan match the differences in prices between the DAM and IDM. Although the total final electricity use by the end of the optimisation horizon was not directly constrained to a certain value, the solver found that the optimum flexibility operation did not cause large deviations in this value. Therefore, multiple intraday shifting can result in financial benefits for flexible consumers participating in sequential markets.

Shifts in exports of the final product are also observed in the figure. Although the product price was constant and shifting its exports do not change the revenues (if the amount of product remains unchanged), the processes required to generate the final product were operated smartly to reduce their associated electricity costs.

In this example, the framework utilisation aimed to optimise the day-ahead and intraday production schedule, using profit as the objective function. The results show a high potential added value by the sequential schedule optimisation approach. This approach decreased the energy costs of the system throughout the day by 10.3.

Although we used profit as an objective in this example, the framework can also identify the amount of feasible flexibility to be offered in reserve markets, minimise polluting emissions resulting from production processes, find re-design opportunities in the processes structure, just to mention a few.

10.6 Conclusions

The superficially studied industrial flexibility shows balancing potential that could generate value for both the energy-intensive industry and the electricity system. Unfortunately, the interdependencies between industrial processes in a plant make the planning and scheduling task complex.

For this reason, we propose a modelling framework to simplify the modelling and optimisation tasks. It contemplates interactions between multiple energy and material resources that allow the integration of production planning and energy management into a single problem. Additionally, the framework intends to be compatible with existing grid operation mechanisms and with other flexibility sources. Thus, this chapter reviews various models used in balancing markets, grid operation, and demand-side management in all consumer sectors. As a result, relevant characteristics were identified and used to design our modelling framework. Grouping multiple equations in intuitive blocks facilitates the interpretation of the models. Consequently, they can be easily represented in a map of blocks and edges to display their topography.

To illustrate the framework, its application on a steel powder production line showcased a successful planning and scheduling procedure that optimises the plant's profit, considering interactions in the day-ahead and intraday electricity markets. Nevertheless, other valorisation mechanisms can be explored by defining a different objective function (e.g., carbon emissions, profit from participation in reserve markets, among others).

Acknowledgements

The authors gratefully acknowledge the financial support of the Flemish Government and Flanders Innovation & Entrepreneurship (VLAIO) through the Moonshot project InduFlexControl (HBC.2019.0113).

References

[1] Roesch M, Bauer D, Haupt L, *et al*. Harnessing the full potential of industrial demand-side flexibility: an end-to-end approach connecting machines with markets through service-oriented IT platforms. Applied Sciences (Switzerland). 2019;9(18):3796–3822.

[2] Valdes J, Poque González AB, Ramirez Camargo L, *et al.* Industry, flexibility, and demand response: applying German energy transition lessons in Chile. Energy Research and Social Science. 2019;54:12–25.

[3] Golmohamadi H. Application of demand response programs to heavy industries: a case study on a regional electric company. In: 32nd Power System Conference. Tehran: Power System Conference (PSC); 2017. pp. 1–6.

[4] Herre L, Tomasini F, Paridari K, *et al.* Simplified model of integrated paper mill for optimal bidding in energy and reserve markets. Applied Energy. 2020;279:115857–115872.

[5] IEA. World energy balances 2020; 2020. Available from: https://www.iea.org/data-and-statistics?country=BELGIUM&fuel=Electricityandheat&indicator=ElecConsBySector.

[6] Geysen D, Six D, Verbeeck J, *et al.* Simplified assessment methodology for optimal valorization of flexible industrial electricity demand. IndustRE – Horizon 2020: 646191; 2016. June. Available from: http://www.industre.eu/downloads/.

[7] Yu M, Lu R, Hong SH. A real-time decision model for industrial load management in a smart grid. Applied Energy. 2016;183:1488–1497.

[8] Pei W, Ma X, Deng W, *et al.* Industrial multi-energy and production management scheme in cyber-physical environments: a case study in a battery manufacturing plant. IET Cyber-Physical Systems: Theory and Applications. 2019;4(1):13–21.

[9] Zhang X, Hug G, Kolter JZ, *et al.* Demand response of ancillary service from industrial loads coordinated with energy storage. IEEE Transactions on Power Systems. 2018;33(1):951–961.

[10] Waldemarsson M, Lidestam H, Karlsson M. How energy price changes can affect production- and supply chain planning – a case study at a pulp company. Applied Energy. 2017;203:333–347.

[11] Unterberger E, Buhl HU, Häfner L, *et al.* The regional and social impact of energy flexible factories. In: Procedia Manufacturing. vol. 21. New York, NY: Elsevier B.V.; 2018. pp. 468–475.

[12] Schott P, Sedlmeir J, Strobel N, *et al.* A generic data model for describing flexibility in power markets. Energies. 2019;12(10):1893–1922.

[13] Gils HC. Assessment of the theoretical demand response potential in Europe. Energy. 2014;67:1–18.

[14] Chen Y, Xu P, Gu J, *et al.* Measures to improve energy demand flexibility in buildings for demand response (DR): a review. Energy and Buildings. 2018;177:125–139.

[15] Parvania M, Fotuhi-Firuzabad M, Shahidehpour M. Optimal demand response aggregation in wholesale electricity markets. IEEE Transactions on Smart Grid. 2013;4(4):1957–1965.

[16] Rodríguez-Sánchez R, Santos Múgica M, García Lázaro S, *et al.* Definition and characterization of services to be provided by flexibility elements. FlexPlan – Horizon 2020: 863819; 2020. D2.1. Available from: https://flexplan-project.eu/publications/.

[17] Barth L, Ludwig N, Mengelkamp E, *et al.* A comprehensive modelling framework for demand side flexibility in smart grids. Computer Science – Research and Development. 2018;33(1):13–23.

[18] Siano P, Sarno D. Assessing the benefits of residential demand response in a real time distribution energy market. Applied Energy. 2016;161:533–551.

[19] Dufter C, Guminski A, Orthofer C, *et al.* Lastflexibilisierung in der Industrie – Metastudienanalyse zur Identifikation relevanter Aspekte bei der Potenzialermittlung. Viena; 2017. Available from: https://www.ffegmbh.de/attachments/article/673/KurzfassungTagungsbeitrag.pdf.

[20] Williams HP. Model Building in Mathematical Programming. 5th ed. New York, NY: John Wiley & Sons, Ltd; 2013.

[21] Van den Bergh K, Bruninx K, Delarue E, *et al.* A Mixed-Integer Linear Formulation of the Unit Commitment Problem; 2014. February. Available from: https://www.mech.kuleuven.be/en/tme/research/energy_environment/Pdf/wpen2014-07.pdf.

[22] ENTSO-E. Balancing Report 2020. ENTSO-E; 2020. Available from: https://eepublicdownloads.entsoe.eu/clean-documents/Publications/MarketCommitteepublications/ENTSO-E_Balancing_Report_2020.pdf.

[23] E-Bridge Consulting, IAEW. Impact of merit order activation of automatic Frequency Restoration Reserves and harmonised Full Activation times. ENTSO-E; 2015.

[24] Pelzer A, Richrter M, Lombardi P, *et al.* Energy-intensive industry as the backbone for demand side flexibility modeling and optimization of industrial flexibility flexgraphs. In: International ETG Congress 2017; 2017. pp. 611–616. Available from: http://ieeexplore.ieee.org/stamp/stamp.jsp?tp=&arnumber=8278789&isnumber=8278714

[25] Petersen M, Hansen LH, Mølbak T. Exploring the value of flexibility: a smart grid discussion. In: IFAC Proceedings Volumes (IFAC-PapersOnline). vol. 45. IFAC Secretariat; 2012. pp. 43–48.

[26] ENTSO-E. Market report 2021. ENTSO-E; 2021. Available from: https://ee-public-nc-downloads.azureedge.net/strapi-test-assets/strapi-assets/ENTSO_E_Market_report_2021_2e499deda8.pdf.

[27] Reza Norouzi M, Ahmadi A, Esmaeel Nezhad A, *et al.* Mixed integer programming of multi-objective security-constrained hydro/thermal unit commitment. Renewable and Sustainable Energy Reviews. 2014;29:911–923.

[28] Heussen K, Koch S, Ulbig A, *et al.* Energy storage in power system operation: the power nodes modelling framework. In: IEEE PES Innovative Smart Grid Technologies Conference Europe, ISGT Europe; 2010. pp. 1–8.

[29] Wang S, Geng G, Ma J, *et al.* Operational bottleneck identification based energy storage investment requirement analysis for renewable energy integration. IEEE Transactions on Sustainable Energy. 2020;12(1):92–102.

[30] Petersen MK, Edlund K, Hansen LH, *et al.* A taxonomy for modeling flexibility and a computationally efficient algorithm for dispatch in Smart Grids. In: Proceedings of the American Control Conference. IEEE; 2013. pp. 1150–1156.

[31] Fink J, Hurink JL, Molderink A. Mathematical modelling of devices and flows in energy systems; 2014. Available from: https://ktiml.mff.cuni.cz/~fink/publication/flow.pdf.

[32] Geysen D, Six D, Verbeeck J, *et al.* Adapted methodology for optimal valorization of flexible industrial electricity demand. IndustRE – Horizon 2020: 646191; 2016. June. Available from: http://www.industre.eu/downloads/.

[33] Ulbig A, Andersson G. Analyzing operational flexibility of electric power systems. International Journal of Electrical Power and Energy Systems. 2015;72:155–164.

[34] Karlsson M. The MIND method: a decision support for optimization of industrial energy systems – principles and case studies. Applied Energy. 2011;88(3):577–589.

[35] Waldemarsson M, Lidestam H, Rudberg M. Including energy in supply chain planning at a pulp company. Applied Energy. 2013;112:1056–1065.

[36] Zeng Y, Xiao X, Li J, *et al.* Multi-objective optimization of the energy system in an iron and steel plant for energy saving and low emissions. In: Computer Aided Chemical Engineering. vol. 43. New York, NY: Elsevier B.V.; 2018. pp. 1141–1146.

[37] Zeng Y, Xiao X, Li J, *et al.* A novel multi-period mixed-integer linear optimization model for optimal distribution of byproduct gases, steam and power in an iron and steel plant. Energy. 2018;143:881–899.

Chapter 11
Industrial demand response: coordination with asset management

Salman Mohagheghi[1] *and Abdulrahman Almazroui*[1]

Demand response (DR) is one of the pillars of the modern distribution system, where consumers would voluntarily reduce consumption in response to financial incentives. While for residential consumers, demand curtailment is mainly a matter of inconvenience, for industrial customers, reduction in electric demand can lead to severe operational ramifications such as a halt in production, a pile-up of inventory, or wasted labor. These challenges have caused industrial DR to remain less explored compared to residential demand-side management. Although the industrial sector may be small by numbers, its energy consumption is the dominant load on most distribution systems. This further underlines the potential benefits gained by involving industrial loads in DR events.

One way to motivate industrial DR is to improve opportunities for indirect cost savings as a result of participation in a DR event. For instance, it has been shown in the literature that demand curtailment can be performed in conjunction with inventory management in order to help the plant operate at or near just-in-time. Another option can be to coordinate DR with asset management, i.e. by shutting down lines and workstations under stress or those that are due for maintenance. This way, the direct financial incentives from participating in DR can be augmented with long-term benefits of optimal asset utilization. Providing one such solution is the goal of the current chapter. A multi-objective optimization framework is proposed here that allows plant operators to balance and optimize different financial, operational, and resource objectives while taking advantage of DR to alleviate operational stress on assets. The approach can further incentivize plant managers to participate in DR events while allowing electric utilities to employ this significant untapped potential.

11.1 Introduction

Asset management is the process of managing resources and financial investments to maximize the reliability of the power system, balance customer needs, and optimize

[1]Department of Electrical Engineering, Colorado School of Mines, Golden, CO, USA

shareholder value [1]. It helps balance performance (efficiency), cost (of assets, replacement, and maintenance), and risk (of failure). Over the years, the industry's approach to asset management has moved from being mere asset operators to asset monitors, asset managers, and more recently, asset optimizers [2]. This was in part inevitable since load growth, aging equipment, rate freezes, and regulatory uncertainty have all forced electric utilities to look for ways to increase earnings [3], which further underlines asset management as a viable solution. The objective of asset management is to maximize the performance of the equipment while minimizing the cost. This requires an analysis that considers both the reliability and the cost of failure. Many utilities increase the loading of their assets with the goal of deferring spending, e.g. by postponing capacity expansion projects. In the short run, this reduces costs, but it also increases risk, which needs to be analyzed and incorporated into the asset utilization strategy. Based on the outcome of the analysis, the utility could decide whether to do nothing, change maintenance policy, overhaul, or replace the component [4]. In general, operating the components within manufacturer-specified limits would be the easiest way to prolong the effective lifetime of assets and avoid premature failure. The latter would require repairs and overhauls, resulting in downtime.

Asset failure in industrial plants can have significant ramifications. First and foremost, failure of a production line or a workstation may result in a halt in production and failure to meet the desired demand, which in addition to lost revenue may trigger contractual penalties. This is especially likely in manufacturing plants with highly interdependent production lines and with little or no backup resources. Further, a failed workstation may directly or indirectly lead to inventory buildup across the plant. Depending on the criticality and shelf-life of the outputs produced by individual workstations, accumulation of pre-process and/or post-process inventory may result in a significant waste of material, capital, and energy. Last but not least, crew members may need to be reassigned to avoid lost person-hours. All this affects the efficiency, sustainability, and productivity of the plant.

To avoid such operational issues, many industrial plant managers are moving from simple corrective or time-based maintenance strategies to preventive ones, where maintenance actions are performed before the (at times inevitable) operational stressors can develop into component failures. This, however, is not always easy to implement because of the potential negative impacts of a downed workstation, as stated above. One solution could be a comprehensive and unified operation strategy that takes into account various aspects related to financial objectives, inventory constraints, crew constraints, and physical health of assets. While such an approach would undoubtedly result in long-term benefits, short-term losses due to interruption in service may still outweigh the gains. To offset the balance between short-term gains and losses, the plant manager may coordinate asset operation and maintenance with potential financial incentives from utility-initiated DR programs.

DR is a demand-side management solution that targets residential, commercial, and industrial customers, and is developed for reduction or shifting of demand at a specific time for a specific duration. DR programs can be roughly classified into three groups according to the party that initiates the demand reduction action

[5]: (a) incentive-based DR, in which load curtailment or demand reduction signals are issued by the electric utility and sent to the participating customers in the form of voluntary demand reduction requests or mandatory commands; (b) rate-based DR, in which variations in the hourly price of electricity is intended to encourage customers to adjust their consumption according to the needs of the electric utility; and (c) demand reduction bids, where customers (mainly larger commercial and industrial ones) initiate and submit bids to the electric utility for payments in exchange for reduced energy consumption. These bids would consist of the amount of proposed power curtailment, the corresponding duration, and the posted price. The electric utility would assess the bids, sort them, and accept the ones that it finds financially viable.

In the past decade, many industrial plants, especially those equipped with local distributed energy resources (DER), have been taking advantage of DR. When the electric utility issues a DR event (for instance, by sending a demand reduction request to the plant), the plant can switch to local generation and reduce its consumption from the viewpoint of the utility. These DER units may consist of small-scale gas turbines, diesel engines, or battery systems, most of which have a short startup time of a few seconds to about 5 min [6, 7]. In particular, battery energy storage systems allow industrial plants to store the energy from on-site solar photovoltaics (PV) and/or small-scale wind turbines (which are generally non-dispatchable resources) to be used at a later time when the DR event is issued. Alternatively, when an on-site generation or energy storage is not available, the plant operator can comply with a DR request by shifting the loads to a future time. Of course, this requires a detailed analysis to ensure that operational constraints, crew constraints, inventory constraints, and/or maintenance requirements are not violated. Real-time scheduling of industrial loads has been the focus of much research in the past few years. For example, El-Metwally *et al.* [8] used sequential ordering of loads to improve load factors through actions such as load shifting and peak shaving. Some authors have adopted solutions that are based on those commonly used for real-time computing. For instance, Subramanian *et al.* [9] modeled deferrable loads as tasks with attributes such as arrival time, departure time, and energy requirements, and proposed algorithms for allocating resources to these tasks based on their energy needs and/or deadlines. A similar approach was proposed in [10]. O'Brien and Rajagopal [11] used a greedy algorithm to shift deferrable loads in such a way that the overall load profile matches a specified target profile. From a different aspect, Dobrin *et al.* [12] combined offline scheduling and fixed-priority scheduling to achieve flexibility while coping with complex timings of different tasks. Others have viewed the problem of load scheduling from the angle of variable electricity prices such as time-of-use (TOU) and real-time pricing (RTP) [13, 14].

An alternative to demand shifting or reverting to local generation would be to simply reduce/curtail demand. Examples of intelligent load shedding algorithms have been proposed in the literature based on the priority ranking of loads [8], expert systems [15, 16], and artificial neural networks [17]. However, when it comes to industrial plants, shedding load is not merely a matter of inconvenience. In the highly interconnected operational environment of modern manufacturing plants, different workstations and production lines are highly interdependent, and the unavailability of each one can have significant operational impacts on the overall plant, i.e. it may lead to the unwanted buildup of inventory in dependent/supplier workstations, a need for reassigning workers to other lines and workstations with a possibility of wasted labor, and under the worst-case conditions, failure to meet the desired production level. Hence, there is a need for unified and comprehensive optimization solutions that will consider these (at times) contradictory objectives before making recommendations for load curtailment. The ultimate goal is to ensure the overall financial viability of the plant operation, which means financial gains from participating in DR must be weighed against potential losses due to waste in material or labor.

If coordinated with asset management, DR is an effective strategy that can enable opportunistic asset maintenance while allowing the plant to take advantage of the program's financial incentives. This can serve as indirect cost savings, which will tip the balance in favor of DR. To do this, the industrial plant can participate in various DR programs for load curtailment or shifting, or alternatively, can proactively take part in the DR bidding market. However, implementing industrial DR would require an optimal solution to incorporate financial, asset, and technical considerations within a unified framework of operation [18]. Devising such a solution is the goal of this chapter. A multi-objective mixed-integer optimization framework is presented that facilitates the sustainable operation of a manufacturing plant by coordinating operational efficiency with asset management. A case study has been presented for a sample industrial plant, a luxury vehicle cockpit assembly line, to further demonstrate the effectiveness of the proposed approach.

11.2 Proposed strategy

11.2.1 General idea

Without DR incentive, the natural goal of the plant operator would be to maximize financial benefits while ensuring wasted labor and inventory buildup are minimized. Such an approach may force some workstations to work continuous hours to produce more end-product, which can at times lead to operational stress. On the other hand, with DR, a new revenue stream will open up for the plant, which can also assist with other operational constraints and objectives such as reducing the stress on assets and/or minimizing inventory buildup. This way, part of the financial losses incurred due to shutting down one or more workstations will be compensated for by the financial gains resulting from participating in DR.

To do this, the plant operator can run a scenario with no DR incentive to find the power consumption level that would maximize profits subject to other operational constraints. A follow-up scenario is then run with a DR incentive. The difference between the two would indicate the energy level that can be offered in the electric utility's DR market as capacity relief. In the most general case, four different objectives can be considered:

- maximizing the net revenue,
- minimizing the inventory buildup,
- minimizing the waste in labor, and
- maximizing the number and duration of operation breaks for various workstations.

The last objective is what leads to stress relief on assets. By reducing the number of consecutive hours during which a workstation has to be operational, not only the operational stress is reduced, but it can also be queued for preventive maintenance. Naturally, the overall problem is subject to a variety of operational constraints, which makes it a constrained multi-objective optimization model as described in the next section.

11.2.2 Problem formulation

The problem has been formulated as a constrained multi-objective optimization model as outlined below.

11.2.2.1 Objective functions

A total of four objective functions have been defined that cover various aspects of the plant's operation.

O_1: Maximizing the financial benefits

This objective ensures that the plant takes advantage of all opportunities to maximize its net revenue. These consist of the financial gain from selling the end product [first term in (11.1)], the cost of electricity consumption [second term in (11.1), with a negative sign], the financial gain from participating in DR bidding and providing capacity relief [third term in (11.1)], and the financial losses associated with unfulfilled orders [last term in (11.1), with a negative sign since it must be minimized]. Note that $p^{\max}$ is a fixed value that represents the energy consumption level of the plant with no DR incentive, i.e. when maximizing profits from production is the main driving force:

$$O_1 = \max \left(\begin{array}{l} \rho \cdot \sum_{t=1}^{\mathrm{T}} \alpha_n \cdot p_{n,t} - c_t \cdot \sum_{t=1}^{\mathrm{T}} \sum_{i=1}^{n} p_{i,t} \\ + r^{\mathrm{DR}} \cdot \left[p^{\max} - \sum_{t=1}^{\mathrm{T}} \sum_{i=1}^{n} p_{i,t} \right] - \pi \cdot \beta \end{array} \right) \tag{11.1}$$

O_2: Minimizing inventory buildup

In an effort to move towards just-in-time (JIT) strategy, it is often desired to minimize the inventory buildup at the input of a workstation (pre-process inventory) and/or its output (post-process inventory). Different materials may have different criticality levels based on their shelf life (e.g. food products or chemicals may expire sooner than other produced outputs). In order to determine the capital value of this inventory buildup, workstation products can be divided into three different categories,

A, B, and C [19]. The products in the A category have the highest value in the organization. The B category also consists of important products but is less strictly monitored, whereas the third category, C, is relatively less important than the other two; although, it usually exists in larger quantities [18]. This categorization can be used when deciding on the operation status of a workstation, i.e. operation strategies that lead to a significant buildup of product categories A and B must be avoided if at all possible. To model this, the inventory buildup of different workstations is weighed here based on a heuristically defined criticality coefficient γ. It should be noted that the post-process inventory of the last workstation (i.e. workstation n) is not included in this equation, since it represents the final product of the plant:

$$O_2 = \min \sum_{t=1}^{T} \sum_{i=1}^{n-1} \gamma_i \cdot s_{i,t} \tag{11.2}$$

O_3: Minimizing wasted labor

When a workstation is shut down to allow the plant to take advantage of DR, crew members working at that workstation need to be reassigned to other units. Unassigned crew indicates a wasted resource that could incur financial losses to the plant (if workers continue to get paid regardless) or the workers (if they lose their pay during the unassigned hours). In either case, this must be minimized as indicated in (11.3):

$$O_3 = \min \sum_{t=1}^{T} \left(W^{\text{total}} - \sum_{i=1}^{n} w_{i,t} \right) \tag{11.3}$$

O_4: Minimizing stress on assets

In this study, asset management is viewed in terms of reducing the total number of consecutive hours during which a workstation may be operational and is formulated as maximizing the number of "operation breaks" a workstation receives during any (H + 1)-hour working period, as indicated in (11.4) – also see (11.12) for more details:

$$O_4 = \max \sum_{t=\mathrm{H}+1}^{T} \sum_{i=1}^{n} b_{i,t} \tag{11.4}$$

11.2.2.2 Constraints

The problem is solved subject to the following constraints:

Workstation production constraints:

The product of a workstation (viewed as its post-process inventory) is assumed to be linearly proportional to its energy consumption. This assumption has been made for proof-of-concept purposes and does not affect the generality of the problem. Nonlinear relationships or stepwise functions can also be assumed, although this could change the structure of the optimization model into a nonlinear one.

At the end of each time-step, the post-process inventory at a workstation is a function of the inventory buildup at the previous time-step, the workstation's production level at that time-step, minus the production level of those workstations that directly use the former's product. This is modeled as in (11.5). The post-process inventory of each workstation must remain within the acceptable range at all times, as indicated in (11.6). Reducing the upper limit in (11.6) moves the operation towards just-in-time. Depending on the units defined for the output of each workstation, the post-process inventory can be considered to be a discrete or continuous variable:

$$\forall i, \forall t : s_{i,t} = s_{i,t-1} + \alpha_i \cdot p_{i,t} - \sum_{j=1,\neq i}^{n} v_{j,i} \cdot \alpha_j \cdot p_{j,t} \tag{11.5}$$

$$\forall i, \forall t : S_i^{\min} \leq s_{i,t} \leq S_i^{\max} \tag{11.6}$$

Workstation operation constraints:

The power consumption level of any workstation is limited by its lower and upper bounds [see (11.7)]. The upper bound is intended to prevent damages to the workstation and/or its components, whereas the lower threshold is more of a financial constraint, i.e. it may not be economically viable for the workstation to operate below a certain level. Further, the total amount of power consumed by the plant must be limited to the maximum allowable level, set by either the contractual agreement with the electric utility or the maximum capacity of the service transformer supplying the plant [see (11.8)]. Finally, if the workstation is operational, the necessary number of workers will be assigned to it [see (11.9)], which naturally cannot be more than the total number of workers available for the entire plant [see (11.10)]:

$$\forall i, \forall t : P_i^{\min} \cdot u_{i,t} \leq p_{i,t} \leq P_i^{\max} \cdot u_{i,t} \tag{11.7}$$

$$\forall t : \sum_{i=1}^{n} p_{i,t} \leq P^{u,\max} \tag{11.8}$$

$$\forall i, \forall t : w_{i,t} = W_i \cdot u_{i,t} \tag{11.9}$$

$$\forall t : \sum_{i=1}^{n} w_{i,t} \leq \mathrm{W}^{\text{total}} \tag{11.10}$$

Workstation interdependence constraints:

For any workstation j, the produced output is limited by the amount of pre-process inventory available. Note that pre-process inventory for workstation j is the same as post-process inventory of the immediately preceding workstation i. Assuming that there are multiple workstations, each providing the required pre-process inventory for workstation j, this constraint can be expressed as in (11.11). The term in the parenthesis consists of the amount of post-process inventory at workstation i at the end of the previous time-step, the amount produced by workstation i during the current time-step, minus the amount used by other workstations other than j that use i's product. The second term is included in the equation to remove the effects of any workstation i that is not connected to workstation j. Without this term, since $v_{j,i}$ is zero, the term inside the minimum function will become zero, thus dominating the constraint. Naturally, the function *min* can be linearized by breaking it into multiple inequalities:

$$\forall j, \forall t : \alpha_j \cdot p_{j,t} \leq \min_i \left\{ \begin{array}{l} v_{j,i} \cdot \left(s_{i,t-1} + \alpha_i \cdot p_{i,t} - \sum\limits_{k=1,\neq i,\neq j}^{n} v_{k,i} \cdot \alpha_k \cdot p_{k,t} \right) \\ +(1 - v_{j,i}) \cdot \mathrm{M} \end{array} \right\} \quad (11.11)$$

Asset management:

The approach to asset management adopted in this model is to ensure that, as much as technically possible, workstations do not consecutively work higher than a predetermined number H of hours. The goal is to allow for operation breaks in between, which could be used simply for operational stress relief or for conducting preventive maintenance. This is modeled as in (11.12). Notice that without operation breaks in the consecutive (H + 1)-hour period, the summation will add up to (H + 1). This indicates that the workstation has worked for more than H consecutive hours, which is not desirable. A nonzero $b_{i,t}$ indicates a break in between, which denotes that the workstation has worked for H or less number of consecutive hours:

$$\forall i, \forall t \in [\mathrm{H}+1, \mathrm{T}] : \sum_{d=0}^{\mathrm{H}} u_{i,t-d} = (\mathrm{H}+1) - b_{i,t} \quad (11.12)$$

Unfulfilled orders:

Ideally, the plant manager would like to meet the target production level during each workday. However, to allow the plant to take advantage of DR, this constraint is relaxed, as shown in (11.13), although subject to a threshold [as shown in (11.14)]. Of course, the shortage in meeting the demand leads to penalties, as indicated in the objective function (11.1):

$$\beta = D^{\mathrm{des}} - \sum_{t=1}^{\mathrm{T}} \alpha_n \cdot p_{n,t} \quad (11.13)$$

$$\beta \leq \beta^{\max} \quad (11.14)$$

Integrality and non-negativity constraints:

$$\forall i, \forall t : p_{i,t} \geq 0 \quad (11.15)$$

$$\forall i, \forall t \in [\mathrm{H}+1, \mathrm{T}] : b_{i,t} \geq 0 \quad (11.16)$$

$$\forall i, \forall t : u_{i,t} \in \{0,\ 1\} \quad (11.17)$$

$$\beta \geq 0 \quad (11.18)$$

11.2.3 Solution methodology

Due to the multi-objective nature of the optimization model, a goal programming approach has been adopted to ensure that no individual objective function would dominate the others, and that Pareto optimal solution is reached. Pareto optimal solution is one in which no individual objective function can improve without worsening others. This is especially important when two or more objective functions are contradictory to one another. For instance, objective O_1 in this study tries to push the workstations to produce more in order to maximize profits, which may result in higher inventory buildup, hence being contradictory to objective O_2.

In this goal programming approach, all objective functions are solved one at a time, i.e. in a single objective framework. Their corresponding optima are then set as the goals (targets) to be achieved in the multi-objective framework. These goals are usually slightly relaxed to ensure that all targets can be met. The problem can be modeled as shown below, where L is the multi-objective function. We wish to minimize the distance of each objective function O_q from its pre-determined goal (target) G_q. This is achieved by defining non-negative surplus or deficiency variables for the minimization and maximization objectives, respectively. These variables will then convert those requirements into soft constraints [see (11.20)–(11.23)]:

$$\min L \tag{11.19}$$

Subject to:

$$O_1 + b_1 \geq G_1 \tag{11.20}$$

$$O_2 - b_2 \leq G_2 \tag{11.21}$$

$$O_3 - b_3 \leq G_3 \tag{11.22}$$

$$O_4 + b_4 \geq G_4 \tag{11.23}$$

$$\forall q : \frac{b_q}{G_q} \leq L \tag{11.24}$$

$$\forall q : b_q \geq 0 \tag{11.25}$$

11.3 Case study

11.3.1 System description

The proposed methodology is applied to a sample industrial manufacturing plant for luxury vehicle cockpit assembly. The schematic diagram of the plant layout and the workstations is illustrated in Figure 11.1. The plant consists of multiple workstations that are arranged in series or parallel configurations. The data for the baseline electric demand consumption of the workstations, the criticality level of their product (i.e. post-process inventory), the number of crew members necessary for operating each one, and the hourly rate of electricity is provided in the Appendix.

It is assumed that at time 0 all workstations have a starting pre-process inventory of 4 units, with the exception of workstations 1, 7 and 8 whose inputs are provided from external resources and considered to have no limitations. The problem has been solved for 10 time-steps, i.e. $T = 10$. It has also been assumed that a total of 20 workers are available during every hour of the workday. These workers are sufficiently skilled to be able to be assigned to any workstation necessary. The desired number of units to be produced by the plant during the workday is 20 with a revenue of $40,000 for

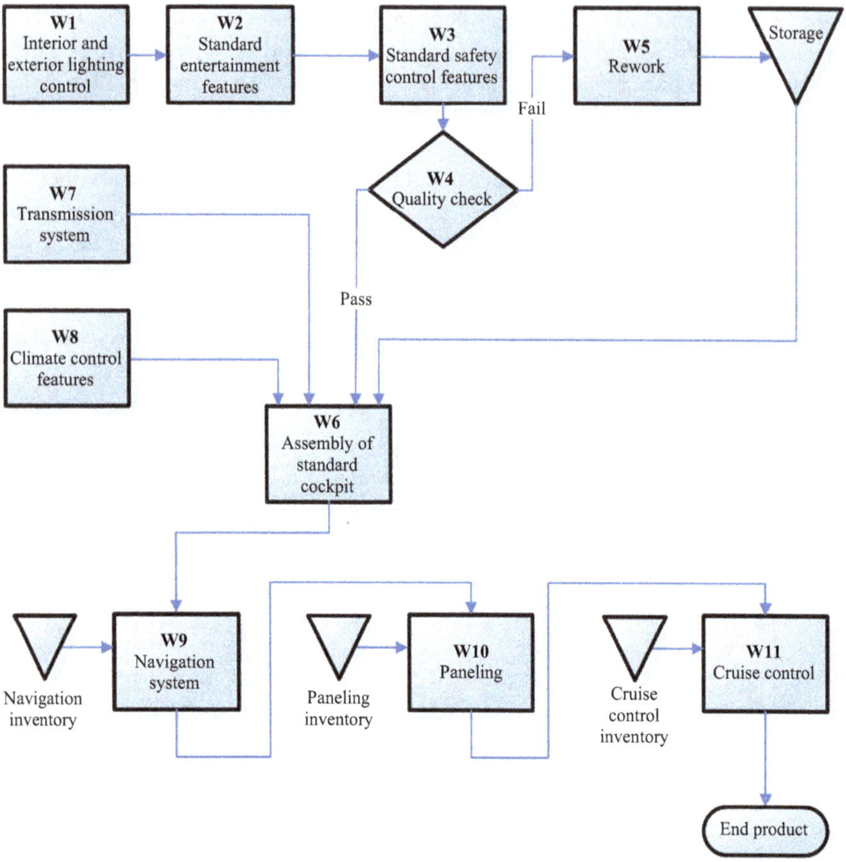

Figure 11.1 Schematic diagram of the manufacturing plant under study

each unit, and the maximum permissible number of unfulfilled orders is eight units with a penalty of \$5,000 per unit. The maximum power to be received from the utility is assumed to be 1,000 kW per hour and the baseline DR incentive rate is \$50/kW of demand reduction. Lastly, it is desired that no workstation operates for more than four consecutive hours. These values are all considered for demonstration purposes and do not affect the generality of the problem.

11.3.2 Results

To solve the multi-objective optimization model, the individual objective functions are first solved one at a time, subject to all relevant constraints. This will provide the global optima for each objective, which will then be used to set the goals (targets) to be met in the multi-objective framework. The results are shown in Table 11.1. With only objective O_1, the optimization model tries to maximize the financial gains, which

Table 11.1 Single objective optima

Obj.	Net revenue ($)	Inventory buildup (units[1])	Wasted labor (person-hours)	No. of operation Breaks	Total energy consumption (kWh)	Unfulfilled orders
O_1	798,815	420	76	115	4,875	0
O_2	562,814	10	96	145	3,120	7
O_3	446,443	352	17	16	4,721	8
O_4	536,645	200	128	263	2,925	8

[1]This is a weighted sum, considering the criticality level of each workstation product.

means that all other limiting factors such as inventory buildup, wasted labor, or stress on workstations are ignored. As a result, the workstations are pushed to operate at or close to full capacity throughout the workday, meeting the desired production level. Despite providing the highest net revenue, it can be seen that this case would result in the highest post-process inventory buildup, since the operation is encouraged. When O_2 is considered as the single objective, the level of post-process buildup reduces significantly, as expected. However, this occurs at the expense of lower net revenue and higher unfulfilled orders because more workstations are expected to be shut down or to work at a lower capacity. Higher instances of workstation shutdown mean that more labor is wasted (since the plant is left with a higher number of unassigned workers) but more substations are receiving operation breaks (due to the higher number of idle hours). With only O_3 to optimize, the optimization model prioritizes wasted labor over other factors. As expected, that number reaches its lowest value compared to other cases. Interestingly, the number of operation breaks decreases significantly (i.e. an indication that workstations are operating more continuously), yet the net revenue is low. This is because many workstations work at less than capacity so as to satisfy maximum worker assignment, but not necessarily to produce more, which is why the number of unfulfilled orders is at the highest permissible level. Finally, when considering only O_4, the goal is to maximize the number of operation breaks, which is achieved by implementing more workstation shutdowns. Naturally, this results in a larger number of workers being left unassigned (i.e. the highest value of wasted labor) and a significantly low power consumption (i.e. 60% of the first case). It can be seen that the inventory buildup is not particularly high, which aligns with the low operation capacity.

What is evident from Table 11.1 is the contradictory nature of the four objective functions. A goal programming approach would make sure that no objective will dominate the others and that each will reach the closest value possible to its own optima. To do this, we define the optimal value of each objective function as the goal (target) to be (ideally) met in the multi-objective framework, i.e. goal values G_q in equations (11.20)–(11.23). In problems such as the one studied here, where the individual objectives cannot simultaneously achieve their individual optima, the goal (target) values are often slightly deteriorated with respect to the true optima in order to provide some flexibility for the multi-objective problem. For minimization objectives, this means slightly increasing the value (typically by 10% or so), whereas the value is slightly decreased for maximization objectives. The multi-objective optimization model is then solved as minimizing (11.19) subject to constraints (11.1)–(11.18) and (11.20)–(11.25). The results are shown in Table 11.2. It can be seen that no objective function manages to achieve its single objective optima, with objectives O_2 and O_3 having the largest percentagewise difference, which indicates that they have the largest impact on the overall solution (and hence on deteriorating other objective functions). For the overall solution, the total number of unfulfilled orders stands at four, with the overall power consumption of 3,900 kWh. Further, the total incentive received from DR amounts to $48,750.

Table 11.3 shows the hourly power consumption level of individual workstations over the 10-hour dispatch period. Many workstations operate at lower than rated capacity (to reduce post-process inventory buildup) or are shut down (to facilitate asset stress relief). Figure 11.2 illustrates the post-process inventory buildup across

Table 11.2 Multi-objective optima

Objective function	Single objective optima	Goal value	Multi-objective optima	Shortage/surplus with respect to single objective optima
O_1	$798,815	$718,933	$667,658	17%
O_2	10	11	25	150%
O_3	17	19	44	158%
O_4	263	236	63	77%

Table 11.3 Hourly power consumption of individual workstations (kW)

Station no.	Time step									
	1	2	3	4	5	6	7	8	9	10
1	15	25	15	25	15	25	15	25	15	25
2	30	50	30	50	30	50	30	50	30	50
3	60	100	60	100	60	100	60	100	60	100
4	6.4	44.8	6.4	44.8	6.4	44.8	6.4	44.8	6.4	44.8
5	6.4	19.2	6.4	19.2	6.4	19.2	6.4	19.2	6.4	19.2
6	0	25.6	0	25.6	0	25.6	0	25.6	0	25.6
7	0	160	20	140	20	140	20	140	0	160
8	20	140	20	140	20	140	20	140	20	140
9	0	20	0	20	0	20	0	20	0	20
10	0	25.6	0	25.6	0	25.6	0	25.6	0	25.6
11	0	32	0	32	0	32	0	32	0	32

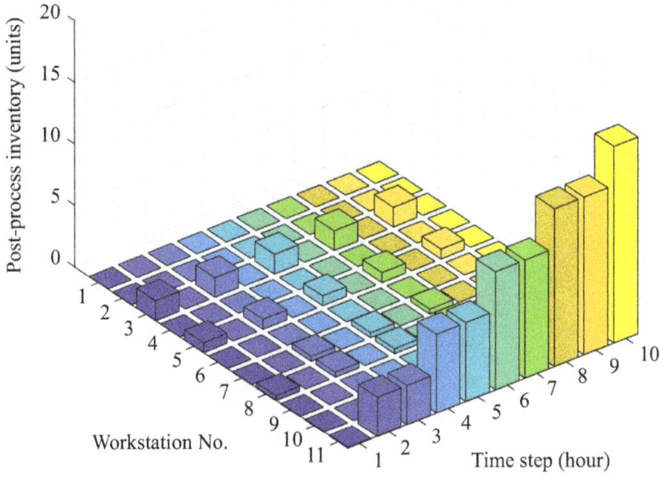

Figure 11.2 Post-process inventory buildup across the plant during the workday

the plant during the workday. It can be seen that the overall operation is reasonably close to being just-in-time, i.e. not accumulating any unnecessary inventory and using almost all units produced by each workstation. Note that the post-process inventory of the last workstation is the final product of the plant and is therefore not penalized [see (11.2)]. Figure 11.3 demonstrates the workstations that undergo operation breaks during the workday. Although these breaks provide opportunities for asset stress relief and possibly preventive maintenance, they do result in wasted labor.

11.3.3 Discussion

Certain aspects of the problem formulation can be changed to make it applicable to more specific cases. For instance, the number of workers assigned to each workstation was assumed to be a fixed number, regardless of the workstation's loading level. However, it is possible that workstations require a smaller number of workers when operating at partial capacity. This change can easily be applied by modifying (11.9), i.e. by making the assigned workers to be a function of discrete levels of consumption power of the workstation, rather than its operating status. Further, it is possible that some workstations can only be operated by especially trained skilled workers. This means that not all workers can be reassigned to all workstations.

Further, the approach to asset health assessment may be extended by considering the loading level of each workstation in addition to the number of consecutive hours it

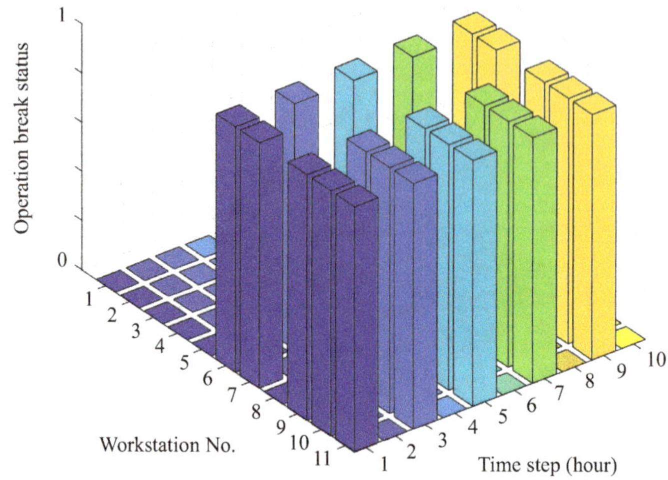

Figure 11.3 Operation break status of different workstations during the workday (1: under operation break, 0: operational). No workers are assigned to the workstations under operation break.

has been operational. This can be easily accomplished by modifying (11.12) in a way that each working hour of a workstation is also weighed based on its loading level, hence converting the equation into a weighted sum.

Lastly, many manufacturing plants may be equipped with the onsite generation, e.g. battery energy storage, diesel generators, or renewable energy resources such as small-scale wind turbines or solar PV systems. In these situations, the available onsite power can be utilized for local consumption, to sell back to the utility (for instance at peak hours), or to charge the battery unit for future use. This will provide the plant operator with a larger number of degrees of freedom to integrate into the plant's operation strategy and is expected to improve financial gains. One such solution was proposed in the first author's earlier work in [20].

11.4 Conclusions

In the recent years, with the push towards sustainability and energy efficiency, asset management has become the focal point of many modern manufacturing plants. Global competition and reduced profit margins often push these plants towards continuously operating close to their full capacity, which can place additional stress on the components and under extreme cases, lead to their premature failure. This, in turn, could result in a halt in operation, unwanted inventory buildup, potential waste, and the possible need for component replacement. Advanced asset management approaches can help manufacturing plants move from being reactive asset users to proactive asset optimizers. Various condition-based or opportunistic maintenance approaches can be adopted to optimize the utilization rate of assets and minimize asset failures and/or degradation. While in the long run, this can lead to significant cost savings, its short-term benefits may be more difficult to justify given the potential temporary lost revenues and/or sub-optimal inventory utilization and buildup.

If coordinated with asset management, DR is an effective strategy that can enable opportunistic asset maintenance while allowing the plant to take advantage of the program's financial incentives. To do this, the industrial plant can participate in various DR programs for load curtailment or shifting, or alternatively, can proactively take part in the DR bidding market. However, implementing industrial DR would require an optimal solution to incorporate financial, asset, and technical considerations within a unified framework of operation. One such solution was proposed in this chapter. A multi-objective mixed-integer optimization framework was presented to facilitate the sustainable operation of a manufacturing plant by coordinating operational efficiency and asset management. A case study was provided to show how various operational objectives can be coordinated with asset management to ensure that the plant operator can reduce the operational stress on assets while taking advantage of DR financial benefits.

11.5 Nomenclature

11.5.1 Indices

i	Index used for workstations
j	Index used for workstations
k	Index used for workstations
n	Number of workstations, and the index for the last workstation that produces the final product
q	Index used for objective functions within the multi-objective framework
t	Time index

11.5.2 Parameters

c_t	Cost of power provided by the utility at time t (\$/kWh)
D^{des}	Desired production level of the plant (number of units, volume, or weight produced); this can be viewed as the post-process inventory of the final workstation
G_q	Goal (target) value for objective function q within the multi-objective optimization framework (units vary)

H Desired a maximum number of consecutive hours during which a workstation can be operational (h)

M A very large value

$P_i^{\max}$ Workstation i maximum (rated) power, i.e. power consumed by workstation i when working at full capacity (kW)

$P_i^{\min}$ Workstation i minimum permissible power (kW). The unit cannot operate at lower capacity and must therefore be shut down

$P^{u,\max}$ Maximum power available from the utility (kW)

r_t^{DR} Incentive rate for participating in DR at time t ($/kWh)

$S_i^{\max}$ Maximum acceptable post-process inventory for workstation i (number of units, volume, or weight)

$S_i^{\min}$ Minimum acceptable post-process inventory for workstation i (number of units, volume, or weight)

$v_{j,i}$ Binary parameter indicating whether workstation j is directly connected to workstation i, i.e. uses its product (=1: if j is connected to i, 0: otherwise)

W_i Number of workers to be assigned to workstation i when it is operational

W^{total} Total number of workers available during the workday

α_i Factor relating the number of units/volume/weight produced by workstation i to the power consumed by it (No. units/kW, m^3/kW, or kg/kW)

$\beta^{\max}$ Maximum permissible number/amount of unfulfilled orders, i.e. maximum permissible shortage in the production level of the last workstation compared to the desired level (number of units, m^3, or kg)

γ_i Criticality level of the product produced by workstation i

π Penalty due to unfulfilled order ($/unit, $/m^3$, or $/kg)

ρ Revenue from selling the end product ($/unit, $/m^3$, or $/kg)

11.5.3 Variables

$b_{i,t}$ Number of break hours that workstation i receives in the (H + 1)-hour period terminating at time t

b_q Surplus or deficiency variable for objective function q within the multi-objective optimization framework (units vary)

O_q Objective function q within the multi-objective optimization framework (units vary)

$p_{i,t}$ Operational level of workstation i at time t (kW)

p_t^u Utility power provided to the plant at time t (kW)

$s_{i,t}$ Post-process inventory at the output of workstation i at time t (number of units, volume, weight)

$u_{i,t}$ Binary variable indicating whether workstation i is operating at time t (=1: if workstation is operating, 0: otherwise)

$w_{i,t}$ Number of workers assigned to workstation i at time t

β Unfulfilled orders, i.e. shortage in the production level of the last workstation compared to the desired level (number of units, m^3, or kg)

11.6 Appendix

Table 11.4 Workstation data

Workstation	P_i^{min} (kW)	P_i^{max} (kW)	α_i	$S_i^{min/max}$ (units)	W_i	γ_i
1	5	50	0.16	0/20	2	1
2	5	50	0.08	0/30	1	3
3	10	100	0.04	0/30	3	1
4	6.4	64	0.125	0/20	1	3
5	6.4	64	0.125	0/20	2	3
6	25.6	256	0.125	0/25	3	5
7	20	200	0.02	0/22	2	3
8	20	200	0.02	0/20	1	1
9	2.5	25	0.16	0/25	2	5
10	12.8	128	0.125	0/25	1	5
11	4	40	0.1	0/–	2	N/A

Table 11.5 Rate of power purchased from the utility

Rate	Time (h)									
	1	2	3	4	5	6	7	8	9	10
c_t	0.2	0.2	0.2	0.2	0.2	0.2	0.4	0.4	0.4	0.4

References

[1] W. Reder, "Asset management – fact or fiction?" *IEEE Power & Energy Magazine*, pp. 93–96, May/June 2005.

[2] Y. Mansour, L. Haffner, V. Vankayala, and E. Vaahedi, "One asset, one view – integrated asset management at British Columbia transmission company," *IEEE Power & Energy Magazine*, pp. 55–61, May/June 2005.

[3] R.E. Brown and B.G. Humphrey, "Asset management for transmission and distribution," *IEEE Power & Energy Magazine*, pp. 39–45, May/June 2005.

[4] G.J. Anders, J. Endrenyi, and C. Yung, "Risk-based planner for asset management," *IEEE Computer Applications in Power*, pp. 20–26, October 2001.

[5] S. Mohagheghi, J. Stoupis, Z. Wang, Z. Li, and H. Kazemzadeh, "Demand response architecture – integration into the distribution management system,"

in Proc. *IEEE International Conference on Smart Grid Communications* (SmartGridComm'10), Gaithersburg, MD, USA, October 4–6, 2010, pp. 501–506.
[6] I. Kuzle, D. Bošnjak, and S. Tešnjak, "An overview of ancillary services in an open market environment," in *Proceedings of the Mediterranean Conference on Control and Automation*, Athens, Greece, July 2007.
[7] X. Yu and L.M. Tolbert, "Ancillary services provided from DER with power electronics interface," in *Proceedings of the IEEE Power and Energy Society General Meeting*, Montreal, Quebec, Canada, October 2006.
[8] M.M. El-Metwally, M.S. El-Sobki, H.A. Attia, and S.A. Wahdan, "Priority ranking of industrial loads and application of demand side management technique," in *Proceedings of the International Middle East Power Systems Conference*, December 2006.
[9] A. Subramanian, M. Garcia, A. Domínguez-García, D. Callaway, K. Poolla, and P. Varaiya, "Real-time scheduling of deferrable electric loads," in *Proceedings of the American Control Conference*, Montreal, Quebec, Canada, June 2012, pp. 3643–50.
[10] M.L. Della Vedova and T. Facchinetti, "Real-time scheduling for industrial load management," in *Proceedings of the IEEE EnergyCon Conference and Exhibition*, Florence, Italy, September 2012.
[11] G. O'Brien and R. Rajagopal, "A method for automatically scheduling notified deferrable loads," in *Proceedings of the American Control Conference*, Washington, DC, USA, June 2013, pp. 5087–92.
[12] R. Dobrin, G. Fohler, and P. Puschner, "Translating off-line schedules into task attributes for fixed priority scheduling," in *Proceedings of the IEEE Real-Time Systems Symposium*, London, UK, December 2001, pp. 225–34.
[13] K. Collins, M. Mallick, G. Volpe, and W.G. Morsi, "Smart energy monitoring and management system for industrial applications," in *Proceedings of the IEEE Electrical Power and Energy Conference*, London, Canada, October 2012.
[14] S. Bahrami, F. Khazaeli, and M. Parniani, "Industrial load scheduling in smart power grids," in *Proceedings of the International Conference on Electricity Distribution*, Stockholm, Sweden, June 2013.
[15] B. Delfino, F. Croce, P.A. Fazzini, S. Massucco, A. Morini, F. Silvestro, and M. Sivieri, "Operation and management of the electric system for industrial plants: an expert system prototype for load-shedding operator assistance," in *Proceedings of the IEEE Industry Applications Society*, Rome, Italy, August 2000, pp. 3260–68.
[16] F. Shokooh, J.J. Dai, S. Shokooh, J. Tastet, H. Castro, T. Khandelwal, and G. Donner, "An intelligent load shedding system application in a large industrial facility," in *Proceedings of the IEEE Industry Applications Society*, October 2005, pp. 417–25.
[17] C.T. Hsu, H.J. Chuang, and C.S. Chen, "Artificial neural network based adaptive load shedding for an industrial cogeneration facility," in *Proceedings of the IEEE Industry Applications Society*, Australia, October 2008.

[18] S. Mohagheghi and N. Raji, "Managing industrial energy intelligently," *IEEE Industry Applications Magazine*, pp. 53–62, March/April 2014.
[19] Y. Chen, K.W. Li, and S.F. Liu, "A comparative study on multicriteria ABC analysis in inventory management," in *Proceedings of the IEEE Intl. Conf. on Man, Machines and Cybernetics*, Singapore, October 2008, pp. 3280–85.
[20] M. Choobineh and S. Mohagheghi, "Optimal energy management in an industrial plant using on-site generation and demand scheduling," *IEEE Transactions on Industry Applications*, vol. 52, no. 3, pp. 1945–1952, May/June 2016.

Chapter 12
A machine learning-based approach for industrial demand response

Ronke M. Ayo-Imoru[1]*, Ahmed A. Ali*[1] *and Pitshou N. Bokoro*[1]

Considerable interest is now being vested in low-carbon energy sources in other to meet the world's ever-growing energy demand without causing damage to the environment, which has given rise to the increasing contributions of renewable energy sources to the energy grid. This development is not without its challenges to modern electric power systems. Due to the intermittent nature of renewable energy resources, its increase has resulted in energy demand–supply mismatch, grid imbalance, or grid instability. A reliable and cost-effective approach is required to address this energy trade imbalance caused by the influx of renewable energy sources. Demand response (DR) is a concept that aims at achieving energy balance in the grid by controlling and adjusting flexible loads.

Industrial DR has the potential for a significant contribution to the operational flexibility of power systems. This is because the industrial sector is one of the major electricity consumers in the world, as many industrial loads consume much electrical energy. Therefore, a proper industrial demand response regime will help in ensuring a safe and secured grid, improve energy balance, promote decarbonization, more grid reliability and cost reduction for customers.

Machine learning is one of the primary drivers of the fourth industrial revolution; it is currently widely applied in many areas, and DR is not left out. This work explores the application of machine learning for industrial DR. Over the years, different tools based on machine learning have been developed for DR applications. This research aims to review and analyze machine learning approaches, the current trends, and innovations in industrial DR.

Some of the different machine learning tools analyzed from this review included the artificial neural network (ANN), adaptive neuro-fuzzy inference system (ANFIS), Markov chain, random forest, and support vector machine (SVM) approaches. The strengths, challenges, and opportunities of the different machine learning approaches were further analyzed and explained.

[1]Department of Electrical Engineering Technology, University of Johannesburg, Johannesburg, South Africa

12.1 Introduction

In tackling the issues of climate change, there is a need for low-carbon emission energy sources. Consequently, there has been a rapid increase in the contribution of renewable energy sources (RES) to the electricity grid in recent times. The intermittent nature of the RES results in instability of the grid. Therefore, there is a need to stabilize the grid by balancing the electricity supply and demand. It is reasonable or easier to adjust the electricity supply by adjusting electricity generation across the different plants, but this is not economical and not good for the health of the generation plant. Generating electricity at a constant load is better for the generating plants and more sustainable. These issues have given rise to the concept of demand response (DR). The DR concept focuses on stabilizing demand and supply from the demand side.

DR is an approach that brings flexibility to power systems by adjusting the demand side. In DR, flattening the peak of the demand curve reduces the instability in the grid. It involves consumers changing their electricity consumption patterns, and this they achieve via different methods. DR helps stabilize the electric grid, helps generation plants to work optimally with minimal interruptions, saves cost for both the electricity supplier and the consumer, and helps consumers be involved in the electricity demand and supply process. Implementing IDR reduces the need for flexibility options like storage, which are mostly more costly [1]. DR is a critical smart grid technology. An intelligent DR implementation could provide 185 GW of system flexibility globally, almost the same amount of electricity currently installed in Australia and Italy combined [2].

The types of (DR) can be classified based on the loads connected. There is the residential DR, which focuses on adjusting electricity consumption patterns in households. At the same time, the Industrial demand response (IDR) focuses on changing electricity consumption patterns in the industries. The type of load used in the industries is heavy. Having a thriving (IDR) will have a significant impact on the power system stability. The focus of this chapter is the IDR. Different approaches have been deployed in the application of IDR. The focus of this work is the use of machine learning in achieving IDR.

Machine learning is a branch of data science that deals with training the computer with data in other to discover patterns that can be used in developing an algorithm for further analysis of other data with which it had not been trained. Machine learning is a fast-growing field that is presently applied widely in almost every field – the health sector, banking, engineering, aerospace, finance, and many others. For DR, machine learning has also been greatly employed. From the search on current trends in demand response, it was discovered that most of the available publications on DR were for residential demand response. Only a few publications are on machine learning applications for industrial DR. Hence, this has prompted the interest in carrying out this research. This research focuses on investigating work done so far on the application of machine learning in industrial DR, the best approaches used, and the results obtained from this approach.

This chapter consists of five sections for a proper exploration of the application of machine learning for industrial response: in Section 12.2, the types of industrial loads are discussed, Section 12.3 discusses DR in the industry; Section 12.4 discusses the machine learning approaches for IDR and Section 12.5 contains the conclusion.

12.2 Industrial load

The term 'load' in electricity refers to any device or component connected to the electricity supply and uses electricity, converting it to other forms of energy [3]. Generally, in power systems, load classifications are residential, agricultural, commercial, and industrial loads [4]. Thus, Industrial load refers to the electrical devices and components that the industries use. These industrial loads include machines, heating appliances, cooling appliances, and other electrical devices that the industries use. Some of these devices can be in operation for days.

In literature, Industrial loads have different classifications, Shoreh *et al.* in [5] classify industrial load into two, namely production and support services. The production loads are involved in the manufacturing/industrial process, while the support loads are used in the production processes. Starke *et al.* in [6] classify industrial load into three types which are mechanical I, mechanical II, and thermal loads. The mechanical I can provide energy and DR capacities; they cannot modulate but can be turned on and off. The mechanical II type of industrial load can modulate and are most suitable for DR. The thermal loads are process equipment that are continuously running and cannot be interrupted except for scheduled maintenance. Lu *et al.* 2021 in [1] also categorized industrial load into three: shiftable, non-shiftable, and controllable loads. The shiftable load has two operation states of on and off. The non-shiftable load must work continuously once the operation has started and the controllable load (can work at different operating levels at different power demands).

12.2.1 Characteristics of industrial load

The power range for industrial load depends on the industry's size. The small-sized/scale industry is between 3 and 20 kW, the medium-scale industry is about 25–100 kW, and the large industry is from 100 to 500 kW. Examples of industrial load are motors, air conditioners, refrigeration systems, defrosting systems, production equipment processing lines [7].

Some of the challenges facing industrial loads include Multiple power sources, different electrical uses, disturbance from grid, complex distribution network [8]. Proper management of the industrial load is of great benefit to the industries as it will help overcome the stated challenges. Therefore, the application of DR is beneficial to both the industry and the electricity supply units.

12.3 Industrial DR

The industry has a significant role in the implementation of DR because of the magnitude of power, industrial loads usually consume, which can provide much flexibility to the grid, and industries also mostly have infrastructures for communication, control, and market participation which enhances DR [9]. Examples of heavy industries are steel, pulp, paper, cement, oxygen generation, oil refinery, and aluminum production. An effective industrial response will reduce the need for alternative flexibility sources like costly storage systems and fast-run power plants. The DR policy for industries is focused on increasing and effectively optimizing the flexibility of production [10]. Some of the major themes of DR in the industry are further explained below.

12.3.1 Industrial load forecasting

Load forecasting is one of the critical aspects of IDR that determines the successful implementation of DR applications. Accurate load forecasting is beneficial to both the power system operators and the consumers [11]. The industry with a proper forecasting tool will be able to optimize its energy management system. Almost every model and approach developed for industrial DR involves load forecasting. Another type of forecasting in IDR is price forecasting.

12.3.2 Role of technology in IDR

For the successful application of DR, technology plays a very vital role. The progress made so far in DR applications has been mainly due to advancements in technology. For IDR implementation, some of the required technologies include a good communication network and protocols, advanced control-device technologies, and smart metering, which must work impeccably to form fully functioning systems [12].

12.3.3 Role of policy in IDR

For effective implementation of IDR, there is a need for policies to be put in place. For example, having regulations that encourage government administrators to use incentives to attract customers' interests in involuntarily engaging in DR in which they vary their load, and also adopt enabling price mechanisms, and which will further improve the energy efficiency [12]. One of the essential drivers of industrial DR is government policy [12].

12.3.4 Incentives and price-based DR

DR is classified into two the price-based and incentive-based DR. The approaches used in the price based are further broken into the day-ahead pricing (DAP) [13] and real-time pricing (RTP) [14]. While for the incentive-based, we have the Time of use TOU approach. In RTP, the actual online condition of the power grid is better reflected. The DAP approach is based on the assumption that the electricity price of the next day is known in advance, which gives rise to fixed scheduling of electricity operations and does not allow for an adequate response to unanticipated variations [1]. TOU involves fluctuating the electricity tariff during the peak and off-peak periods. In this DR type, the price of electricity is increased during high electricity demand periods and reduced during off-peak periods [15]. The DR types are shown in Figure 12.1.

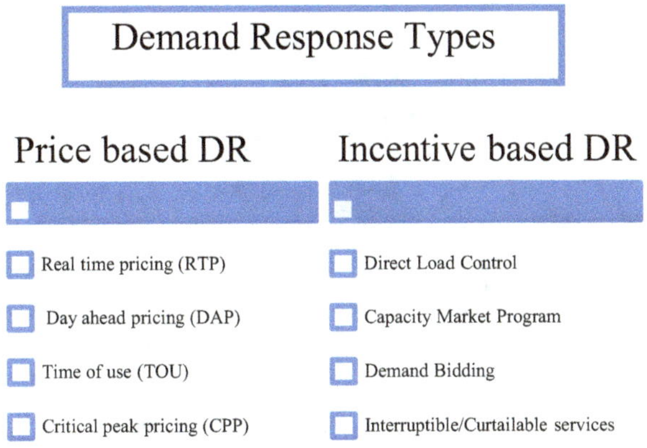

Figure 12.1 DR classification

12.3.5 Ancillary services

Besides DR, the industrial load also serves for ancillary services. Ancillary services are services provided to help maintain a stable and secure operation of the power system. These services help to maintain grid reliability. These services are provided to ensure voltage stability, frequency stability, power quality, and transient stability [16]. Examples of ancillary services include spinning reserves, reactive power control, speed regulation control, voltage control, automatic generation, load shedding and automatic islanding.

12.4 Machine learning in IDR

Machine learning is described as a branch of Artificial intelligence that involves the use of data and algorithms in learning progressively similar to how humans learn until accuracy is attained [17]. Machine learning uses algorithms to find patterns in historical data and for many other applications like prediction, diagnosis, and estimation. Machine learning can be classified as supervised, unsupervised, and reinforcement learning which is shown in Figure 12.2.

A. *Supervised learning* involves using an algorithm that has both the input data and the target data. It is like having a data set with a teacher where the target data is the teacher like the supervisor. So the algorithm gradually learns in the presence of the target data (the supervisor). In DR, supervised learning has been employed by [18] for DR applications in industrial and commercial buildings. Examples of supervised learning are Neural Networks, Support Vector Machine (SVM), Linear Regression, Random Forest and Naïve Bayes, as shown in Figure 12.2.

B. *Unsupervised learning:* It involves training the algorithm without target data (teacher/supervisor). This method is majorly used for finding patterns in a dataset,

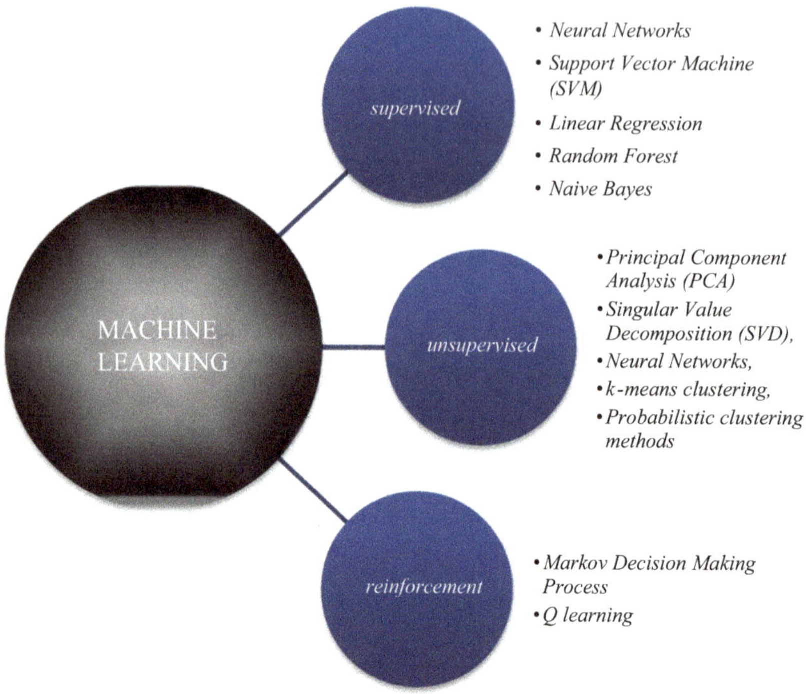

Figure 12.2 Machine learning classification

clustering, and dimension reduction. Examples of unsupervised learning are principal component analysis (PCA), singular value decomposition (SVD), neural networks, k-means clustering, probabilistic clustering methods.

C. *Reinforced learning:* In reinforced learning, learning is done through an agent that is able to interact with its environment and learn through trial and error. A comprehensive review of the applications of reinforced learning for DR was done by [19]. Examples of reinforced learning are the Markov decision-making process and the Q learning process.

Machine learning is widely used in almost every field today, and DR is not left out. However, machine learning has been chiefly used more for residential DR and has very few applications in this area for IDR, which has necessitated this work. From five different academic search engines employed, the research output for industrial DR and in machine learning (ML) applications for IDR are minimal compared to other types of DR. Table 12.1 shows the result from these search engines.

From Table 12.1, we can see that very little has been done in the application of machine learning in the industrial DR. The few research output on machine learning applications on IDR found in the literature is shown in Table 12.2.

Table 12.1 Articles for machine learning in Industrial DR

Search engine	DR	IDR	ML
Scopus	14,415	142	32
Core	333,348	177	158
Mendeley	16,381	43	12
Google Scholar	151,000	625	148
Microsoft Academic	17,192	266	31

Table 12.2 Machine learning application in Industrial DR

Articles	Machine learning approach	Industry	Application
[1] [20]	Multi-agent deep reinforcement learning	Lithium-ion battery manufacturing industry	Minimize electricity cost Maintain production task
[1]	Recurrent neural network	Steel powder manufacturing industry	Peak shifting to off-peak Electricity cost reduction
[21]	Artificial neural networks	Desalination Air separation plant	Load forecasting and price forecasting
[18]	Deep learning reinforcement algorithm	Generic	Load forecasting
[6]	GA	Assessing IDR potential	Load curve modelling
[6]	Linear regression	Assessing IDR potential	Relationship between electricity consumption and price
[19]	Reinforcement learning	Generic	Review of algorithms for DR
[22]	Fuzzy logic	Generic	Electricity price forecasting
[23]	GA nondominated sorting GA II (NSGA-II)	Multimachine shop floor and humans	Scheduling of jobs for machines
[24]	Support vector machine	Generic	Demand elasticity estimation
[25]	particle swamp optimization	Ball mills in a slurry shop floor	Energy cost reduction

Some of the machine learning that have been mostly used in literature for industrial DR include the genetic algorithm (GA), neural networks, SVM and fuzzy logic which are further described in the following sections.

12.4.1 Genetic algorithm (GA)

GA is a heuristic machine learning approach based on Charles Darwin's theory of evolution [6]. The basic principle behind GA is the principle of selection of the fittest,

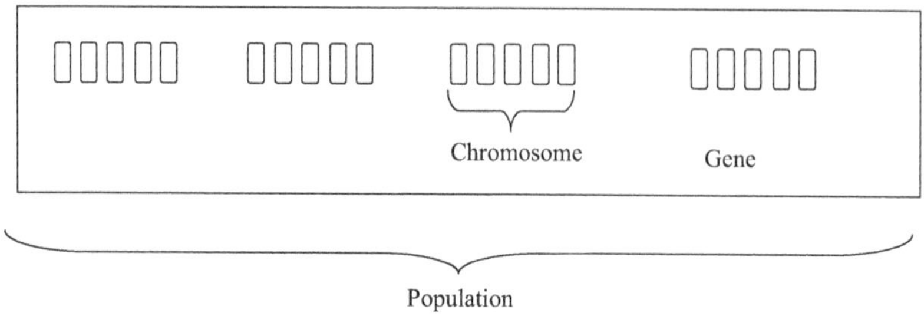

Figure 12.3 Some genetic algorithm parameters

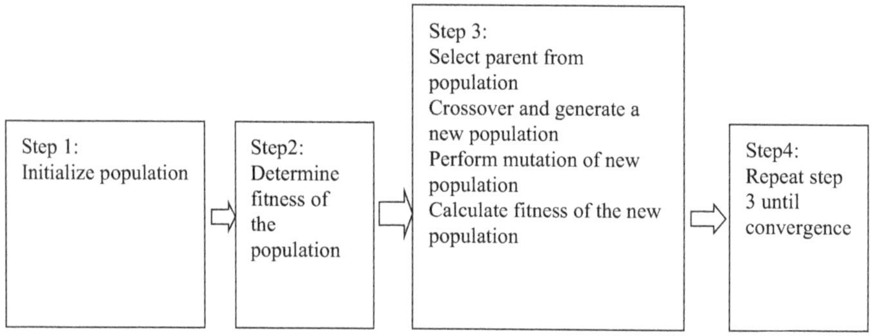

Figure 12.4 Steps involved in GA process

like natural selection, where the fittest genes are selected for the reproduction of the next generation. GA uses majorly five parameters in its operation: population size, cross over, fitness function, search space, and mutation probabilities [23]. The GA process always starts with a set of solutions called population. The individual solution in a population is called a gene. The genes are presented in a string called chromosomes. Figure 12.3 shows the relationship between the genes, chromosomes, and population.

The fitness function is characterized by the fitness score, which indicates how to fit each individual is. The selection probability, which is the possibility of being chosen, is determined by the fitness score. Cross over involves the two selected genes being paired to produce the offspring. The process involved in GA is shown in Figure 12.4.

12.4.2 Support vector machine (SVM)

SVM is a supervised learning algorithm that learns by assigning labels to objects. They are primarily used for solving classification problems, regression problems and for outliers detection [24]. SVM is used for non-linear problems and performs well with a limited amount of data. The setback with SVM is that they can be computationally expensive. This is why in computation, SVM classifiers are done alongside a kernel function to reduce computational price.

A hybrid system of least square support vector machine (LS-SVM) was developed to determine the demand elasticity by assessing the response of demands to different influence signals for a DR system [24]. SVM is used in identifying miscellaneous electric loads in [26].

12.4.3 Artificial neural network (ANN)

ANN is a supervised learning algorithm that is inspired by the operation of human learning. It is patterned after how the brain works. The human brain is a highly complex, non-linear, and parallel information processing system. It can perform functions like pattern recognition, perception, and motor control faster than any computer that ever existed. ANN can be called a virtual brain – the ANN components are similar to that of the brain, which includes a collection of neurons that are interconnected through synapses [27]. The neurons are simple processing units with the ability to store knowledge and make it available for later use by other components of the code. Neuron interconnections have a strength called synaptic weight, which is used to store the acquired knowledge. A simple ANN structure is shown in Figure 12.5.

ANN has different procedures for learning, which are called learning algorithms. The learning algorithms modify the synaptic weight of the network in an orderly fashion to attain the desired design objective [27]. A neural network can learn and generalize to produce output from inputs it did not encounter during training. The other strengths of the neural network are that it can: work well in non-linear systems; do input–output mapping; adapt to changes in the environment, and give confidence level on decisions that it made. ANN was used in forecasting load and locational marginal price for an inverse DR process developed by [21]. The inverse DR process was used in demonstrating IDR in desalination and air separation plants. Load and price prediction was made using deep learning neural networks for IDR by [28].

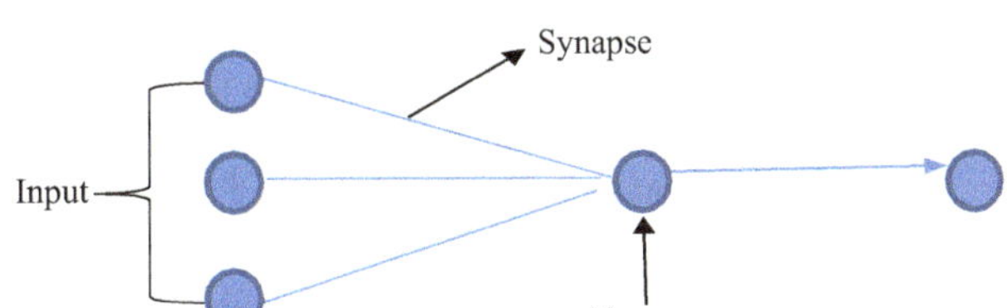

Figure 12.5 A simple neural network

ANN was used in forecasting the temperature and power demand perturbation for cold storage that uses DR [29].

12.4.4 Fuzzy logic

Fuzzy logic uses human expertise in interpreting incomplete and imprecise information in IF-THEN rules to solve problems. It performs numerical computation by using linguistic labels stipulated by membership functions. The fuzzy system incorporates human knowledge and performs inferencing and decision-making [30]. Fuzzy logic may be viewed as a methodology for computing words rather than numbers. Although words are inherently less precise than numbers, their use is closer to human intuition. Furthermore, computing with words exploits the tolerance for imprecision and thereby lowers the cost of the solution. Fuzzy logic relies on age-old skills of human reasoning. The strengths of fuzzy logic are approximation capability; ability to compute with words; tolerance for imprecision, and the ability to model non-linear functions. A fuzzy-based method is used to model the uncertain electricity price for industries to make decisions for maximum profit for the consumer. The fuzzy α-cuts method developed by [22] gives the consumer upper and lower limits price, giving them a price range from which to make their decisions. Reference [31] developed a Fuzzy system that is used in deriving priority factors for the difference in other to make appropriate DR decisions in the industry.

12.4.5 Adaptive neuro-fuzzy inference system (ANFIS)

The adaptive neuro-fuzzy inference system (ANFIS) is the neuro-fuzzy method of interest in this research. ANFIS was proposed by [32]. ANFIS combines the strength of fuzzy logic approximation with the adaptive capability of the ANN. The membership functions are generated using a neural network approach and, therefore, it does not require expert knowledge, which is usually a prerequisite for a standard FL system. In ANFIS, the membership function (MF) parameters change through a learning process. ANFIS uses either back-propagation or a combination of least squares estimation and back-propagation for MF parameter estimation. The ANFIS approach and applications are well described in [32].

12.4.6 Linear regressions

Linear regressions are a supervised learning algorithm mainly used for trend forecasting, predicting effects, and determining the strength of predictors. Linear regressions could be simple linear regression or multiple linear regression. In assessing the IDR potential at the western interconnects of the united states of America, Linear regression analysis was used in analyzing the relationship between electricity consumption and peak in relation to the sales of the industries [6]. A summary of the different machine learning algorithms, showing their strengths and weaknesses, is given in Table 12.3.

Table 12.3 Machine learning algorithms, applications, and weaknesses

Machine learning algorithms	Application	Strengths	Weaknesses
Genetic algorithm	Optimization	Good at parallel implementation Works for discrete and continuous functions Solutions improve over time	Give different output results Not good in dealing with complexities Slow speed of convergence
Artificial neural network	Forecasting/ prediction	Good non-linear problems Can do a parallel implementation	Computationally expensive Requires a large amount of data
Fuzzy logic	Decision making/control	Approximation capability; Ability to compute with words; Tolerance for imprecision; Ability to model non-linear functions	Completely dependent on human expertise Inaccuracy of output due to the inaccuracy of data
Adaptive neuro-fuzzy inference system	Prediction control	Fast learning capacity High generalization capacity	Computationally experience
Support vector machine	Classification and regression problems	Performs well with a limited amount of data High speed of performance Good at generalization High accuracy in classification	Computationally expensive Mathematically complex
Linear regression	Forecasting Estimating predictors strength	Simple to interpret Easy to implement Sensitive to poor quality data	Sensitive to outliers It works with a non-linear data set
Particle swarm optimization	Optimization	Easy implementation	Slow convergence speed
Principal component analysis	Dimension reduction	Improves visualization Reduces overfitting	Information loss Difficult to interpret
Random forest	Classification Regression	Reduces overfitting Can work with missing data	Computationally expensive Large training time

12.5 Conclusion

Industrial DR is a vast resource for proper and efficient energy management that can bring stability to the grid in an economical manner. This chapter has discussed the different aspects of industrial DR while focusing on machine learning applications in IDR.

The different machine learning algorithms from the literature were covered. The ANN is good at prediction and its primarily used for prediction in IDR, forecasting load, and price. The GA is mainly used for optimization problems.

In this chapter, the strength and weaknesses of some selected machine learning algorithms were explored. It can be concluded that all the machine learning algorithms have their strength and weaknesses. To obtain the best algorithm for IDR, It is best to understand what to trade-off with the different algorithms and how to combine more than one algorithm to get the best result.

References

[1] R. Lu, R. Bai, Y. Huang, Y. Li, J. Jiang, and Y. Ding, "Data-driven real-time price-based demand response for industrial facilities energy management," *Applied Energy,* vol. 283, p. 116291, 2021.

[2] IEA. "Promoting digital demand-driven electricity networks: digital solutions to support power systems in transition." https://www.iea.org/areas-of-work/promoting-digital-demand-driven-electricity-networks [accessed 9 October 2021].

[3] K. Daware. "Industrial loads." https://www.electricaleasy.com/2016/06/types-of-electrical-loads.html [accessed 1 October 2021].

[4] E. Fact. "Types of industrial loads." https://www.electricportal.info/2019/01/what-types-electrical-loads-curves.html [accessed 1 October 2021].

[5] M. H. Shoreh, P. Siano, M. Shafie-khah, V. Loia, and J. P. Catalão, "A survey of industrial applications of Demand Response," *Electric Power Systems Research,* vol. 141, pp. 31–49, 2016.

[6] M. Starke, N. Alkadi, and O. Ma, "Assessment of industrial load for demand response across US regions of the western interconnect," Oak Ridge National Lab. (ORNL), Oak Ridge, TN, USA, 2013.

[7] E. Lee, K. Baek, and J. Kim, "Evaluation of demand response potential flexibility in the Industry based on a data-driven approach," *Energies,* vol. 13, no. 23, p. 6355, 2020.

[8] Honeywell. *Industrial Load Management System*. (2015). [Online]. Available: www.honeywellprocess.com/library/marketing/brochures/brochure-on-industrial-load-management-system.pdf

[9] X. Zhang, G. Hug, Z. Kolter, and I. Harjunkoski, "Industrial demand response by steel plants with spinning reserve provision," in *2015 North American Power Symposium (NAPS)*, IEEE, 2015, pp. 1–6.

[10] H. Golmohamadi, R. Keypour, B. Bak-Jensen, J. R. Pillai, and M. H. Khooban, "Robust self-scheduling of operational processes for industrial demand response aggregators," *IEEE Transactions on Industrial Electronics,* vol. 67, no. 2, pp. 1387–95, 2019.

[11] A. Bracale, P. Caramia, P. De Falco, and T. Hong, "A multivariate approach to probabilistic industrial load forecasting," *Electric Power Systems Research,* vol. 187, p. 106430, 2020.

[12] B. Shen, C. C. Ni, G. Ghatikar, and L. Price, "What China can learn from international experiences in developing a demand response program," Lawrence Berkeley National Lab. (LBNL), Berkeley, CA USA, 2012.

[13] M. T. Kelley, R. C. Pattison, R. Baldick, and M. Baldea, "An MILP framework for optimizing demand response operation of air separation units," *Applied energy,* vol. 222, pp. 951–66, 2018.

[14] J. M. Yusta and J. A. Dominguez, "Measuring and modeling of industrial demand response to alternative prices of the electricity," in *Proceedings of the. Power Systems Computation Conference***,** *Session*, 2002, vol. 15.

[15] K. Jessoe and D. Rapson, "Commercial and industrial demand response under mandatory time-of-use electricity pricing," *The Journal of Industrial Economics,* vol. 63, no. 3, pp. 397–421, 2015.

[16] J. Wang and W. Xu, "Present situation and problems of large-scale wind power transmission and accommodation policy," in *Large-Scale Wind Power Grid Integration*, Elsevier, 2016, pp. 257–76.

[17] IBM. *What is Machine learning?* [Online]. 2020. Available: https://www.ibm.com/cloud/learn/machine-learning [accessed 9 October 2021].

[18] G. Krishnadas and A. Kiprakis, "A machine learning pipeline for demand response capacity scheduling," *Energies,* vol. 13, no. 7, p. 1848, 2020.

[19] J. R. Vázquez-Canteli and Z. Nagy, "Reinforcement learning for demand response: a review of algorithms and modeling techniques," *Applied Energy,* vol. 235, pp. 1072–89, 2019.

[20] R. Lu, Y.-C. Li, Y. Li, J. Jiang, and Y. Ding, "Multi-agent deep reinforcement learning based demand response for discrete manufacturing systems energy management," *Applied Energy,* vol. 276, p. 115473, 2020.

[21] M. Buchman, C. Mascarenhas, A.Trueworthy, and S. Watson, *Inverse Demand Response for Energy Intensive Industrial Processes* [Online]. 2017. Available: http://mbuchman.scripts.mit.edu/projects/files/Final%20Paper.pdf [accessed 9 October 2021].

[22] M. Zarif, M. Javidi, and M. Ghazizadeh, "Self-scheduling approach for large consumers in competitive electricity markets based on a probabilistic fuzzy system," *IET Generation, Transmission & Distribution,* vol. 6, no. 1, pp. 50–58, 2012.

[23] X. Gong, Y. Liu, N. Lohse, T. De Pessemier, L. Martens, and W. Joseph, "Energy-and labor-aware production scheduling for industrial demand response using adaptive multiobjective memetic algorithm," *IEEE Transactions on Industrial Informatics,* vol. 15, no. 2, pp. 942–53, 2018.

[24] L. Xie and H. Zheng, "Demand elasticity analysis by least squares support vector machine," in *2013 6th International Congress on Image and Signal Processing (CISP)*, IEEE, 2013, vol. 2, pp. 1085–89.

[25] S. Ma, Y. Zhang, Y. Liu, H. Yang, J. Lv, and S. Ren, "Data-driven sustainable intelligent manufacturing based on demand response for energy-intensive industries," *Journal of Cleaner Production,* vol. 274, p. 123155, 2020.

[26] L. Du, Y. Yang, D. He, R. G. Harley, T. G. Habetler, and B. Lu, "Support vector machine based methods for non-intrusive identification of miscellaneous electric loads," in *IECON 2012 – 38th Annual Conference on IEEE Industrial Electronics Society*, IEEE, 2012, pp. 4866–71.

[27] S. S. Haykin, "*Neural networks and learning machines/Simon Haykin*," New York, NY: Prentice Hall, 2009.

[28] R. Lu and S. H. Hong, "Incentive-based demand response for smart grid with reinforcement learning and deep neural network," *Applied Energy,* vol. 236, pp. 937–49, 2019.

[29] H.-M. Hoang, M. Akerma, N. Mellouli, A. Le Montagner, D. Leducq, and A. Delahaye, "Development of deep learning artificial neural networks models to predict temperature and power demand variation for demand response application in cold storage," *International Journal of Refrigeration,* 2021.

[30] R. R. Yager and L. A. Zadeh, *An introduction to fuzzy logic applications in intelligent systems*, Springer Science & Business Media, 2012.

[31] S. Mohagheghi and N. Raji, "Managing industrial energy intelligently: Demand response scheme," *IEEE Industry Applications Magazine,* vol. 20, no. 2, pp. 53–62, 2013.

[32] J.-S. Jang, "Input selection for ANFIS learning," in *Proceedings of IEEE 5th International Fuzzy Systems*, IEEE, 1996, vol. 2, pp. 1493–99.

Chapter 13

Feasibility assessment of industrial demand response

Jose-Fernando Forero-Quintero[1], Roberto Villafáfila-Robles[1] and Daniel Montesinos-Miracle[1]

Industrial Demand Response (IDR) has been used for regulation and balance purposes for many years [1]. Large energy intensity manufactures have responded to external signals, generally from System Operator (SO) to shift or shutdown loads according to emergence or unexpected situations in power system [1,2]. Historically, such ancillary services were very constrained and only remunerated by means of payments according to an individual contract agreement between SO and industrial customer [3]. Nowadays, new IDR programs have been investigated, which have received a great impulse because of the development of new control and communication systems, Distributed Energy Resources (DER) based on Renewable Energy Sources (RES) [4], energy storage capacity, alongside market liberalization and pricing diversification. Such new IDR programs tend to focus on controllable, deferrable and interruptible loads aggregation [2], as well as on DER and Energy Storage Systems (ESS) aggregation through Virtual Power Plants (VPP) [5] or the energy market participation of an aggregator on behalf of the prosumers [6].

IDR programs generate economic, social and environmental benefits for whole stakeholders (consumers, utilities, regulatory entities, system operators and governments, among others) [2,8], where IDR impacts meaningfully in the frequency regulation, reserve management [6] (primary reserve, contingency, replacement and operational), as well as in a decrease of the overall costs. Additionally, benefits in the energy market have been identified in terms of reducing energy prices and volatility [7,9]. Studies on flexibility services have been carried out, in where an attractive service portfolio has been found for IDR companies (usually referred to as aggregators). Likewise, several reforms to the energy markets and tariff structure have been identified to fit the new flexibility services. All above technical concerns have been tackled by several studies, but feasibility assessment of Industrial Demand Response applications has not been received enough attention [1,10]. In some works, the feasibility

[1]Centre d'Innovació Tecnològica en Convertidors Estàtics i Accionaments (CITCEA-UPC), Department d'Enginyeria Elèctrica, Universitat Politècnica de Catalunya, Barcelona, Spain

has been corroborated by means of case studies where cost–benefits analysis (CBA) was ill-defined and incomplete.

Accordingly, this chapter addresses the problem of how to accomplish a suitable CBA for IDR applications, reviewing the available theories, standpoints, concepts and indicators in literature. On the other hand, this chapter also aims to provide an overview about the cost and benefit structures of the devices and processes involved in IDR, considering standpoint (industrial customer, SO and aggregator), timelines (short, medium or long terms), energy markets (wholesale, retail, balancing and flexibility), pricing features and profitability indicators (Net Present Value, Capital expenditure, savings, net revenues, etc.). This chapter is organized as follows. Section 13.1 describes the IDR development process, where, together with Life Cycle Cost (LCC) approach, it defines the main costs associated with each development step. In Section 13.2, IDR benefits are exposed, taking particular care in benefit sources, electricity tariff and energy markets. Profitability indicators involved in industrial demand response projects are discussed in Section 13.3. Finally, case studies, conclusions and final considerations are drawn in Sections 13.4 and 13.5, respectively.

13.1 Cost assessment of IDR

CBA is defined as the process of profitability assessment, which measures the net cost and benefits of a project along its useful life, including the residual value of the facilities and devices. CBA has multiplex components, among them the life cycle cost (LCC) [11]. LCC concept is expected to be used in most CBA for industrial projects. The LCC approach addresses not only operational cost, but also pre-feasibility study expenses, flexibility potential assessment, design and installation of the IDR programs, as well as communication and control expenditures. Such costs are relevant, given that the complexity of the industrial production process is greater. Accordingly, this chapter applies the LCC approach in the CBA and therefore, in the development cycle of the IDR programs.

Figure 13.1 shows the development cycle of the IDR programs, which are composed of five steps [12], which will be remarked in the next subsections. Every step encompasses a set of costs, which will be examined according to weight in the total cost analysis of each one. The development cycle of the IDR programs is a closed non-stop process, where a process of a continuous improvement process, at the end of the cycle, identifies possible design modifications and changes of the operation and management regime, resulting in an increasing efficiency each time.

13.1.1 Measurement of flexibility potential

Industrial demand response is linked strongly with the flexibility of electricity consumption. Certainly, not all flexible devices can participate in demand response programs, but they can be used as a bandwidth where demand response sets up its actuations. Such bandwidth has two boundaries: baseline and expected line scenarios. The flexibility potential is exactly the space between these borders. Both baseline and

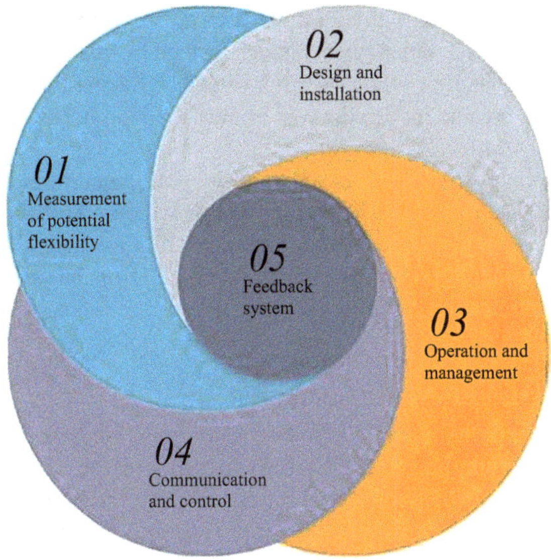

Figure 13.1 Development process of the IDR programs. Source: Own preparation based on [10]

expected line scenarios are established through optimization process and modelling to match the flexible capacity of the loads, processes and devices with operation and management strategy [6,8].

13.1.1.1 Baseline

The main starting point for finding IDR potential is to consolidate a baseline scenario [2,7]. The baseline scenario describes the energy flows through the production process, as well as electricity consumption of cross-sectional technology (air-conditioned, lighting, office equipment, among others) [13]. Furthermore, baseline also condenses the existing cost, not only associated with the energy consumption, but also raw material cost, payroll and administrative expenses, including a precise study on each stakeholder interaction [7].

Estimating the baseline scenario implies a prioritization of all costs with respect to their weight within the total cost [7]. On the one hand, average calculations about historical records [7], daily consumption over recent years, monthly payments to the suppliers among others factors are used by models, which seek to estimate a baseline scenario where the flexibility potential of the IDR is reflected. On the other hand, manufacturing execution system (MES) has emerged as a novel tool to control and manage the manufacturing process, where energy management and manufacturing process are still addressed separately. Surely, a unified platform will be raised in the future [14]. Currently, designing and analysing the IDR programs can be accomplished by means of the energy management system (EMS) [12]. Baseline also is gauged through consumption patterns regards to normal and emergency situations, future scenarios or related with a given pattern formed by quarter-hourly load curves with lower standard deviations than a certain value [1]. Each estimation method aims to be the most accurate and reduce the cost of measurement of flexibility potential of the IDR at the same time.

All above estimation methods, based on EMS, MES, monitoring system or modelling strategies, often entails a cost that can be high, which are not properly described in the literature. Generally, these costs grow in relation to energy intensity level of the industrial manufacturers, as well as the complexity of the production process, the existing communication and control systems, IDR goals and flexibility services [2,7].

13.1.1.2 Analysis of flexibility potential

Measurement of flexibility potential of the IDR programs consists not only of finding minimum and maximum power and energy capacities but also bears within itself a more specific analysis, which must be done considering certain variables.

Generally, measurement of flexibility potential can be seen as a process divided into four stages or four categories depending on measurement scope. Such a process starts with a theoretical potential related to a certain industrial sector. There are technical boundaries to control industrial loads and hence, the demand response execution or any flexibility services [9]. For instance, for participating in ancillary markets is required a minimum interruptible load (capacity power) able to shut down in a short period of time [2], whereas for offering strategic reserve is needed to store energy during a larger time than for balancing the frequency or voltage. The flexibility potential must also incorporate the flexibility of the production process, the probable storage capacity of the final and intermediate products [9], the flexibility of an operation permeable regime, as well as energy generation strategy of the DG units [6].

CBA must be carried out to filter the flexibility potential according to the expected profitability of possible DR actions. In this same way, DR actions should be accepted by stakeholders, which limits the flexibility potential previously estimated [4]. The stages and categories of the measurement of flexibility potential can be summarised as follows.

- *Theoretical*: Flexibility potential refers to all manufacturing facilities, loads and processes that could be an object of demand response.
- *Technical*: Subset theoretical potential, which is excluded from uncontrollable or hard-controllable devices.
- *Economic*: Profitable potential resulting from the application of a CBA on the technical potential.
- Practical: Usable potential composes of the interventions on loads and processes accepted by stakeholders [4], as well as the restrictions involved in the contact between utility and end-user, Quality of Services (QoS) and privacy considerations and other boundaries.

Due to the above and as an overall rule, energy audit may be needed mainly for small and medium industrial manufacturers, which are not aware of their flexibility potential and capacity [1]. Finally, computing capacity and specialized staff have been required to obtain relevant results in the measurement of flexibility potential [4]. Usually, these costs are dismissed because the technical concerns are the centre of attention in the literature, but it is clear that industrial companies will need proper agencies (aggregator or specified company department) to design, execute and manage an IDR program [15, 16].

13.1.2 Design and deployment

A challenging phase of the development process of the IDR programs is the design and commissioning. The design reflects an expected scenario able to achieve the IDR objectives. Several optimization algorithms are used to define a suitable design, where multiple variables are included in both objective functions and configuration constraints. The usage of optimization algorithms implies some costs associated with the design. Therefore, demand response commissioning involves the required investments to implement the final design, as well as the modifications of electrical installations regarding with original electrical diagrams and also the expenses to adapt the production process [17].

13.1.2.1 Design cost

Once the flexibility potential is known, the designing process aims to squeeze such potential, putting forward an IDR program with flexible loads, devices and process along with an operation scheme [17]. To do this, information on the energy market, flexibility services, tariff structures, production costs, as well as available operation and management strategies should be considered. These factors can strongly determine the kind and features of the IDR program design [2]. For instance, the melting process of scrap steel allows only disruptions below half an hour due to its start-up cost, as reducing the load can provoke the thermal equilibrium break-up of the melting pots [2]. Such preparation and recuperation times after interruption should be also considered, as they are usually very distinct for each production process [1]. Likewise, designing IDR programs must be coordinated with the maintenance cycles of the machinery to make more efficient flexibility services [1]. In addition, designing IDR programs involves also retail and spot electricity price forecasting, as well as generation and consumption planning of the energy [8]. The energy cost calculation (accordingly with electricity, gas, oil or any raw material price), as well as a bidding strategy of the DG, when they participate in ancillary and balancing markets, are also relevant aspects in the designing process.

Finally, several amounts of flexibility can subtract from the manufacturing process. Changing the lineal or hierarchical processes for independent or interdependence ones allow increase the flexibility of the industrial manufacturer [2]. Within the designing process, some adjustment measures of the manufacturing processes can be found. In the same direction, communication and control systems must be selected accordingly with the features and capacities needed to guarantee suitable and feasible operation. This will be discussed in Section 13.1.4.

Given the amount of information and complexity of the design calculations, advanced software and specialized algorithms could be needed to obtain a proper IDR program. Therefore, computational capacity and performance must be considered for both storing pertinent information and resolving large number of scenarios. Consequently, specific staff (workforce) may also be required to manage all flexible facilities and analyse the results of the adopted strategies, as well as propose improvement actions.

13.1.2.2 Capital cost

All investments incurred on equipment, infrastructure and others one-time expenses to bring an IDR project to operable status can be seen as capital costs [5]. In the literature, design and pre-feasibility costs use to be included as capital expenditures, but they are shown separately in this chapter with the aim to identify some neglected costs of each development cycle of the IDR program. Capital cost involves expenses from the design stage ends to the IDR project begins the operation, including commissioning and start-up costs.

The existing control and communication system (CCS) [2] weigh heavily within required investments [2], given that small and medium industrial customers must invest in CCS for cross-sectional technologies and adapt the CCS of the existing manufacturing process technologies [9,13]. Conversely, large manufacturers have already installed CCS throughout the production process and cross-sectorial technologies, therefore, investments will be focused on optimal production scheduling and storage capacity [5]. With respect to the production process refitting, investments to integrate parallel machines at the manufacturing process [15], reduce the interdependence between production stages and create storage capacity in intermediate processes, i.e. raw materials, non-finished and final products [5], are commonly found in the literature.

All above investments are modelled depending on the financing method, which could be either debt, equity or a mixed between them, in where it is common to observe different interest rates. Each interest rate must be contemplated in CBA, together with future expenses for replacing assets at the end of their useful life. Capital cost is related to the hurdle rate of the company, in other words, the minimum rate of return must be achieved to be profitable, and therefore, calculating capital cost is a relevant task for the financial managers [18]. The relation (13.1)–(13.3) defines the cost of debt (C_{c-d}), equity (C_{c-e}) and total capital cost:

$$C_{c-d} = \sum_{n=1}^{N} I_n^d * K_n^d \tag{13.1}$$

$$C_{c-e} = \sum_{m=1}^{N} I_m^e * K_m^e \tag{13.2}$$

$$C_{cT} = \frac{1}{(C_{c-d} + C_{c-e})} * (C_{c-d} + C_{c-e}) * 100 \tag{13.3}$$

where I_n^d represents the interest rate and K_n^d the debts for each company's debt (n). Likewise, I_m^e is the dividend rate and K_m^e the debts for each company's investor (m). Normally, the existing smallest dividend rate is used in investment without risk, as well as a return for bearing extra risk. Such rate is common for whole investors depending on their shareholding structure [18]. Finally, there is a set of costs, which must be counted to increase the accuracy of the CBA. All acceptance testing such as user, factory or operational acceptance testing are usually included in commissioning cost, as well as, expenses incurred in cases where electrical connections have not been faithfully done with respect to electrical diagrams, issues that cannot be envisaged previously [19]. Thus, besides of commissioning time in normal conditions, extra time could be contemplated for Information and Telecommunications (IT) failure, time zone adjustments, software incompatibilities, organization problems, among others [19].

13.1.2.3 Opportunity cost

Within economic metrics, opportunity cost is commonly used to take decisions in situations with multiple solutions, as well as supporting in the design stage of the IDR program. Basically, opportunity cost measures the total potential missed due to choosing one alternative over another [5]. Therefore, opportunity cost can be a guide to define the minimum and maximum tariff of a certain service and consequently if these could be commercially exploitable [9]. For instance, when energy surplus appears, an opportunity cost analysis is carried out to decide whether such energy is employed for self-consumption or balancing services to the grid. In other words, the possible savings and profits generated by self-consumption are the minimum expected revenue levels by balancing services, therefore, in this case, opportunity cost influences in minimum tariff of the balancing services. This approach can be put in practice in different case studies, becoming a useful analysis tool. The following formula depicts the opportunity cost:

$$\mathrm{OC} = \mathrm{FO} - \mathrm{CO} \tag{13.4}$$

When opportunity cost (OC) corresponds with the difference between the return of the best foregone option (FO) and return of chosen option (CO). Authors have underlined the relevance of opportunity cost in demand response programs regards reserve management, as such reserve capacity cannot take part in other markets, therefore, the reserve price is mainly driven by OC. Opportunity cost goes together with risk analysis, which should be comparable to allow pinpoint conclusions [8]. Investors and decision makers always aim to keep the opportunity cost at its lowest levels and controlled.

13.1.3 Operation and management

The next stage in the IDR Program development is the operation and management (O&M), which possesses costs that could affect ostensibly the CBAs. From measurement of the flexibility potential and design stage, several operation strategies are raised with the aim to maximize profits or minimize total cost. Such operation mode, not only corresponds with the IDR program, but also with changes in the production schedule. Hence, this section describes the O&M cost of the IDR program, as well as the assumable cost due to the impacts on the production process.

13.1.3.1 Direct and indirect cost

The costs generated by operation and maintenance of the IDR program can be, directly or indirectly, related to either its functioning or the collateral impacts on the production process. Such costs are classified into direct and indirect [1,10]. Direct costs relate to the technical capacity required to carry out the demand response program without considering demand response actions. Whereas, indirect cost is linked with the expenses incurred during demand response is executed. Indirect cost is considered more complex than direct cost, and in consequence, suitable penalties are employed to model them. Table 13.1 lists the direct and indirect costs commonly used in CBA of the Industrial Demand Response programs.

Each cost is defined as follows:

- *Lost and delayed revenues in the production process*: Modifications on the Production process caused by regulation services (down or up) could provoke possible defaults in the clients' orders, given that these can limit production and logistics schedules, reduce production flexibility, loss of productivity or increase the intrinsic risk of the value chain.
- *Increased workforce*: Increasing stop and start-up frequency of the machinery could require greater workforce, not only in quantity, but also qualified personnel able to take technical decisions or analyse information about performance of the system. Likewise, the control and monitoring levels could be augmented, creating the need for further staff or investments in automation systems.
- *Increased power consumption*: Machinery as electric motors have energy consumption peak within the star-up and pre-heating periods, therefore, demand response actuations could increase this energy requirement. For other hand, a greater backup power system could be needed to guarantee security and reliability in the preparation and recovery period before and after interruption, respectively.
- *Wear and tear costs*: Each asset and device, both value chain and cross-sectional technologies, have maintenance periods according with theoretical nominal performance. Such normal performance can be modified by IDR actions, either

Table 13.1 Direct and indirect cost of O&M stage of IDR

Direct c	Indirect cost
Lost and delayed revenues [7]	Wear and tear cost
Increased power consumption	Ramp-up delay cost [7]
Shut down and Start-up cost [8]	Increased workforce cost [7]
Increasing maintenance cost [7]	

Source: Own preparation.

frequency and intensity of the work pace. It exits a threshold where changes in performance can either extend or damage the useful life of equipment depending on the frequency of interruptions [1].

- *Ramp-up delay cost*: After regulation, machinery should restore nominal performance within of recovery period time. Such time could be elongated due to, for example, raw material deficit or coordination failures, given that the start-ups are unreliable and susceptible to the contingencies.
- *Shut down and start-up cost*: Costs involved in turn machinery on and off when the demand response program is executed. Such costs exclude the ramp-up expenses which are linked to operational constrains.
- *Increasing maintenance cost*: Industrial plants with demand response programs must consider an increasing maintenance cost due to the increase in the number of interruptions and regulation activations, as well as changes in the duration of interruptions, not only in controlled assets but also in external devices to the IDR program.

13.1.3.2 Energy purchasing

In the CBA framework, energy purchases are categorized as a cost for executing both demand response actions and flexibility services [7,9]. Energy purchases include electricity, gas, heat, and among other energy sources. Undoubtedly, energy interactions will augment in the next years, in both quantity and number of stakeholders, as well as markets and tariffs [9,10]. The industrial manufacturers can participate in balancing and flexibility markets as individuals, as well as a conglomerate through an aggregator. Particularly, for aggregation mode, the aggregator role is relevant due to its knowledge and expertise in the energy sector, as well as its capacity of managing, coordinating and monitoring end-user's flexibility.

There are flexible services such as energy trading and price arbitrage, where in industrial manufacturers not only react to the wholesale price passively but also act to purchase or sell energy, generating additional benefits. This activity can be carried out also with an aggregation model through a marketer entity. Currently, activities such as price arbitrage are studied by research works although they are not allowed in most EU-members.

13.1.3.3 Penalties

Industrial consumers, involved in the IDR program, could be penalized because of non-compliance with certain demand response commitments such as reaction times, power reduction, or amount of load shift [20]. Such penalties are included as costs in CBA, together with sanctions for reactive power, harmonic injection, among other factors attributed to the manufacturers that affect the Quality of Service (QoS). Additionally, as mentioned before, indirect costs are modelled as penalties for CBA, given that these costs are not easily measurable [8].

Likewise, penalty cost is generally neglected, but they are needed to guarantee the consistency of the models and comparison of the results. Finally, within the optimization process, penalties have also been used to set up unpleasant decisions or performances such as abrupt changes of the steam pressure or mechanical mass, which are not relevant in the model but can be considered inefficient or undesirable [21].

13.1.4 Communication and control

The IDR program's success is going ever hand-to-hand with a proper level of communication, control, and automation [4]. Demand response and flexibility services require the compliance of technical conditions. Minimum continuous energy delivery time and minimum response time are examples of these requirements, which can be guaranteed by the communication and control system (CCS) [2]. CCS is also needed for observing and monitoring the responsive load performance to assure the production process [9]. Likewise, participating in balancing markets requires communication among SO and consumers to offer energy bids and receive the corresponding acceptance, as well as the later billing process.

Regarding cost of CCS, investment in control, communication, and monitoring infrastructures must be done, as was aforementioned in Section 13.1.2.2. To active demand response actions, information, transaction, and control costs should be incurred in order to gather, consolidate and process the information of the IDR program operation. Such costs are mostly in terms of computing burden and qualified personnel [5]. The computational burden is an additional issue in both the design and operation stages of the demand response programs. Discrete-time models cannot represent adequately events without slots of width, which could impact, for instance, the makespan minimization results. Conversely, the usage of the continuous-time models could delay tasks without enough gain of accuracy in other estimations, such as cost minimization [22]. Therefore, there is a trade-off between accurate model and computational burden, which should be adequately managed, taking into account the loss of accuracy, delays, costs of data collection, transaction, and control.

Finally, the IDR program is tended towards a completely automated operation, namely Auto-IDR [23], which uses signals between SO and consumer's equipment to arrange energy exchanges, tariffs and delivery time without any manual labour. Likewise, standards such as Open ADR [23], Green Button or SEP are being used to unify the communication protocols and allow major interoperability grades between diverse types of devices and systems [2].

13.1.5 Feedback system

Given that IDR programs are composed of continuous actions during long period times, implementing a feedback system could be the convenience to evaluate the performance in short terms in order to correct possible deviations, as well as proposing improvement measures to increase the DR program efficiency [12]. Besides, with the help of Internet of Things and database technologies, a feedback system safeguards valuable information related to the energy performance and profitability of the DR programs, which will be available for the whole stakeholders (end-users, aggregators or SO) to refine operation schemes, production scheduling, and energy uses.

Additionally, a feedback system could realize interventions on IDR actuations in order to react to unexpected events immediately. Within Auto-IDR programs, the feedback system allows to automate various interactions among SO, aggregators and end-users, where excessive controls and authorizations can be avoided. The main characteristics of a feedback system can be as follows:

- Assessment frequency (critical days, typical days, weeks or years).
- Input and output modules.
- Communication protocols.

Overall, expenditure in the feedback systems is included either in EMS investments or in CCS, since the feedback systems are commonly shown as a part of these systems. This chapter was independently depicted to highlight its role in the CBA. The costs incurred in the feedback system can be seen in the same terms of the CCS costs, depending on IDR complexity, existing communication and control infrastructure, and CBA assessment period.

13.2 IDR benefits

The rollout of the IDR program seeks essentially to obtain benefits for both industrial consumers, aggregators, system operators and the whole power system. Energy bill savings, revenues, productivity increase and efficiency improvements are the most common benefits in the literature [2]. There are additional benefits such as reduction in greenhouse gas emissions (GHG) or the increase of RES penetration level that require a study more deepen to account properly for its positive impacts. On the other hand, benefits such as the reduction in wholesale price electricity cost or positive effects regarding the national economy are partial or completely neglected due to a lack of suitable measurement methods. Considering the time frame to profitability (short, medium and long terms) and CBA standpoints, the demand response benefits are defined and discussed in this section. Some mathematical expressions are included in order to guide the numerical calculations.

13.2.1 Regulation services

To guarantee safe and reliable power system operation, system operators employ regulation services, which aim to reduce the unbalance among generation and demand electricity, as well as handle both expected and unexpected situations. Such regulation services are mostly supplied by conventional generators, together with local devices such as on-load tap-changer of the transformer and demand-side response [6]. In the past, some large industrial consumers provided regulation services when certain contingencies in the network could affect the continuity of the electricity services. With the evolving of the control and communication technologies and the expansion of the DERs and ESS, all high energy intensity plants with DR programs can participate in the ancillary energy markets, providing not only regulation services but also others balancing services to the grid. For that reason, the provision of regulation and balancing services have become a new revenues source for industrial manufacturers.

To measure the profits for regulation services, industrial consumers with IDR programs must consider the tariff models and remuneration schemes in each balancing market. As known, the price of the regulation service is based on the marginal cost of the conventional generation units in spite of the cost of industrial consumers and conventional generators are very diverging. For instance, the cost of regulation supplied by the pulp industry with CHP is composed of 58% by additional start-up costs, 33% by intra-day energy purchases and 9% by increased fuel costs [8]. Additionally, the cost of regulation raises in relation to the activation time, reaction time and number of actuations [8]. For the sake of simplicity, case studies use the existing prices to estimate the profits for regulation services, which could underestimate or overestimate the profitability. Besides, the acceptance mechanism of the regulation bids must also be represented, some factors are used in CBA calculations [15].

In addition, bidding strategy can also affect the profits for regulation services. Such strategies are based on either current regulation service price or cost analysis, both providing a minimum viable price regulation. For instance, a strongly dependent relation between the cost of regulation and spot price on the bid size can be found, even could estimate that regulation price should be over three times the spot price to compensate the cost for supplying regulation services [8].

13.2.2 Reserves

Reserves play a relevant role in the electrical power grids. They are designed both by replacing the consumed secondary (or tertiary) regulation and facing contingencies as when the largest generator in the system is tripped [1]. Reserve pricing is linked to numerous factors, which are discussed in the next subsections. In the context of industrial demand response (IDR), contingency and replacement reserves have been studied, given their capacity and time response. Both regulation and reserve services supplied by the IDR program are generally suitable for many industries [6].

13.2.2.1 Contingency reserve

Contingency reserve (known as spinning and no-spinning reserve) is currently a part of the capacity of generation aimed to deliver energy as a response to contingency events, which could be needed in a small period of time and remain for minutes or hours. Part of these reserves is synchronised with the electrical power grid. In Germany, contingency reserve is managed through an ordinance on interruptible load agreements (AbLaV*), which is a market structure for interruptible loads [9]. Each provider must submit offers with minimum and maximum capacity, as well as minimum and maximum duration of switch-off. Profit calculations are described in the relation (13.5):

$$P_{cr} = \sum_{t=1}^{N} T^t_{cr-up} * E^t_{cr-up} + \sum_{t=1}^{N} T^t_{cr-down} * E^t_{cr-down} \tag{13.5}$$

*Verordnung zu abschaltbaren Lasten – AbLaV.

where T^t_{cr-up} and E^t_{cr-up} are contingency reserve price and energy delivered for up-regulation, respectively. N is analysed period time and t minimum time unit (e.g. hours, weeks, months or years). In respect of down-regulation, profits are assumed as savings of the energy commitment, mainly in terms of operational cost since down-regulation price ($T^t_{cr-down}$) is zero for certain countries, independently of the quantity of energy ($E^t_{cr-down}$).

13.2.2.2 Replacement reserve

During scheduled shutdowns or contingencies with tolerance range in terms of reaction speed, replacement reserve can be employed for smoothing the negative impacts of these events. Besides, such reserve is useful to replace other reserves, preserving the security and reliability of the operation for long periods. Unlike contingency reserve, replacement reserve providers are not forced to a certain reaction speed but must be available and be able to supply the reserve for long periods (e.g. hours). As stated before, reserves are capacities, which must not be used in other services, therefore, prices are established according to the auction process and the opportunity cost of the market participants. For instance, in the Spanish market, replacement reserve is settled within of balancing auctions in the Intraday market, which use tariff every quarter-hour both for up-regulation and down-regulation, as can be seen in relation (13.6) [16]:

$$P_{rr} = \sum_{t=1}^{N} T^t_{rr-up} * E^t_{rr-up} + \sum_{t=1}^{N} T^t_{rr-down} * E^t_{rr-down} \tag{13.6}$$

where T^t_{rr-up} and $T^t_{rr-down}$ are replacement reserve prices for up- and down-regulation actions. Meanwhile, E^t_{rr-up} and $E^t_{rr-down}$ corresponds with energy delivered for up- and down-regulation, respectively. N is a certain analysed period time and t minimum time unit (e.g. hours, weeks, months or years).

13.2.3 *Self-consumption*

Joint with energy efficiency measures, demand response has proved to be able to obtain savings in the production plant's electricity bill due to smart manage their consumption [2]. Currently, IDR not only produces electricity bill savings but also reductions in diverse expenses such as raw materials purchases, improved gas and vapour consumption, among other positive impacts caused by a smart self-consumption. Basically, revenues from self-consumption are estimated based on case reference (baseline), which are previously defined in the design stage and measurement of flexibility potential. Savings for smart self-consumption reached by the IDR program are gauged both in terms of saved energy (e.g. MWh) or saved money as shown in relation (13.7) [7]:

$$P_{self} = \sum_{t=1}^{S} T_t * (E^{sb}_t - E^{sIDR}_t) \tag{13.7}$$

Considering that T_t means the price of each energy resources (S) (electricity, gas, heat, vapour, and etc.). E_t^{sb} and E_t^{sIDR} correspond with energy or resources demand for baseline scenario and IDR application scenario, respectively.

13.2.4 Changes in energy purchasing and flexibility trade

Benefits from the IDR program are also present in the energy exchanges cost between whole stakeholders. This section will deal with possible benefits from changes in energy purchasing and the provision of flexibility services.

13.2.4.1 Changes in energy purchasing

At the baseline scenario, energy purchases are measured and classified. IDR programs modify this baseline scenario to respond to flexibility requirements or market signals. These changes in energy purchases generate benefits in multiple ways. Initially, industrial manufacturers could acquire electricity at low prices and hourly bands where security and reliability are high. On the other hand, energy purchases are diversified towards various energy sources, improving further the reliability of the energy supply. Finally, better agreement conditions between utility and industrial customers can be expected, obtaining advantageous tariff or extra services due to the reduction of the contracted load and demand peak.

13.2.4.2 Flexibility trade

As is predicted, flexibility trade will be a plenty developed in the next years, due to its relevant role in the increase of the RES penetration level and electric vehicle (EV) inclusion, as well as because the number of flexible stakeholders in the grid are augmenting continuously [13]. Basically, flexibility trade involves the sales and purchases of flexibility services offered by flexible end-users, aggregators or energy capacity owners (i.e. EV, distributed generators (DGs) or energy storage system (ESS)) on the one hand, and acquired by system operators, utilities, balancing party responsible (BRP), marketers or even interconnectors or foreign generators. Existing energy markets could incorporate the flexibility trade, but given its complexity, some flexibility markets have been discussed in the literature (local energy markets, capacity market, ramping product market, among others). Each flexibility market possesses a set of prices for each flexibility service, which are based on aspects such as the marginal cost, return rate, tariff policy, subsidies and penalties, among others.

For industrial consumers, the role of the aggregator is relevant, which allows the participation in the flexibility markets of the small and medium manufacturers, as well as others stakeholders (EV or ESS) without the minimum power capacity to enter the market. An overall model of a flexibility market is displayed in Figure 13.2. As shown, small and medium-sized enterprises, EV fleet and small and medium smart buildings facilities owners use aggregators to take part in the flexibility markets [24, 25]. Aggregators manage the end-user's flexibility to offer it into flexible markets through a bidding strategy. Prices, duration and location should be accorded, not only with the end-users but also with flexibility consumers [2, 26]. Bid acceptances T_{bid-ac}, $T_{bid-ag-ac}$ are transmitted from flexibility market to flexibility providers, activating the demand response actions.

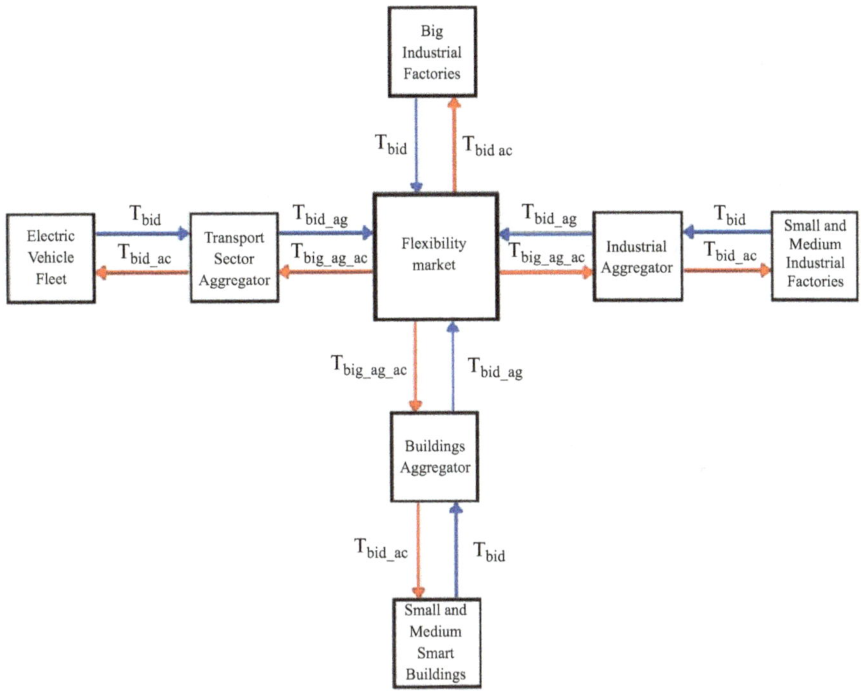

Figure 13.2 Flexibility market and bidding process. Own preparation.

According to the overall above scheme, the profits produced by flexibility trade represents the difference between the prices that flexible consumer are willing to pay and the prices that ensure a reasonable reward by providing flexibility services. Prices and tariffs into flexibility markets have been only contemplated theoretically and can be fused depending on the market development and regulatory framework.

13.2.5 Transmission and distribution network support

Transmission system operator (TSO) and distribution system operator (DSO) receive benefits from IDR programs, which could be useful to reduce technical losses [6], relieve congested lines and defer reinforcement investments. These benefits have been neglected in most case studies in the literature, due to the vast and complex required information from utility to calculate possible rewards. Besides, privacy issues have not been tackled yet for transferring sensible information adequately.

Investment deferral and congestion management, which have received more attention recently, are discussed in the next subsections. There are other benefits such as loss reduction and positive effects on competitiveness, which could mention but deeper studies are required.

13.2.5.1 Investment deferral

Commonly, system operators (SO) control the network expansion through grid planning with investment schedules, which pretend to assure a reliable, efficient and economic operation for next year's [27]. Such investments encompass reinforcement electrical lines to upgrade assets such as substations, transformers, moreover the pipe enlargement or electrical infrastructure upgrades. In addition, it not only should consider the capital cost but also the whole cost associated with the expansion and upgrades, such as pre-feasibility studies, design and commission cost, fixed and variable costs, etcetera. All above costs and investments can be deferred and avoided by IDR programs, generating positive impacts within a finite planning horizon.

Benefits from investment deferral are related to the minimization of the net present value of the cost of expansions and upgrades. In relation (13.8), avoided costs are calculated as the difference between the net present value of the expansion plan investments and the value when the investments have been deferral for the same planning horizon. Such horizon must match with the IDR program useful life [27].

Mathematically, benefits can be formulated as follows:

$$P_{id} = \sum_{n=1}^{N} \left(\sum_{t=1}^{T} \frac{C_{t,n}}{(1+d)^t} - \sum_{t=1}^{N} \frac{C_{t,n} * (1+i)^{\delta}}{(1+d)^{t+\delta}} \right) \tag{13.8}$$

where $C_{t,i}^{bl}$ and $C_{t,i}^{IDR}$ are the investments for each expansion and upgrade required (N), d is the discount rate, T is the assessment period and δ is the deferral length. It is worth mentioning that the above relation corresponds to the radial structure of the network [27].

13.2.5.2 Congestion management

Currently, congestion in transmission and distribution lines is becoming a more and more important problem for system operators [28]. Increasing demand and penetration of renewable energy sources are boosting changes in the electrical infrastructure. Said that, TSO and DSO attempt to maximize the usage of the network without harming the Quality of Service (QoS) [28]. Congestion management seeks suitable energy flows that represent the minimum costs possible. However, there are nodes with low influence on the distribution and transmission network total load, therefore such nodes and customers connected there cannot be selected for successful IDR programs. DSO and TSO are in charge of selecting proper nodes for congestion management, given that they possess enough technical information. The chosen nodes and their load both before (do_{cm}) and after (d_{cm}) of the IDR programs are employed in relation (13.9) to measure the benefits of congestion management service [29]:

$$P_{cm} = T_{inc-cm} * (do_{cm} - d_{cm}) - Pe_{cm} * (LR_{cm} - do_{cm} + d_{cm}) \tag{13.9}$$

being T_{inc-cm} the tariffs or incentives for each unit load reduction, Pe_{cm} the penalty for disregarding the minimum required load and LR_{cm} is the load reduction requested by DSO or TSO.

13.2.6 Other benefits

IDR programs, not only have positive impacts on consumers, utilities and system operators, but also on social welfare and the environment [6]. Such benefits have been mentioned in multiple cost–benefits analyses, with the desire of quantifying and in consequence, remunerating adequately these impacts. Besides, CBA could be evaluated more accurately whether all benefits and costs are examined.

13.2.6.1 Greenhouse gases emission savings

Reducing pollution emissions is already a global challenge due to climate change evidence of the last years, therefore, multiple policies have been raised around energetic efficiency, renewable energy sources, demand response, electrification of the transportation sector, among others. Industrial Demand Respond has positive effects on the carbon footprint of existing manufacturing processes, besides replacing the existing energy generation with other lower polluting emissions energy sources [10]. IDR programs can reduce the emissions of the conventional generators (top-down approach), as well as of the industrial factories (bottom-up approach).

According to the above approaches, benefits could be gauged using two methods. First, a count of the emissions (RE_{IDR}^{top}) produced by involved power plants (N) through pollution coefficients (*bp* and *cp*) is done and expressed mathematically in the relation 13.10. Such savings are in terms of power output both before ($P_{g,n}$) and after ($P_{g,n}^{IDR}$) of IDR program application:

$$RE_{IDR}^{top} = \sum_{n=1}^{N} b_p * \left(P_{g,n} - P_{g,n}^{IDR}\right) + c_p * \left\{\left(P_{g,n}\right)^2 - \left(P_{g,n}^{IDR}\right)^2\right\} \tag{13.10}$$

In regards to the second method, industrial factories must estimate the total emissions of their production activities by hourly (preferably), considering that there is an existing lineal relation between electricity consumption and production rate, besides, any excess product will be stored without relevant costs. Both baseline and IDR scenario, the second method can be realised as follows:

$$RE_{IDR}^{bot} = \sum_{n=1}^{T} \left(E_{p,t} * P_{rate,t} - E_{p,t}^{IDR} * P_{rate,t}^{IDR}\right) \tag{13.11}$$

where $E_{p,t}$ and $E_{p,t}^{IDR}$ are the total emissions before and after the IDR execution. Likewise, $P_{rate,t}$ and $P_{rate,t}^{IDR}$ are the production rates before and after of the IDR program. For simplifying calculations, hourly CO_2 factors of the electricity generation mix are used to measure equivalent emissions, considering the factory location [1]. Besides, the evaluation period (T) should involve from the preparation stage to the recuperation period of the actions scheduled by IDR programs [1]. Finally, these emission reductions must be adequately converted to CO_2 equivalent emissions and later in terms of savings with the help of the carbon emission price.

13.2.6.2 RES penetration level

Undoubtedly, the reduction of Greenhouse Gasses (GHG) emissions is related to the Renewable Energy Resources (RES) expansion. The increase of the RES Penetration level is prompted with a set of policies and programs, where industrial demand response can be found [2,6]. Augmenting the RES penetration level is a relevant and challenging task at a time [1,4]. Shifting loads of the IDR programs could allow the matching between renewable energy generation and certain electricity usage. For instance, wind farm produces a high amount of energy at night, precisely when most EVs are charging [27]. Moving a significant amount of demand towards hours of the day where PV generation has an output power peak could also bolster the investment on RES [1]. Given that, increasing the RES penetration level could not be significantly caused by a unique factory, necessarily it should be tackled from an aggregation point of view.

According to the above and due to the positive effects of both electrical networks and consumer installations, two different approaches to gauge the impacts of a high-RES penetration level due to IDR programs can be discussed. First, as mentioned before, IDR programs prompt the new renewable generation units and maximize the use of existing units. Usage rate and efficiency indicator of RES are used as an indicator of profits. On the other side, consumers can be motivated to apply techniques of demand-side management [30] and adjust adequately the demand curve with the generation behaviour. The installation of new energy renewable facilities during the IDR Program execution could indicate a certain avoided cost that had been invested in environmental policies to obtain the same RES penetration level.

13.2.6.3 Positive effects in national economy

Finally, IDR programs have also positive effects on the national economy. As mentioned before, the IDR program allows to increase RES penetration level and reduce the GHG emissions. Given that renewable generation units work at a lower Levelized Cost of Energy (LCOE), the higher amount of renewable energy, the lower the whole electricity price (Spot price). Additionally, RES receives subsidies or reduced taxes, which are not related to the power effective usage, therefore, IDR programs could increase their utilization and hence their efficiency [6]. Likewise, the power network stability and security will be improved due to the generation diversification

13.2.6.4 Machinery useful life

As known, the IDR program implies wear and tear costs in the machinery due to changes in the nominal operation. It has proved that IDR programs could be also beneficial to the machinery's useful life [1], only if interruption frequency is lower than nominal frequency or if power during IDR execution is lower than nominal power but higher than inefficiency operation of the performance curve. Such an operational regime could extend maintenance cycles, as well as the useful life of refrigeration, lubrication and filtration devices.

13.3 Feasibility assessment

To fill out CBA, it is necessary to define profitability indicators that allow comparing diverse economic and transversal information about projects. Such indexes help to deduce whether an IDR program action and thereby the whole IDR program are profitable or conversely requires a financial stimulus. In this section, indicators most commonly used are defined mathematically. Besides, a review of the relevant information they provide is carried out.

13.3.1 Indicators

Whenever a profitability analysis is carried out, several indicators should consider, mainly so that good decisions can be taken regard to investments or energy policies. Basically, such indexes involve valuable information, which depends on the way they are calculated and presented. First, the selected indicators are usually related to the flexibility objectives such as savings in electricity bill and energy cost, as well as the minimization of the environmental impacts, among others.

Leaving aside the above, there are specific economic indicators (e.g. net present value, capital and operational expenditures, among others) that can reflect the good health of a given investment. Frequently, these indexes are shown either in terms relative with respect to baseline case or in terms of absolutes. In the end, it should be remembered that each indicator must be related to a flexible source and the services it can provide. In this section, will be shown a review of the indicators that are more applied in the literature.

13.3.1.1 Net present value

Net present value (NPV) is an economic decision metric used to compare investments in projects with different revenues and costs streams within their lifetime [1,4]. According to relation (13.12), the difference between total revenues and cost is related to discount rate, which is chosen by investors or holders as a minimum profitable rate of return. Positive and negative values of NPV indicate the profitability level of certain projects. Broadly, investors prefer projects with the highest NPV, as well as with other economic indexes. NPV is defined mathematically as follows:

$$\text{NPV} = \sum_{t=1}^{T} \frac{C_t}{(1+r)^t} - C_o \tag{13.12}$$

Being C_t and C_o are the net cash inflow during the period t and initial investment costs, respectively. Also, r is the discount rate (%) and T is the total time periods.

13.3.1.2 Payback period and internal rate of return

Continuing with project decision indexes, payback period (PP) and internal rate of return (IRR) are indicators of the length of time required to recoup original investment and financial equilibrium point in a given project [1,7]. The payback period becomes relevant as it shows the information on the liquidity of the project and management of possible risks in projects with high uncertainty. Meanwhile, IRR shows the point in which the project generates revenues for the holders or investors. In the following relations, PP and IRR are described mathematically:

$$\mathrm{NPV}_c = \sum_{t=1}^{T} \frac{C_{ct}}{(1+r_c)^t} - C_o \tag{13.13}$$

$$\mathrm{NPV}_b = \sum_{t=1}^{T} \frac{C_{bt}}{(1+r_b)^t} - C_o \tag{13.14}$$

$$\mathrm{PP} = N_{ul} * \frac{NPV_c}{NPV_b} * 100 \tag{13.15}$$

As defined above, the payback period is calculated based on net present value both cost (NPV_c) and benefits (NPV_b), where C_{ct} and r_c are the net cash inflow and discount rate for costs, respectively. For the benefits side, C_{bt} and r_b are the net cash inflow and discount rate for benefits. Besides, N_{ul} is the IDR program useful life and T is the total time period. On the other side, IRR corresponds to the discount rate when the payback period is achieved. Both discount rates could be streamlined at the same rate or remain as independent rates, depending on IDR program complexity.

The application of PP and IRR in profitability analysis has drawbacks. The lack of the salvage (residual) value could diminish the effectiveness of the results. Besides, inaccuracies could be presented in an IDR program where continuous investments are done, as well as diversified lifespans of numerous assets. Equally, IRR is unappropriated for mutually exclusive projects.

13.3.1.3 Savings

Benefits expressed in savings are commonly employed by IDR projects [2,10]. Savings with respect to a baseline scenario in which the costs were previously estimated [7]. For example, IDR programs reach savings in the electricity bill for industrial end-users. Such savings are calculated in percentage terms per year, hour or during all IDR program lifespans. The relation (13.16) proposes an efficiency parameter (η) for electricity cost, which is defined as follows:

$$\eta = \frac{C_b^o - C_b^r}{C_b^o} * 100 \tag{13.16}$$

where C_b^o and C_b^r are the original electricity cost and rescheduled electricity cost due to IDR program application, respectively. Besides, savings are also found in cost reductions such as gas and heat consumption, raw material losses, etc. Generally, these savings are expressed in the reduction of cost in both sub-process and total cost of the plant. Similarly, power peak reduction rate can be mentioned as an indicator to analyse the profitability, as well the influence of the tariff on savings [10]. Finally, savings can be located during preparation, execution and recuperative stages of load interruptions, given that the average consumption of IDR action is lower than the original average demand [1].

13.3.1.4 Profits

Many CBA, not only for IDR programs but also for diverse types of flexibility applications, use the net profits as a profitability indicator, which are usually the difference between revenues and cost of the project during a time horizon [5,24]. The profit indicator is a simple view of how many revenues are obtained of certain flexibility services, as well as of a set of services seen as a unique assemblage. For instance, an optimization algorithm for maximizing the profits of providing operation services in electric power systems is developed [5]. In the case study, profits are shown as the final count of the whole revenues generated by providing reserves, capacity and energy delivery and net profit for electricity consumption reduction. Conversely, profits are also used to quantify revenues via optimal production scheduling [15]. Considering the aforementioned, profits become a flexible indicator but at the same time generic and trivial.

Profits per year or per power are also applied in other case studies, which could be insufficient to define the IDR program profitability. This approach generally is related to benefits for main stakeholders, aggregators or industrial customers, given that the investor role and its benefits are put at the second level. Finally, not only the profit index but also the remainder profitability indicators are employed for identifying the profitability of adjusts in both energy markets, pricing and energetic policies.

13.4 Case studies

In the literature, there are hundreds of case studies about multiple industry sectors, each of them with particularities in both size and technologies, as well as services, production processes, among others. In addition, each of the energy markets, as well as pricing schemes and subsidies and penalties, creates a wide diversity of technical situations, therefore, finding common patterns is a complex task. For this reason, this section shows case studies where interesting results were obtained. Such case studies made important findings on the IDR program's performance.

13.4.1 Chlor-alkali production industry

A medium chlor-alkali production industry is taken as a case study to analyse the effects of various types of markets and tariffs at Nordic electrical networks [9]. An intermediary storage capacity is modelled to increase the flexibility potential and provision. In addition, the effects of excess capacity on feasibility were studied both numerical models and flexibility metrics. The Chlor-alkali production factory participates in reserve markets through primary reserve and reserve via interruptible loads.

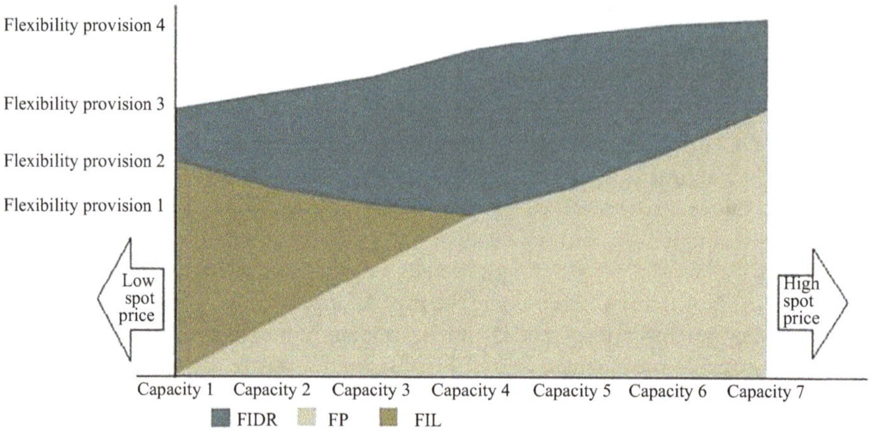

Figure 13.3 Spot price influence on CBA medium chlor-alkali production industry at Nordic electric system

The feasibility study concluded that when the spot price is low enough and the industry has small excess capacity, the most suitable reserve is generated by interruptible loads (FIL). Alternatively, when a factory possesses the medium or high excess capacity and the electricity price is low, the feasible flexibility is caused by IDR program (FIDR). In the same sense but increasing the electricity price, the higher the electricity price, the higher the primary reserve (in comparison with IDR and interruptible loads reserve). Finally, in a scenario with high prices and excess capacity, only primary reserve (FP) will produce the biggest profits and advantages [9]. The above conclusions can be seen in Figure 13.3.

13.4.2 Paper industry in Germany

A paper industry in Germany is considered as a case study [1]. The paper plant has three main demand response actions within the DR program, which was found in an initial flexibility audit. Each demand response action implies a set of costs and benefits, which are defined as both minimum revenues received by the industrial customer's due to the DR program and the minimum payment offered by TSO/DSO. For simplifying, the historical reserve price was used to reward the industrial participants in demand response actions [9]. Finally, the production is affected by two maintenance periods with monthly and semi-annual frequency, respectively. As a result of the optimization tool, an initial investment was estimated at 130 k€ and an extra pulp storage tank was installed to guarantee the duration of interruptions. Table 13.2 shows economic evaluation results:

As seen in Table 13.2, the 'stock preparation' demand response action was the most profitable, since it has the highest NPV and IRR, with a payback period of 2 years and 2 months. Likewise, this demand response action counted 397 tons CO_2 avoided.

Table 13.2 CBA results for paper industry in Germany

Demand response action	**Npv (€)**			**IRR (%)**	**DPP (years)**
	Discount rate (5%)	**Discount rate (10%)**	**Discount rate (15%)**		
Stock Preparation	64,307	47,822	33,602	30.6%	2.2
Winder	272	75	–95	12.1%	3
Storage	–1,978	–1 989	–1,982	–83.2%	>5

Source: Own preparation.

13.5 Conclusions and final considerations

As seen in this chapter, a cost–benefits analysis (CBA) is needed to measure IDR profitability. Such CBA should incorporate a LCC approach, which describe all cost involved in each stage of the IDR development cycle. Together with the cost, a specific analysis of the benefits should be conducted. As mentioned, such benefits are diverse and depend on technical requirements, standpoint (industrial customer, utility or aggregator), timelines (short, medium or long terms), market (wholesale, retail, balancing and flexibility) and pricing features. Finally, a proper choice of profitability indicators must be done, which allows to compare different IDR actions and select the most suitable and profitable option.

Final considerations are featured below:

- Simulations and case studies include a historical database on regulation and reserve prices in the current energy markets. Such prices are settled according to the marginal cost level of the conventional generators. Units with high marginal cost are usually used for regulation or reserve capacity, therefore, the feasibility of IDR programs could have a wide margin to invest without posing any risk to its economic viability. Otherwise, IDR programs with limited internal rates of return could be affected by regulatory changes or with the application of efficient and modern technologies (lower marginal cost).
- According to the opportunity cost concept, the energy savings index is more suitable than money saving one, as energy could be employed in flexibility services generating higher profits in comparison with self-consumption [7].
- Measurement of flexibility potential is an essential part of the IDR program. This stage allows to define the assets and processes that will participate in the IDR program design on one hand, and find investments (batteries, endogenous or exogenous capacity, changes in process and materials) must be done to obtain the biggest profits possible on the other hand. Figuring out adequately IDR potential generates expenses, which are not included in the majority of the case studies and their CBA in the literature.

- Risk management is still a lack in CBA applied to IDR programs. This chapter has been included some IDR development cycle stages, but it must be researched in greater depth in the future.
- Electro-intensive industries should plan their expansion programs considering the future flexibility services that could be rendered since the higher is energy intensity, the higher is IDR potential and the higher are probable profits.

References

[1] Rodríguez-García J, Álvarez-Bel C, Carbonell-Carretero JF, Alcázar-Ortega M, Peñalvo-López E. 'A novel tool for the evaluation and assessment of demand response activities in the industrial sector'. *Journal of Energy*. 2016;**113**:1136–46.

[2] Shoreh MH, Siano P, Shafie-khah M, Loia V, Catalao JPS. 'A survey of industrial applications of demand response'. *Journal of Electric Power Systems Research* 2016;**141**:31–49. http://dx.doi.org/10.1016/j.epsr.2016.07.008.

[3] Ma J, Silva V, Ochoa LF, Kirschen DS, Belhomme R. 'Evaluating the profitability of flexibility'. *IEEE Power and Energy Society General Meeting*, 2012, pp. 1–8. doi:10.1109/PESGM.2012.6344848.

[4] Gils HC. 'Assessment of the theoretical demand response potential in Europe'. *Journal of Energy*. 2014;**67**:1–18. http://dx.doi.org/10.1016/j.energy.2014.02.019.11

[5] Kreuder L, Gruber A, von Roon A. 'Quantifying the costs of demand response for industrial businesses'. In *IECON 2013 – 39th Annual Conference of the IEEE Industrial Electronics Society*, 2013, pp. 8046–51. doi: 10.1109/IECON.2013.6700478.

[6] Khripko D, Morioka SN, Evans S, Hesselbach J, de Carvalho MM. 'Demand side management within industry: a case study for sustainable business models'. *Journal of Procedia Manufacturing*, 2017;**8**:270–77. http://dx.doi.org/10.1016/j.promfg.2017.02.034.

[7] Eldali F, Hardy T, Corbin C, Pinney D, Javid M. 'Cost-benefit analysis of Demand response programs incorporated in open modelling framework'. In *2016 IEEE Power Energy Society General Meeting Conference*, Boston, MA, USA, July 2016; IEEE November 2016, pp. 1–5.

[8] Helin K, Kaki A, Zakeri B, Lahdelma R, Syri S. 'Economic potential of industrial demand side management in pulp and paper industry'. *Journal of Energy* 2017;**141**:1681–94.

[9] Richstein JC, Hosseinioun SS. 'Industrial demand response: how network tariffs and regulation (do not) impact flexibility provision in electricity markets and reserves'. *Journal of Applied Energy*. 2020;**278**:115431. https://doi.org/10.1016/j.apenergy.2020.115431.

[10] Alcázar-Ortega M, Álvarez-Bel C, Escrivá-Escrivá G, Domijan A. 'Evaluation and assessment of demand response potential applied to the meat industry'. *Journal of Applied Energy*. 2012;**92**:84–91.

[11] Ingrid M-C, Aprà FM, Olivella-Rosell P, Villafáfila-Robles R. 'The potential role of flexibility during peak hours on greenhouse gas emissions: a life cycle assessment of five targeted national electricity grid mixes'. *Journal Energies* 2019;**12**(23):4443. https://doi.org/10.3390/en12234443.

[12] Kiliccote S, Piette MA, Hansen D. 'Advanced controls and communications for demand response and energy efficiency in commercial buildings'. In *Second Carnegie Mellon Conference in Electric Power Systems: Monitoring, Sensing, Software and Its Valuation for the Changing Electric Power Industry*. Pittsburgh, PA, January 12, 2006, 150, pp. 1–11.

[13] Mutule A, Obushev A, Grebesh E, Lvovs A. 'Feasibility study for demand response in commercial buildings'. In *The 9th International Scientific Symposium ELEKTROENERGETIKA 2017 Conference*. Stará Lesná. Slovak Republic September 2017, vol. 718, pp. 696–701.

[14] Beier J, Thiede S, Herrmann C. 'Energy flexibility of manufacturing systems for variable renewable energy supply integration: real-time control method and simulation'. *Journal of Cleaner Production*. 2017;**141**:648–61, ISSN 0959-6526. https://doi.org/10.1016/j.jclepro.2016.09.040.

[15] Ramin D, Spinelli S, Brusaferri A.. 'Demand-side management via optimal production scheduling in power-intensive industries: the case of metal casting process'. *Applied Energy*, Elsevier, 2018;**225**(C):622–36. https://ideas.repec.org/a/eee/appene/v225y2018icp622-636.html

[16] Rodríguez-García J, Ribó-Pérez D, Álvarez-Bel C, Peñalvo-López E. 'Maximizing the profit for industrial customers of providing operation services in electric power systems via a parallel particle swarm optimization algorithm'. *IEEE Access* 2020;**8**:24721–733. https://doi: 10.1109/ACCESS.2020.2970478.

[17] Liu Y, Fan Y, Palazoglu A, El-Farra NH. 'A flexible design framework for process systems under demand-side management'. *AIChE Journal* 2020;**66**:e16249. https://doi.org/10.1002/aic.16249.

[18] Chava S. 'Environmental externalities and cost of capital'. *Journal of Management Science* 2014;**60**(9):2223–47. https://doi-org.recursos.biblioteca.upc.edu/10.1287/mnsc.2013.1863.

[19] Piette M, Watson D, Sezgen O, Motegi N. 'Results and commissioning issues from an automated demand response pilot'. In *National Conference on Building Commissioning*. May 2004. https://escholarship.org/uc/item/2fv1x8fr.

[20] Wang Y, Li L. 'Time-of-use electricity pricing for industrial customers: a survey of U.S. utilities'. *Applied Energy* 2015;**149**:89–103, ISSN 0306-2619. https://doi.org/10.1016/j.apenergy.2015.03.118.

[21] Biel K, Glock CH. 'Systematic literature review of decision support models for energy-efficient production planning'. *Computers and Industrial Engineering*, 2016;**101**:243–59, ISSN 0360-8352. https://doi.org/10.1016/j.cie.2016.08.021.

[22] Sun L, Harjunkoski I, Castro P. 'Resource-task network based approach for industrial demand side management of steel production'. In *Computer Aided*

Chemical Engineering, Elsevier, 2013;**32**:259–64, ISSN 1570-7946, ISBN 9780444632340. https://doi.org/10.1016/B978-0-444-63234-0.50044-0.

[23] Samad T, Koch E. 'Automated demand response for energy efficiency and emissions reduction'. In *PES T&D 2012 Conference*, Orlando, FL, USA, May 2012, August 2012, pp. 12–14.

[24] Ocana JC, Sarker M, Ortega-Vazquez M. 'Decentralized coordination of a building manager and an electric vehicle aggregator'. In *2017 IEEE Power & Energy Society General Meeting*, 2017, pp. 1–1. https://doi: 10.1109/PESGM.2017.8273847.

[25] Safari A, Tahmasebi M, Pasupuleti J. 'Smart buildings aggregator bidding strategy as a megawatt demand response resources in the spinning reserve electricity market'. In *2019 IEEE PES Innovative Smart Grid Technologies Europe (ISGT-Europe)*, 2019, pp. 1–4. https://doi: 10.1109/ISGTEurope.2019.8905471.

[26] Zhao W, Wang J, Jiang D. 'Demand response potential of customer-side integrated energy system'. In *IOP Conference Series: Earth and Environmental Science*. 2021;**632**:042026. https://doi.org/10.1088/1755-1315/632/4/042026.

[27] Gutierrez-Alcaraz G, Lu C. 'Demand response and network reconfiguration on distribution system investment deferment'. In *21st International Conference on Electricity Distribution*, Frankfurt June 2011 (January). Paper No. 0743.

[28] Linna NI, Wen F, Liu W, Meng J, Lin G, Dang S. 'Congestion management with demand response considering uncertainties of distributed generation outputs and market prices'. *Journal of Modern Power Systems and Clean Energy* 2017;**5**:66–78.

[29] Zaeim-Kohan F, Razmi H, Doagou-Mojarrad H. 'Multi-objective transmission congestion management considering demand response programs and generation rescheduling'. *Applied Soft Computing*, 2018;**70**:169–81, ISSN 1568-4946. https://doi.org/10.1016/j.asoc.2018.05.028.

[30] Hadi AA, Silva CAS, Hossain E, Challoo R. 'Algorithm for demand response to maximize the penetration of renewable energy'. *Journal IEEE Access*. 2020; **8**:55279–288.

Chapter 14

Measurement and verification of demand response: the customer load baseline

A. Gabaldon[1]*, A. García-Garre*[1]*, M.C. Ruiz-Abellón*[2]*, C. Álvarez-Bel*[3]*, L.A. Fernandez-Jimenez*[4]*, J.L. Martínez-Ramos*[5]*, S. Valero-Verdú*[6]*, and J. Rodríguez-García*[3]

Demand response (DR) is a basic tool to achieve power systems flexibility in the short and medium terms. The effective deployment of DR and the engagement of new resources need knowledge about how DR performs and how to evaluate their flexibility to give a correct economic feedback to customers and aggregators. DR verification requires a reference in the absence of control: the customer baseline load (CBL). The aim of this chapter is to describe several baselines that provide an acceptable evaluation of load response as well as the use of different adjustment methods to improve the CBL. Some of these adjustment factors can be justified through the simulation of physical-based load models (PBLMs), which are also used in DR for planning and operational tasks. The chapter discusses some issues reported by grid operators: detection of abnormal responses (before and after DR) that can be due to gaming or are reactions to maintain load service such as pre-heating, pre-cooling or the change of tasks timeline. All these approaches have been illustrated using real data of an industrial customer. Results show that the adjustment of CBLs can improve several conventional approaches described in the literature.

Keywords: Customer Baseline Load, Physical-Based Load Modelling, Demand Response, Measurement and Verification, Electricity Markets

[1]Power Systems Group, Technical University of Cartagena, Cartagena, Spain
[2]Department of Applied Mathematics and Statistics, Technical University of Cartagena, Cartagena, Spain
[3]Institute for Energy Engineering, Universitat Politècnica de València, Valencia, Spain
[4]Department of Electrical Engineering, Universidad de La Rioja, Logroño, Spain
[5]Department of Electrical Engineering, Universidad de Sevilla, Sevilla, Spain
[6]Department of Mechanics and Energy, Universidad Miguel Hernández, Elche, Spain

14.1 Introduction

Future power systems must be much more flexible than in the past because foreseen energy policies will involve a more significant participation of renewable resources in the generation mix. This new scenario means that the supply side will decrease its controllability, which requires an increasing participation of demand side. For these reasons, the design of new markets is more "customer-centred" [1] and encourages the participation of demand side to recover the lack of flexibility of power systems. A valuable option is developing the portfolio of demand response (DR) including the adequate payment for a resource's performance. Unfortunately, this is a significant concern because this flexibility is difficult to be measured and verified.

The engagement of customers in DR is difficult due to several barriers (regulation, education, costs, etc.). To overcome some of these barriers, right and correct remuneration of resources based on the reduction of power and energy consumption is needed. In last decades, a main concern for regulators has been the definition of economic incentives for participants. For this purpose, the US regulator issued Order 745 [2]. Order 745 stated that DR providers must be remunerated at the same price paid to generators, that is, at full locational marginal pricing (LMP). Some stakeholders raise the issue that paying LMP in all hours presents a significant challenge to the accurate measurement and verification (M&V) of DR.

The future scenario in the European Union (EU) envisages the need to enhance and increase DR. The European Directive 2019/944 establishes that "States shall allow final customers, including those offering DR through aggregation, to participate alongside producers in a non-discriminatory manner in all electricity markets" (article 17, [1]).

An adequate and understandable economic flow is necessary: the estimation of the customer baseline load (CBL) is a cornerstone for defining the performance, cost and revenue of different DR programs. Customers should receive credit according to the actual flexibility they provide, which requires an accurate evaluation of the amount of electricity that would have been consumed by customers in the absence of the DR event. The evaluation of this flexibility needs a forecast of demand considering loads and customers, whose physical behaviour can change due to several parameters: weather, type of day, end-use shares or the frequency in DR calls (i.e. the change of customer behaviour due to frequent DR requests). In "real time", the baseline can be used to show customer and aggregator whether they are meeting DR targets. In the medium term, the baseline enables the evaluation of credits to customers as well as the performance and potential of DR resources by system operators (SOs).

Several factors should be considered in the development of baselines according to the literature [3]: accuracy, simplicity and integrity. The accuracy of the achieved flexibility is important to avoid paying too high an incentive for DR while still encouraging the participation of customers and aggregators. It is also important, from the customer point of view, to recognize his/her contribution in DR and to avoid non-performance penalties or the underestimation of DR achievement.

Another factor is that the economic feedback to customers opens up the possibility of manipulation attempts. Some SOs and consultants think that customers could shift the load to different times to affect the calculation of actual flexibility. These changes in customer pattern should be detected to ensure DR integrity. Finally, methodologies should be as straightforward as possible and should consider the characteristics of customers and markets where DR resources are deployed. This justifies the use of less accurate but understandable baselines in some industrial segments [4] to avoid a lack of interest by aggregators and customers. Note that there must be multiple baselines to cover different types of DR products on a range of different sites [5]. Likewise, the improvement of the CBL through the adjustment coefficients must be adapted to the context, for instance, the use of aggregated and elemental physical-based load model (PBLM).

The rest of the chapter is organized as follows. In Section 14.2, a revision of the significance of CBL is presented. In Section 14.3, traditional baselines and their adjustment are revisited. Section 14.4 depicts the customer example and some considerations to apply a more accurate adjustment of CBL, while Section 14.5 outlines the performance of different CBL methods and the improvement achieved through different coefficients. Finally, in Section 14.6, some conclusions are presented.

14.2 Literature review

Baseline methods have grown in interest due to the forthcoming role of the DR policies in wholesale and retail markets worldwide (UE [6], United States, Australia or South Korea [7]), both to balance renewable and to mitigate environmental emissions of power systems over the 2030–50 horizon. Moreover, aggregators have gained relevance during last decades, and opportunities for them and customers will arise in future decarbonized scenarios [8]. The growth of DR means that payments and revenues are rising [9] and the accuracy of these economic flows should be revisited. For this reason, different laboratories [10], SO [11, 12], aggregators [3], marketers and utilities [13, 14], and energy and environmental agencies [15] have analysed different types of baselines and have proposed new methods or adjustments to improve their accuracy. This proliferation of methodologies makes the management of DR more difficult [16] but also allows the choice of different CBL according to the customer characteristics. In some cases, the CBL methodologies must adapt to DR alternatives. This happens in some of the US SOs [9, 12], and also in other regions, KPX (the SO in South Korea [7]).

The issue of CBL harmonization was first recognized by the North American Energy Standard Board (NAESB) as a barrier [17] and later in the EU [18]. Consequently, NAESB developed a series of definitions that have been acknowledged by US authorities [2, 19].

CBL methods outlined in the literature depend on customer segments and regions, and some reports include a comparison of diverse baselines in the same or even among different countries [7]. For instance, Lawrence Berkeley reported [10] that adjustments usually improve the performance of CBL, but the use of a different model for different groups of loads, due to weather sensitivity of some loads, is suggested. Con Edison presents similar results in [14], and they state that simple baselines perform well in most customer segments.

San Diego Gas & Electric obtained similar results in [13] using 21 different CBLs. The authors conclude that "traditional" baselines reasonably estimate across all customers and all event days, but none of the "traditional" methods are the most accurate for single customers on individual event days. In conclusion, more complex baselines sometimes provide marginal improvements in accuracy but with a higher computational cost. In [20] the authors propose the clustering of customers to reduce the randomness of each individual demand. This improves the performance of CBL, but unlike industrial and commercial customers, the morning adjustment produces an adverse impact in their "overall performance index" for residential customers. This demonstrates the need to make a classification of segments and end-uses into "homogeneous" groups. The same conclusion was previously established in DR planning and management [21–23].

Baseline estimation and short-term load forecasting (STLF) share common methodologies because they provide demand forecasts in the short term (<48 h). STLF comprises multiple methodologies, for example, linear regression and artificial neural networks (ANN) which can also be used to calculate the baseline [24,25]. Some other machine learning methods such as support vector regression (SVR) and support vector machine (SVM) have been employed to forecast demand in [26–28]. Hybrid parameter optimization [26] and ant colony optimization [27] have been proposed to find the optimal parameters for SVR, whereas SVM with simulated annealing has been presented in [28]. The efficiency of ensemble methods based on regression trees, such as random forest or boosting, has been analysed in [29]. Nevertheless, classical methods like ARIMA models still perform well for demand forecasts. For this reason, hybrid models that combine two or more different methodologies (ARIMA, SVM or ANN) outlined good results. For instance, SVM and ARIMA are proposed in [30, 31], and the combination of ANN and SVM has been developed in [32]. A machine learning approach to disaggregate load and photovoltaic (PV) generation from net load data is analysed in [33] to obtain CBLs in prosumers. Authors conclude that reducing errors in the PV output power estimation can improve the performance of the CBLs.

Several approaches have considered CBL for industrial segments. Large industrial customers are considered in [34] using historical data to define a baseline. In [4], CADMUS consulting presents the evaluation of CBL for small, medium and large commercial and industrial customers (C&I) engaged in DR peak shaving programs in Pennsylvania (USA). Authors used regression analysis and day-matching CBLs for large C&I customers, and provide evidence that some of these customers adjusted their loads in response to the advance notifications. Again, authors state similar conclusions than other reports: regression methods can give the best accuracy for some facilities and specific customers, but day-matching methods provided better accuracy on average.

The performance of several CBL approaches to quantify the effects of DR in the demand of a university campus building is presented in [35]. The interest is based in the large share of heating, ventilation and air conditioning (HVAC) in the overall demand of buildings. Specifically, authors propose a simple linear interpolation method that uses the least squares to fit a linear baseline to the fan data in HVAC over a 5-min event just before DR event and 5-min period after the settling time. As a conclusion of the proposed method, it is possible to develop more accurate baselines if customers (or aggregators) use sub-metered data from flexible loads (fans in HVAC systems), rather than the overall building load data. The analysis of the responsive demand in buildings before and after DR through the use of PBLM is proposed in [36]. In this case, simulation and sub-metering are applied for the adjustment of baselines.

Control groups are also suitable for the calculation of baselines. They include customers without DR control to compare their behaviour with those who are responding to DR. Because this approach intends to reflect the response of weather-sensitive (WS) loads, the customer in the control group should experience the same weather conditions as the responsive one. Two alternatives are used in [12]: randomized control trial and matched control groups. The randomized trial involves customers that are able to participate in DR actions but that are randomly selected in advance, and their flexibility is withdrawn during the target period. In matched control groups, the control consists of customers who do not participate in DR but have similar characteristics (segment, end-uses, income) to DR participants. According to [12], this specific option outperforms traditional baselines.

In the context of residential customers, Ref. [24] proposes linear regression, using historical demand and weather as independent variables. Also in this context, [7] compares Korean and French matching-day baselines and it establishes that weighted baselines (South Korea) outperform French CBL. However, the author does not use real data for residential segments, taking estimations from the total consumption instead. In [37], a synchronous pattern-matching principle-based residential CBL estimation approach without historical data requirement is proposed. Customers are split into two groups with and without control. The control group is clustered and then each DR participant is matched to the most similar cluster to predict their CBLs. The main problem of this approach is that it works better with large-enough sample sizes (between 200 and 400 participants) and this reduces DR potential and income for non-participants. In [23], three different baselines (exponential moving average, regression and HighXofY) are analysed for residential customers and the authors discuss their performance. They conclude that the type of CBL will affect customers' decision and participation in future DR events and that bias is more important than accuracy.

14.3 Customer baseline load, non-intrusive load monitoring and physical-based load models

14.3.1 The necessary linkage between DR methodologies

Four tasks are critical for the success of DR policies and the increase of flexibility in power systems: planning, operation, measurement and verification. Figure 14.1

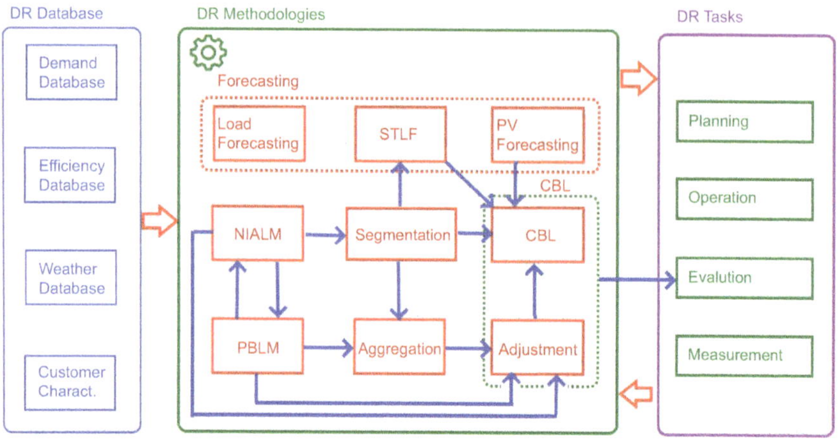

Figure 14.1 Interaction among PBLM, CBL, NIALM and STLF tools according to [21]

depicts a layout of the Spanish Research network REDYD2050 [21] issued from past and present research developments of this group. According to REDYD2050, different methodologies can be used for more than a single task, and this can make the management of DR easier while overcoming some barriers. For example, the statement of some authors about the need to develop a specific tool for each customer and DR service. This idea makes the development of DR more complex, as well as the understanding of these methodologies by customers.

DR actors should be aware that the evaluation and deployment of the DR potential need some end-use and load models, especially "grey" or "white" ones, like PBLM. The aggregated response [38] is another important issue to determine the flexibility potential of resources (e.g. reduction levels, loss of load service, rate of change, energy recovery/snapback, etc.). As DR is basically achieved by responsive loads with some kind of energy or product storage [22] for maintaining service (e.g. HVAC and water heaters, ice or heat storage tanks), aggregators and large customers need some tools to perform DR simulations before the event (planning/operation).

Non-intrusive load monitoring (NIALM) plays an important role in this context. Elemental end-use demand can be obtained to tune the parameters of PBLMs, to validate them or even to define homogeneous and heterogeneous load groups and customers. Moreover, NIALM allows the definition of average end-use patterns (i.e. the calculation of elemental end-use load baselines [39] in customers where sub-metering is not available), all of them from Smart Meter measurement at a lower cost. Another example of linkage between methodologies can be found in market participation: the aggregator needs load forecasts to define the energy requirements in day-ahead markets and avoid penalties in balance markets. This can be done through specific forecasts [29], but CBLs can also be used. Weather or gains in efficiency can be evaluated from modelling, if these models are physical-based. Finally, NIALM should contribute to the verification of the performance of responsive loads (e.g. through a representative sample) to verify the change of use of some devices before DR (to detect potential gaming) or during DR (for the verification of load flexibility).

As discussed in previous paragraphs, segmentation is crucial for achieving greater performance in the CBL, but it is also a fundamental step in other DR tasks. For instance, segmentation is important to perform load aggregation, evaluate their DR potential and adjust their CBL through PBLM [36].

STLF and CBL also provide inputs for PBLM toolboxes (e.g. the change in customer behaviour due to market prices) and are common tools for day-to-day operation of aggregators. It is also worth mentioning the linkage in the opposite sense (from PBLM to CBL) which is further considered in the adjustment of CBL.

14.3.2 Physical-based load models

PBLMs are considered as grey-box or white-box models for end-use loads (heating, cooling, ventilation, water heating, etc.). The advantage of these models is that they consider the physical laws of the loads and their environment to determine the behaviour and changes of the system they serve. The model defines an energy and flow balance between heat gains, heat losses and generation, the storage capacity, the energy conversion and thermal or product flows. The model is tuned through real data measurements or NIALM, including performance tests with control and response, to cover energy responses of actual or load capabilities through DR policies.

The advantage of this kind of models is that they can evaluate the effect of changes in inputs, parameters or state variables. For example, weather sensitivity can be evaluated with more accuracy than some adjustment methods. This gets a more accurate and flexible response than other adjustment methods (i.e. black-box models). More detailed information about the formulation of equations, the thermal-electric equivalent and the physical meaning of the parameters of these models, as well as research papers and scripts, can be found in [21]. Such reference also includes simulation examples, coefficient values and data for several end-uses such as electric heating, water heating or HVAC.

14.3.3 Unadjusted customer baseline load: a review of the main methodologies

There are many works dealing with different CBL approaches. Some methodologies can provide excellent results for specific segments, DR products, markets and situations. However, they may fail in other scenarios [12] such as small/medium customers, prosumers, customers with a high share of WS loads, loads that vary from day to day in a consistent pattern (some industrial facilities) or customers that usually respond to DR calls. Literature agrees that "traditional" baseline methods are a sufficient approach for obtaining a good and simple basis to develop CBLs. Although it might not be the most suitable one for a specific customer, it can be recommended for a set or segment of customers providing high accuracy. Therefore, it is important that regulators provide different methods to compute the CBL to increase customer motivation to engage DR policies. This possibility is usually deployed by operators and utilities (e.g. block exchange notification of DR mechanisms, NEBEF 3.1, in France [40]).

US power systems have a significant experience in DR and consequently with CBL methodologies. For this reason, NAESB has defined five types of methodologies [17]: maximum base-load (MBL); baseline type-I; baseline type-II; meter before/meter after and metering generation output.

Baselines type-I and type-II have been adopted as default methodologies by several SOs [11,12]. Baseline type-I is based on historical demand meter data, which may also include other variables such as weather and calendar data. Baseline type-II assumes the same idea, but it uses statistical sampling to estimate the aggregated consumption. With the increase in the deployment rate of smart meters [5,18], this last type of methodology lacks its practical interest. Other methodologies like MBL are more straightforward since they attempt to reduce the consumption of a demand-side resource to a specific level, regardless of its demand before the deployment. Below, some of the main CBLs in the literature are described.

14.3.3.1 Maximum base-load

It is also known as "non-baseline" or firm service level in some systems [9]. MBL is based on a demand resource's ability to reduce its demand to a specified level: the firm service level. The operator defines the annual peak of demand and the customer contribution to this peak. The difference between these values (contribution to peak and MBL) is the flexibility of demand. The method is used for emergency DR. The customer should keep its demand below this MBL level to avoid some penalties. The advantage is that revenue is easily understandable and DR objectives are fixed in advance.

14.3.3.2 Baselines based on historical data (type I)

This type is the most usual and understandable approach in most systems. The main methods are as follows:

(a) *Y-day simple average method:* To predict the CBL, it uses the average demand over the *Y* most recent non-DR days immediately before the DR event being considered. Usually 5-day and 10-day basis are used for this estimation [35].
(b) *Comparable day method:* This considers historical demand data to compute the CBL. In this case, the method only takes one day that is selected for its similar conditions with the event day (weather forecast, day of the week, etc.). If sufficient relevant factors are not taken into account, the forecast will be erroneous.
(c) *High (Middle/Low) XofY baseline:* The baseline is obtained by averaging recent historical data for a specific time interval. It considers the consumption of *Y* non-DR days preceding the DR event and uses the average of the *X* days with the highest (or middle/lowest) demand within those *Y* days. Some days are excluded, by so-called exclusion rules [19], because operators assume that some variables can modify the pattern of demand (e.g. day before a DR event, when SO calls aggregators and customers). Up to 30 or 60 days (look-back window) can be used to define *Y* [3], but shorter periods such as 5 or 10 days are more common in the literature because long periods could include changes in the consumption

pattern. By far, HighXofY is the most common baseline if DR events are due to peak load periods, whereas Middle/LowXofY baselines are more suitable in spring or fall seasons with no high peaks and for other markets and services like ancillary markets. Some practical examples of these baselines (for peak load shaving on peak days) are High5of10 in California SO [12] and in New York SO [41] or High15of20 in IESO, Canada [41]. These unadjusted CBLs are calculated as follows:

$$\mathrm{CBL}_{\mathrm{XofY}}(d,h) = \frac{1}{X}\sum_{i=1}^{X} A(i,h) \tag{14.1}$$

where $\mathrm{CBL}_{XofY}(d,h)$ denotes the baseline at time h of day d; $A(i,h)$ is the actual load for the ith highest (middle/lowest) energy day, at time h, among the previous Y non-event days, and X is the number of the highest (middle/lowest) days to be averaged in Y after exclusions. Some examples of exclusion are weekends, holidays, event notification days, facility closures, economic participation days [4] or previous DR event days.

(d) *Nearest XofY baseline:* The peculiarity of this method is that it focuses on the total consumption outside the DR event window to determine which of the Y previous days are more similar to the event day. The X days (among the original Y previous days) having the closest total demand to the DR day under consideration are selected. Then, the baseline is computed as the average load of these X days, now including the DR event window. In this method, exclusion days are also applied, for instance days with less or more than a percentage (e.g. 25%) of average load during the window of DR periods.

(e) *Weighted average method:* This is based on a weighted average of the previous day's CBL (some of the above CBL_{XofY}). During DR event days, CBL is defined as the previous day one. In cases where there is not a computed CBL, CBL is the simple average hourly load calculated for each hour of the day from the five more recent workdays with complete meter data with different weights for each day [6]. This method is used by KPX [7]. In this case, among the 10 reference days, the highest two days and lowest two days are excluded and the remainder are used but with different weights. For example, KPX defines weights such as 0.10, 0.15, 0.15, 0.15, 0.20 and 0.25 ranging from the older demand to the most recent demand.

(f) *Exponential moving average:* This is a weighted average of customer historical demand, where the weight decreases exponentially with time, that is it is the same procedure as of that CBL_{XofY} but the days before DR deployment have different weights, and it considers a broader spectrum for "X" days.

(g) *Control group methods:* This method considers the possibility that the aggregator (distribution system operators, DSOs, or SOs) has a database with load curves of other non-responsive customers (in the same segment of residential, commercial or industrial customers) from the DR event day. The customers are clustered into homogeneous groups and the DR customer's load curve is matched to one of these groups. Then, the CBL is calculated by averaging the load curves in the selected cluster in per unit. More complex methods can use a weighted combination of load curves in the cluster, or the load curves of the same customer and other customers on non-DR event days [37].

(h) *Short-term load forecasting methods:* These types of baselines comprise a wide range of methods. For example, the CBL can be built using a customer-specific regression model, which usually considers historical loads, weather conditions (temperature, humidity, etc.) and calendar features (holidays, season, day of the week, etc.). In [35], the baseline is estimated for campus buildings and with a high penetration of HVAC loads. For that, a linear model is fitted over the 5-minute period just before the DR event and the 5-minute period immediately following the set time. Another example is the use of ANN to estimate the CBL, as in [25, 42], where a back propagation neural network is adapted to establish baselines in public buildings (South Korea and China), taking into account meteorological indices. Some other techniques such as self-organizing maps and K-means can be used to find the most similar load patterns to the day of the DR event. As we mentioned before in this chapter, the close relationship between baseline estimation and STLF models allows that many others methods, like random forest, could be used as an approach to compute the CBL.

14.3.3.3 Meter before–meter after

This baseline is an evaluation method where electricity demand over a prescribed period prior to deployment is compared to similar readings during the sustained response period in previous days (X). It not only uses historical data but also uses the current load (day d, time $h-1$) to predict the next values (d, h) for the baseline. An example of this CBL is used in [6]. The CBL is calculated as

$$\mathrm{CBL}_{\mathrm{XofY}}(d,h) = A(d,h-1) + \frac{1}{X}\sum_{i=1;i\neq d}^{X}\left(A(i,h) - A(i,h-1)\right) \tag{14.2}$$

14.3.3.4 Metering generation output

According to NAESB definition [17], this "baseline" is "a performance evaluation method, used when a generation asset is located behind the demand resource's revenue meter, in which the demand reduction value is based on the output of the generation asset". If there is a meter in the generation "behind the meter", the baseline is zero and the flexibility is the reading of the generators which reduces the overall demand. Another problem is the consideration of prosumers with renewable generation, mainly PV and wind, which can be problematic due to volatility and weather dependency of their generation. This generation "behind the meter" is very volatile because it depends on high-variability weather factors such as cloud cover or wind speed and direction. This makes it more dependent on the weather than the own prosumers' load.

PV systems are the most widespread in the residential sector. In these systems, the movement of clouds can cause the power generated to vary by more than 50% in a few minutes or even seconds. The generation values in large PV facilities are usually measured directly, but this is not the case for PV residential systems. Since digital smart meters only measure net load (or net generation when it exceeds load), the value of the energy generated by a residential PV system at any given time is not accessible for the aggregator (nor for the DSO), which complicates the establishment of the CBL, as it requires an estimate or disaggregation of PV generation.

The disaggregation has been addressed with model-based and data-driven methods. Model-based methods usually require data information about the characteristics of the customer PV installation and weather forecasts in order to estimate PV power generation, although [43] proposes a model capable of estimating the PV system characteristics from net load measurements. Reference [39] uses NIALM data to disaggregate PV generation, identify turned-on appliances and estimate critical and non-critical loads. However, some of the aforementioned baselines could be used but the volatility of renewable sources needs to be considered (i.e. the selection of day-matching series must include days which did not suffer great variations in climatic conditions, wind and solar radiation, with respect to DR event days). The use of a PV generation disaggregation method significantly improves the results in the CBL estimation over those obtained with methods based on historical data if the days have not been carefully selected [44].

As a recommendable procedure, any of the previous approaches could be applied as a first stage to obtain a baseline and later, in a second stage, some adjustment coefficients can be computed by means of PBLM to improve the final CBL. In the rest of the chapter, some of those adjustment coefficients are presented and the complete procedure is applied to real data.

14.3.4 Adjustment coefficients for CBL

Baseline can be improved using adjustment methods. There are two main methods: multiplicative and additive adjustment. The objective of these adjustments is to modify the preliminary CBL to adapt it to weather and demand conditions on the DR event day. The easiest way to evaluate these factors is the use of pre-event DR data and then calibrates the baseline using the observed non-event hours prior to DR periods. In [10], the adjustment factor is defined by

$$\mathrm{amf}\,(d) = \frac{\sum_{k=1}^{a1} A(d, h_0 - (b1 + k))}{\sum_{k=1}^{a1} P(d, h_0 - (b1 + k))} \tag{14.3}$$

where amf(d) denotes the adjusted multiplicative factor for the day d; $A(d, h)$ is again the actual load of day d at time h; $P(d, h)$ is the predicted load (from unadjusted baseline or STLF methods [29]) of day d at time h; h_0 is the start time of the DR event; $b1$ is the buffer time and $a1$ is the length of the pre-adjustment band (Figure 14.2). Then, the new CBL is evaluated by

$$\mathrm{CBL_{adj}}\,(d, h) = \mathrm{amf}\,(d) \times * \,\mathrm{CBL}\,(d, h) \tag{14.4}$$

Alternatively, an addition adjustment factor (adf), reflecting the same concept, can be used instead of a multiplicative one. For example, the so called symmetric additive adjustment (SAA; for instance in South Korea and Australia [15], $b1 = 1$ h and $a1 = 3$ h):

$$\mathrm{SAA}(d) = \mathrm{adf}(d) = \max\left\{\frac{\sum_{k=1}^{a1}(A(d, h_0 - (b1 + k)) - P(d, h_0 - (b1 + k)))}{a1}, 0\right\};$$
$$\mathrm{CBL}_{\mathrm{adj}}(d, h) = \mathrm{adf}(d) + \mathrm{CBL}(d, h) \tag{14.5}$$

Some SOs use pre- and post-DR adjustment factors combined in the same baseline [12]. Other approaches [6] use adjustment factors which consider temperature ratios. In this way the weight of WS loads is included. The idea is that the post-event factor gives additional information about the boundary conditions throughout the DR event day (e.g. weather changes that modify demand). CAISO Baseline Accuracy Work Group (BAWG) justifies this approach to avoid contamination of baseline for both pre-cooling and snapback periods to occur in the hours directly before and after the DR event (Figure 14.2, pre- and post-DR buffer period). BAWG recommends a two-hour buffer before and after DR. The problem is that the duration of this buffer is not justified from the point of view of end-use dynamics. BAWG reports changes in the adjustment coefficient of around 3–4% using both periods [12].

Figure 14.2 depicts this idea of using pre- and post-adjustment periods: the DR period ranges from 8:00 to 13:00 and the pre-adjustment period uses data from 5:00 to 7:00 ($a1 = 2$ h) while the post-adjustment period uses data from 15:00 to 18:00 ($a2 = 3$ h). Note that these periods do not overlap. Besides, the consideration of two periods (pre-buffer $b1 = 1$ h and post-buffer $b2 = 2$ h) limits the possibility of perturbations like gaming just before the baseline. Some of these buffers are applied in several systems in the United States (for instance, NYISO [45] which uses a two hours buffer $b1$). It seems necessary that the definition and the duration of both "buffers" should be justified by load behaviour, through load modelling, which is the approach proposed in [36]. In Figure 14.2, n represents the period in which the consumption is affected by the DR event, that is, the sum of pre-DR period and the post-buffer period.

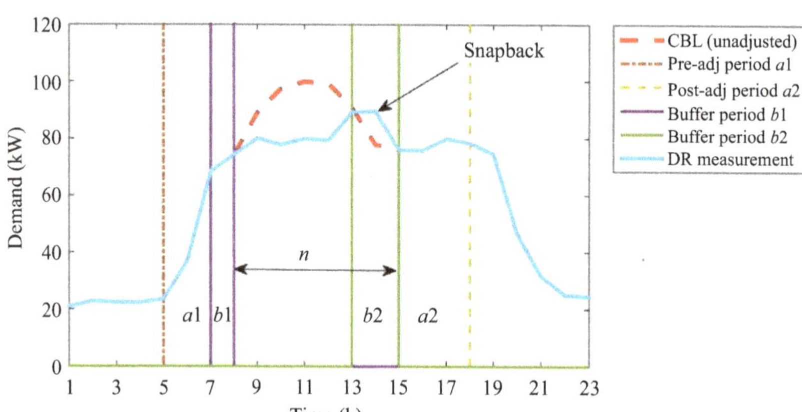

Figure 14.2 Example of baseline adjustment and periods being used in (14.3)

14.3.4.1 Gaming issues

The process of estimating a CBL can be open to manipulation. A change in demand during the adjustment period determines a variation in baseline results, so there can be economic motivations for some DR participants to change their load patterns to generate greater revenue. Some authors report [46] that CBLs are subject to manipulation because participants (customers, aggregators) should theoretically have a greater awareness of their end-uses and consumptions than the power system agents responsible for defining the baselines. For this reason, some CBL approaches include a cap in the adjustment factors from ±20 to ±40%. This involves a limit on the amount by which the CBL can be adjusted to account for differences in patterns of consumption in the very short term. Some authors argue [15] that the use of larger factors can create enough incentives and opportunities for gaming.

Despite gaming possibility, it is reported that customers increase or decrease their loads in response to advance notifications [4] with the aim of maintaining a minimum load service. For example: pre-heating and pre-cooling policies, or demand price elasticity, should be considered in the definition of CBL [38]. In this scenario, it is interesting to analyse the length and proximity of the pre-adjustment ($a1$, Figure 14.2) and the buffer period ($b1$, Figure 14.2) but based on the physical response of loads. PBLM can help to distinguish between gaming and pre-heating or pre-cooling policies [36]. NIALM (or sub-metering) can help in this task, even more precisely, because it can extract elemental load patterns in the aggregated demand, defining which load has changed and how this change happens during the event day. PBLM can classify load changes appointed by NIALM and attribute these changes to a normal or abnormal customer reaction.

14.4 Case study

To test and describe the methodologies presented in Section 14.3, a representative customer has been chosen in the food industry sector. Two main reasons support this choice. First, the food industry is a recurrent sector that has applied DR policies in the past with some degree of success [22]. Second, customer demand in this segment includes a significant share of WS loads, such as electric defrost, refrigerated warehouses, HVAC in offices, cooling production and distribution. Specifically, they represent more than 30% of the overall demand in this type of customers.

Figure 14.3 shows two examples of the weekly demand: the overall load of the facility and a controllable load selected for DR simulation and CBL evaluation, which corresponds to offices, product processing and cut-up packaging.

14.4.1 Detecting pre-heating and gaming through PBLM and NIALM

The choice of adjustment values in (14.3) can be improved through the justification of both the adjustment and buffer periods (Figure 14.2) using PBLM. Moreover, the use of other DR methodologies (Figure 14.1) helps in searching and filtering

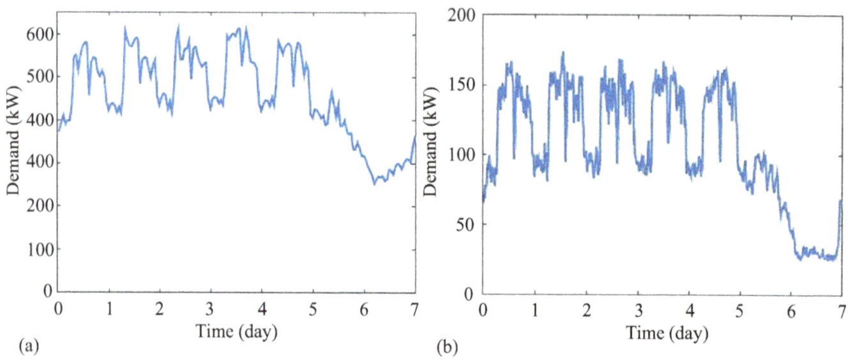

Figure 14.3 Weekly demand of the customer: (a) overall load of the facility and (b) controllable load selected for simulation purposes

some potential disturbances in the evaluation of CBL whereas improving conventional adjustment coefficients, that is a normal pattern to maintain load service during DR such as pre-heating, can be distinguished from "gaming" attempts.

To exemplify the method, some simulations have been performed and the obtained results are shown in Figure 14.4. A group of homogeneous HVAC loads of administration and engineering offices under different weather conditions was simulated (Figure 14.4a), which is the main reason for the adjustment in the demand of WS loads. The choice of this group is based on the relevant share of HVAC loads for many other customer segments.

The load group performs during the winter period and, to consider weather fluctuations, three scenarios have been simulated: the outdoor average temperature, the average minus 3°C and average minus 6°C. A "service function" of HVACs starting at 6:30 and finishing at 13:30, and restarting from 14:30 to 20:30, has been chosen. Figure 14.4a represents the results for the group of loads after running PBLM scripts, see [21] for additional information.

It can be observed that demand is similar at the start of duty cycle of these loads (from 6:30 to 8:00) because all loads are near their full demand, irrespective of the weather, to reach the temperature setpoint and indoor comfort parameters; however, there is difference in a second period (from 8:00 to 13:30 and 14:30 to 20:30) when some loads reach their "steady-state" service, that is thermostat setpoints, depending on weather, service and internal thermal loads (e.g. lighting, machinery and occupancy). These variations in demand justify the definition of periods taken in (14.3) for the evaluation of adjustment factors and explain the failures of adjustment periods for some WS-CBL approaches [45]. The reason is that traditional adjustment may consider a wrong daily period where demand is less dependent on weather. Figure 14.4a also depicts the changes in demand due to pre-heating in morning and afternoon periods. This pre-heating period can be detected by a demand decrease after the pre-heating period with respect to "normal" patterns.

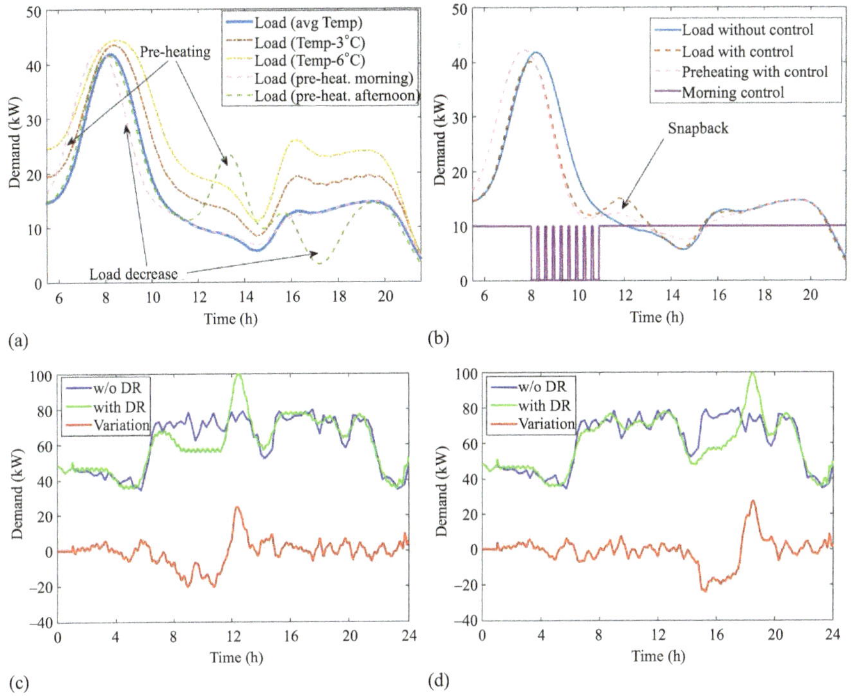

Figure 14.4 Simulation of flexible loads: (a) homogeneous group with different boundary conditions; (b) homogeneous group under control; (c) heterogeneous group with DR in the morning; and (d) heterogeneous group with DR in the afternoon

Figure 14.4b shows a control in the same group from 8:00 to 11:00, considering pre-heating and normal conditions. The necessity of a buffer is justified by some utilities because the buffer period (two hours before event operators [9,12]) is a potential "gaming period" and distorts DR potential, but sometimes the buffer hides actual load changes and service for the customer. The size of snapback helps in the detection of pre-heating, but obviously NIALM or sub-metering is needed for the purposes of checking PBLM results (Figure 14.1).

The response of a heterogeneous group of flexible loads given in Figure 14.3b has also been simulated through PBLM, selecting two different events: one in the morning (from 8:00 to 12:00) and one in the afternoon (from 14:00 to 18:00). During these periods, a forced control cycle to reduce around 20% of the HVAC demand has been considered, a reasonable percentage for avoiding a lack of comfort or to comply with thermal limits for manufacturing and conservation processes. Figure 14.4c and Figure 14.4d presents the results for the two events considered.

After the DR event (see Figure 14.4), there is again an increase in the demand as a response to the control. In the case of the morning event, the increase occurs from 12:00 to 14:00 and in the evening period from 18:00 to 20:00. During the snapback, loads try to recover the previous reduction of demand by increasing their consumption to reach their "normal" operating point. PBLM allows the load behaviour of a DR event to be understood, improving the performance of CBLs. At the same time, PBLM helps to define the duration and magnitude of the recovery periods and potential secondary peaks that can be a concern for distributors and balance responsible parties (BRPs).

14.5 Results and discussion

14.5.1 Comparisons of unadjusted CBLs based on historical data

To illustrate the differences of several unadjusted methods, six different baselines among those described in Section 14.3.3 have been selected: the High5of10, the Low5of10, the Mid4of6, the Nearest5of10, the mid6of10-weighted-average (labelled Weighted average) and, finally, a baseline (labelled Random forest) obtained through STLF (random forest [29]).

A database for an industrial load has been used, with hourly consumptions from 1 July 2019 to 30 June 2021. The random forest method requires a long-enough historical dataset for the training stage (in this case, all data from 2019 and 2020), so the common period for evaluating the six CBL methods was 2021.

Regarding the error metrics, the mean percent error (MPE) has been selected to describe the magnitude and direction of the estimation bias. MPE reflects the percentage by which the baseline, on average, over- or underestimates the "true demand" in the absence of a DR event. To evaluate the precision, both the mean percent average error (MAPE) and the normalized root mean squared error (nRMSE) have been selected. The lower MAPE and nRMSE are, the more precise the baseline is. Note that metrics are defined through relative errors, so they can be used to compare accuracy and precision of CBLs measured in different scales. Mathematically, these metrics are defined as follows [12]:

$$\text{MPE} = \frac{100}{n}\frac{\sum_{i=1}^{n}(y_i - \hat{y}_i)}{\bar{y}} \tag{14.6}$$

$$\text{MAPE} = \frac{100}{n}\sum_{i=1}^{n}\left|\frac{y_i - \hat{y}_i}{\bar{y}}\right|; \qquad \text{nRMSE} = \frac{\sqrt{\frac{1}{n}\sum_{i=1}^{n}(y_i - \hat{y}_i)^2}}{\bar{y}} \tag{14.7}$$

where y_i is the real demand at time i, $\hat{y}_i$ is the CBL (forecasted demand) at time i and $\bar{y}$ is the mean of the real demand for the n values. Note that n refers to the length of the DR evaluation period.

These error metrics were obtained for the whole evaluation period (from 1 January 2021 to 30 June 2021) and all the methodologies (Table 14.1). As it can be deduced from Table 14.1, the two methods that provide more accurate baselines during the whole period analysed are the Nearest5of10 and the Random forest, but with small differences with respect to the rest of the methods. It is also remarkable that all the baselines overestimate the load (they have positive values of MPE) except the Low 5 of 10 that underestimate the load.

Table 14.1 Error metrics of unadjusted CBLs in the whole evaluation period

CBL	MAPE (%)	MPE (%)	nRMSE (pu)
High5of10	11.1	6.5	0.1345
Low5of10	12.2	−6.1	0.1569
Mid4of6	9.8	1.1	0.1263
Nearest5of10	7.2	1.1	0.0992
Weighted average	9.8	1.0	0.1268
Random forest	8.8	2.5	0.1163

Table 14.2 Error metrics of unadjusted CBLs in DR periods

CBL	MAPE (%)		MPE (%)		nRMSE (pu)	
	Morning	Afternoon	Morning	Afternoon	Morning	Afternoon
High5of10	13.5	10.9	8.9	6.5	0.1377	0.1124
Low5of10	14.2	12.6	−6.3	−6.2	0.1458	0.1300
Mid4of6	11.6	9.6	2.7	1.3	0.1194	0.1006
Nearest5of10	8.6	7.2	2.3	1.2	0.0905	0.0764
Weighted average	11.2	9.5	2.3	1.3	0.1156	0.0989
Random forest	10.9	9.0	3.5	2.1	0.1130	0.0958

Although Table 14.1 refers to the whole period under study, the main common and practical use of CBLs arises in peak-shaving events. The customer/aggregator revenue is directly related to this measurement, so CBLs need to be accurate especially in these peak periods. In our study, we have considered two different DR periods for every single day in the evaluation stretch: one in the morning (from 8:00 to 12:00, usually a peak price period in the Spanish energy market) and one in the afternoon (from 14:00 to 18:00). Table 14.2 shows the metrics for DR periods of the unadjusted baselines (data from 1 January 2021 to 30 June 2021). In the case of the morning DR period, the errors slightly increase compared to the global metrics (Table 14.1) whereas in the afternoon period the metrics are practically the same. In both cases (morning and afternoon DR periods), the most accurate CBL methodologies are Nearest5of10 and Random forest, results that correspond to those obtained in the whole period.

Figure 14.5 gives an example of how the different methodologies can work properly in one part of the day (afternoon period) but not another part (morning period). This problem can be reduced with the use of the adjustment coefficients proposed in Section 14.3.4.

14.5.2 Adjustment coefficients: weather sensitive (WS) and PBLM

To improve the performance of the CBLs, SOs use adjustment coefficients that are calculated with data from the previous hours to the DR event. These coefficients

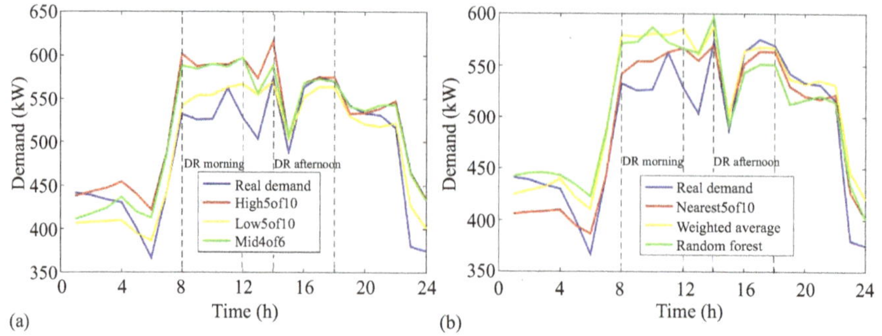

Figure 14.5 Unadjusted CBLs vs. real demand on a specific day (30 April 2021) and DR periods

Table 14.3 Definition of the adjustment coefficients for the CBLs analysed

Adjustment coefficient	Data	Period
Weather sensitive (WS)	Pre-DR	First 2 hours of the 4-hour period pre-DR event
Physically based (PBLM)	Pre-DR	Last 2 hours of the 4-hour period pre-DR event (defined by PBLM)
Backward (BW)	Post-DR	Last 2 hours of the 4-hour period post-DR event (defined by PBLM)

have been explained in Section 14.3.4. In our case, two different multiplicative pre-adjustment coefficients have been evaluated: WS and PBLM. In addition, another adjustment; using data from the hours after the DR event, is proposed, that is a post-adjustment coefficient, known as backward (BW) adjustment. Table 14.3 specifies the periods where the adjustment coefficients are calculated.

14.5.2.1 WS (NYISO)

The WS coefficient is commonly used by NYISO. The two first hours of the four-hour period before the start of the DR event are used. That is, in Figure 14.2, the buffer $b1$ has a 2-h duration and the period $a1$ in which the coefficient is calculated also lasts 2 h.

In the case of the defined DR event in the morning, whose duration is from 8:00 to 12:00, the coefficient is calculated in the period from 4:00 to 6:00. Table 14.4 depicts the error indexes for the two DR periods. Note that the adjustment coefficient slightly improves the errors for the morning period and that the reduction of MAPE is higher when the method is more imprecise. The High5of10 and the Low5of10 are the two methods demonstrating a greater improvement.

Figure 14.6 plots the WS-adjusted baselines, the real demand for two representative days of the 2021 database (21 January 2021 and 30 April 2021) and the two DR

Table 14.4 Error metrics of adjusted WS-CBLs in DR periods

CBL	MAPE (%)		MPE (%)		nRMSE (pu)	
	Morning	Afternoon	Morning	Afternoon	Morning	Afternoon
High5of10	9.3	5.7	1.7	−1.2	0.0980	0.0618
Low5of10	10.2	6.8	1.4	1.4	0.1067	0.0729
Mid4of6	9.8	5.8	2.5	−0.1	0.1026	0.0632
Nearest5of10	9.1	5.6	2.1	−0.1	0.0964	0.0615
Weighted average	9.6	5.6	2.5	0.2	0.1009	0.0603
Random forest	9.7	6.0	1.0	−0.4	0.1018	0.0654

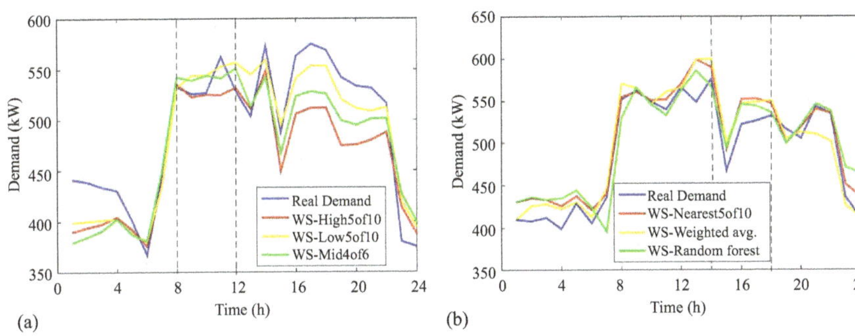

Figure 14.6 WS-adjusted CBLs vs. real demand (a) in a DR morning event for a specific day (30 April 2021) and (b) in a DR afternoon event for a specific day (21 January 2021)

events (morning and afternoon). It can be observed that the WS adjustment coefficient improves the precision of the unadjusted baselines (Figure 14.5a).

In the case of the defined DR event in the afternoon, whose duration is from 14:00 to 18:00, the adjustment coefficient is calculated in the period from 10:00 to 12:00. As can be seen in Figure 14.6, the load presents more variability in the afternoon DR period than in the morning one, therefore the adjustment coefficient works better for the improvement of the CBL curves. Note that, in the afternoon DR period, the error is reduced from 7 to 12% to around a 6% of MAPE for all the methods (Table 14.4).

14.5.2.2 PBLM

In this case, the duration of the period $a1$ and the buffer $b1$ are determined from the analysis of the physical behaviour of the load. In our study, we have selected the 2 hours before the start of the DR event, without any buffer period. Although the use of a buffer period is sometimes necessary to avoid gaming, in other cases the buffer

Table 14.5 Error metrics of adjusted PBLM-CBLs in DR periods

CBL	MAPE (%)		MPE (%)		nRMSE (pu)	
	Morning	Afternoon	Morning	Afternoon	Morning	Afternoon
High5of10	5.5	4.3	1.5	−0.1	0.0601	0.0476
Low5of10	6.3	4.5	0.1	0.2	0.0679	0.0502
Mid4of6	5.6	4.3	1.4	0.1	0.0611	0.0483
Nearest5of10	5.5	4.2	1.0	−0.1	0.0602	0.0471
Weighted average	5.5	4.2	1.4	0.2	0.0598	0.0472
Random forest	6.1	4.9	1.1	−0.1	0.0666	0.0549

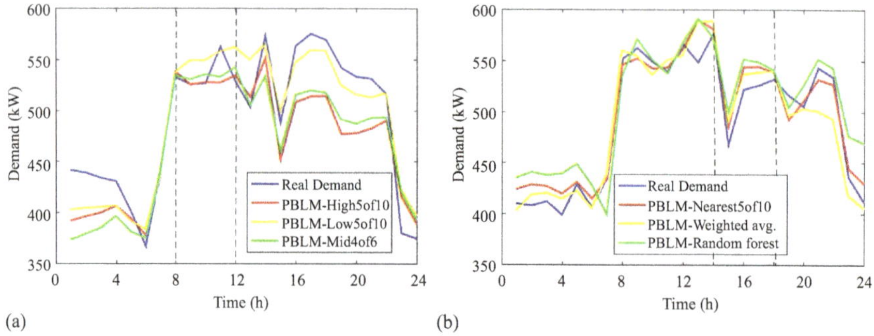

Figure 14.7 PBLM-adjusted CBLs vs. real demand (a) in a DR morning event for a specific day (30 April 2021) and (b) in a DR afternoon event for a specific day (21 January 2021)

hides the real behaviour of the loads. As it has been explained in Section 14.3.4.1, gaming and pre-heating or pre-cooling can be detected through PBLM simulations, so the $a1$ period and the buffer $b1$ could be changed if any of these behaviours are detected.

The PBLM adjustment coefficient for the DR morning event is consequently calculated in the period from 6:00 to 8:00. Table 14.5 shows that PBLM adjustment coefficient performs better than the traditional WS coefficient, reducing the MAPE error to approximately 6% in all the traditional CBL methodologies, a percentage of error lower than the one obtained with the WS adjustment coefficient. For the afternoon event, the PBLM coefficient is calculated in the period from 12:00 to 14:00. In this case, the improvement in the precision of the CBLs is similar to those obtained with the WS coefficient, because the periods for calculating both coefficients adequately reflect the changes in the load behaviour. However, the PBLM still improves the MAPE error slightly, from around 6% to 4.5%.

Figure 14.7 plots the PBLM-adjusted baselines and the real demand. The PBLM adjustment improves the precision of unadjusted CBLs (Figure 14.5a).

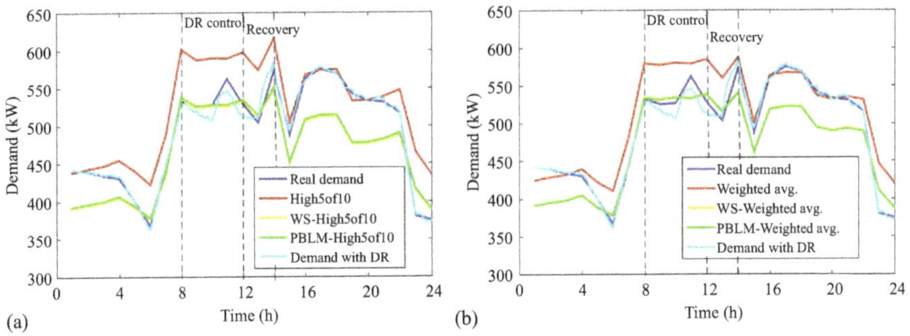

Figure 14.8 Comparison of unadjusted and adjusted CBLs in a DR morning event for a specific day (30 April 2021): (a) High5of10 and (b) weighted average

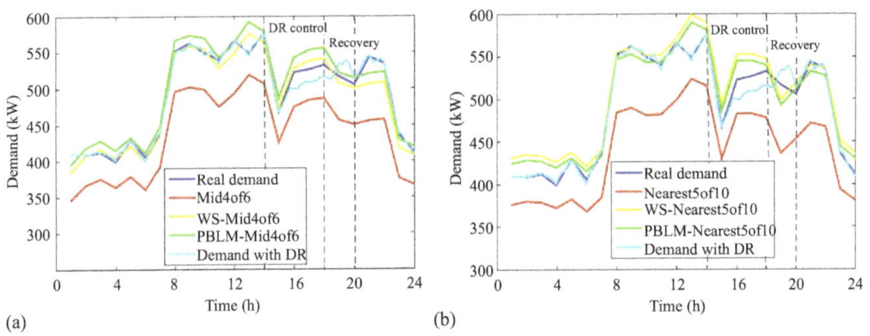

Figure 14.9 Comparison of unadjusted and adjusted CBLs in a DR afternoon event for a specific day (21 January 2021): (a) Mid4of6 and (b) Nearest5of10

Figures 14.8 and 14.9 show the comparative among the unadjusted, WS and PBLM baselines with respect to the real demand for two specific days and two DR periods. The benefits of using the adjusted factors are demonstrated.

14.5.2.3 Backward adjustment coefficient (BW)

The BW adjustment is a post-adjustment coefficient that makes sense for energy recovery periods that occurs after a DR event. It is important to consider that energy savings reported during DR control periods are partially recovered after control (payback time: from 12:00 to 14:00 in a morning DR event and from 18:00 to 20:00 in an afternoon DR event). Thus, it is necessary to study this "energy recovery period".

Hence, PBLM provides another important insight to define the BW adjustment coefficient. As the load does not recover its "steady-state" demand until some hours after the control, PBLMs help to define the duration of the recovery period (see Section 14.4).

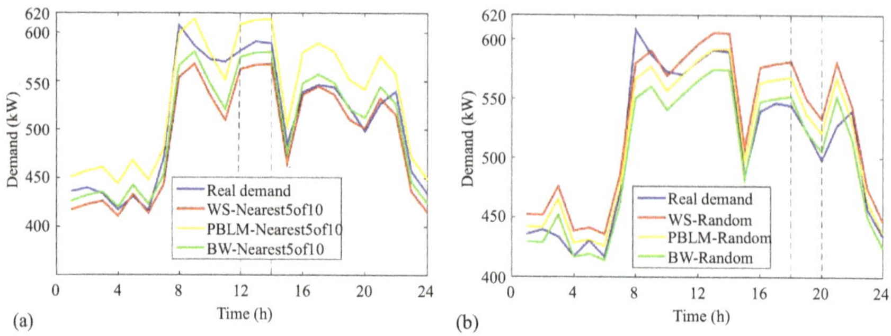

Figure 14.10 BW-adjusted CBLs vs. other adjusted CBLs in DR events for a specific day (26 January 2021): (a) Nearest5of10, morning event; (b) random forest, afternoon event

Table 14.6 Error metrics of adjusted BW-CBLs in DR recovery periods

CBL	MAPE (%)		MPE (%)		nRMSE (pu)	
	Morning	**Afternoon**	**Morning**	**Afternoon**	**Morning**	**Afternoon**
High5of10	3.1	3.0	0.3	0.4	0.0379	0.0361
Low5of10	3.2	2.8	0.1	0.1	0.0385	0.0346
Mid4of6	3.1	2.9	0.1	0.1	0.0377	0.0348
Nearest5of10	3.0	2.8	0.2	0.2	0.0372	0.0342
Weighted average	3.1	2.7	0.2	0.1	0.0369	0.0336
Random forest	3.8	3.2	0.4	0.1	0.0447	0.0391

In our study, adjustment values are taken not at the end of control, but in the 2-h period that starts 2 h after the end of the event, that is, at the end of the recovery period defined by PBLM. Notice that the adjusted CBLs obtained with this BW adjustment are only valid from the end of the DR event to the end of the day, and therefore, calculating the energy savings during DR periods with these BW-adjusted CBLs does not make sense. The main goal of these adjusted CBLs is to calculate the energy payback, because the increase in demand can be important for SOs, producing new peaks after events.

In the case of the DR morning event (Figure 14.10a, from 8:00 to 12:00), the BW adjustment coefficient is calculated in the period from 14:00 to 16:00 and these CBLs are only valid from 12:00 to 24:00. For the DR afternoon event (Figure 14.10b, from 14:00 to 18:00), the BW adjustment coefficient is calculated in the period from 20:00 to 22:00, and these CBLs are only valid from 18:00 to 24:00. Table 14.6 shows the error indexes for all the CBL methodologies in the energy recovery period (dashed lines in Figure 14.10). As it can be seen, the BW adjustment significantly reduces the MAPE index to less than 4 and 3% in the DR morning and afternoon events, respectively.

14.5.3 DR control events: effects on energy calculations

It is worthy to note the results presented in Sections 14.5.1 and 14.5.2 are values in terms of power and energy, but market prices can change from one hour to another during DR periods (control and snapback). For this reason, it is necessary to consider an economic evaluation of error, weighting the error based on market prices. This problem also deals with compensation mechanisms to demand-side (e.g. hourly LMP prices) which is an issue and a complex problem for regulators [47].

For instance, the energy recovery analysed in baselines increases its interest because DR also involves changes in the balance of energy which affects aggregators and third parties like BRPs. The assumption of several roles by aggregators can be an issue with regard to market regulations (e.g. article 102 of Treaty on the Functioning of the EU on dominant positions in internal markets [48]). For instance, France has decided that the aggregator should pay BRP/suppliers for these changes [48], and this impacts cost-effectiveness. Appropriate CBL designs that can consider these effects are needed. Changes in demand during those periods are also a concern for associations such as Eurelectric [8], which recommends improving aggregation development. PBLMs, BW adjustment and NIALM tools can contribute to solving some of these issues.

14.6 Conclusions

DR policies are necessary to make an increase of the renewable share in the future power generation mix credible. The verification of flexibility becomes a key issue. Baselines are a mainstay to provide the verification of DR and the subsequent revenue and payment. An accurate and precise CBL also emerges as an important tool for engaging and empowering customers in markets, for example for decision making in electricity markets, recognizing and giving credit for their real flexibility.

From the point of view of the demand-side, demand resources need an accurate and fair verification of their flexibility. In this way, the accuracy of demand forecasts with DR (PBLM modelling) and without DR (CBL), with incomes and penalties associated, is a crucial factor for the development of DR and subsequently the integration of distributed energy resources. Aggregators, suppliers, operators or BRP need accurate demand models since most of their decisions and their economic viability are based on demand forecasting and its flexibility. Specific and complex methodologies are proven to define accurate CBLs, but this option usually increases the complexity of DR and sometimes requires a different model for each customer. Moreover, these models can present problems if DR performs periodically and the customer changes its behaviour, or simply if the aggregator develops more complex products like the participation in several markets and services. Literature shows that unadjusted CBLs are not the best option, but they can improve their performance through adjustment factors. Until now, these factors have been proposed based on experience. This chapter highlights the convenience of using adjustment factors explained by PBLM, and that a double-adjusted CBL displays an even better performance. Thus, this methodology arises as an adequate and simple baseline estimator.

This work also states the benefits of synergies associated with the use of other aggregator tools, such as NIALM, segmentation and ICT resources to verify load response. In this case, the adjustment period can be justified and improved both before and after the period of DR events to improve the evaluation of DR and the changes for balance compensations among the different agents involved. In this way, different DR actors can obtain necessary feedback to perform a better evaluation of the DR potential and develop distributed energy resources (DER) in the 2050 horizon.

Acknowledgements

This work has been supported by the Agencia Estatal de Investigación, Ministerio de Ciencia e Innovación (Project RED2018-102618-T funded by MCIN/AEI/10.13039/501100011033) and the Ministerio de Educación (Spanish Government) under grant FPU17/02753.

References

[1] European Commission: *Directive (EU) 2019/944 of the European Parliament and of the Council of 5 June 2019 on Common Rules for the Internal Market for Electricity and Amending Directive 2012/27/EU*. Available from https://eur-lex.europa.eu/legal-content/EN/TXT/?uri=CELEX%3A32019L0944, accessed October 2021.

[2] FERC: *Demand Response Compensation in Organized Wholesale Energy Markets (Final Rule), Order 745, June 2011*. Available from https://www.ferc.gov/legal/maj-ord-reg.asp. accessed November 2021.

[3] EnerNOC: *The Demand Response Baseline. White Paper* (2009).

[4] The CADMUS Group: *Demand Response Program Annual Evaluation: Phase III of Act 129* (2018).

[5] Smart Energy Demand Coalition (SEDC): *Mapping Demand Response in Europe Today*. Available from http://www.smarten.eu/wp-content/uploads/2015/09/Mapping-Demand-Response-in-Europe-Today-2015.pdf, accessed October 2021.

[6] Segovia-Leon, E.J., Hunter, B.: 'Recommendations for baseline load calculations in DR programs' *eDREAM H2020 Proj. D3.2*, no date.

[7] Lee, S.: 'Comparing methods for customer baseline load estimation for residential demand response in south korea and france: predictive power and policy implications chaire european electricity markets'. *Work. Pap. 39*, no date.

[8] EURELECTRIC: 'Designing Fair and Equitable Market Rules for Demand Response Aggregation'. http://www.eurelectric.org, accessed July 2021.

[9] PJM: 'Demand response in PJM markets and operations', https://www.pjm.com/markets-and-operations/demand-response.aspx, accessed January 2020.

[10] Coughlin, K., Piette, M.A., Goldman, C., Kiliccote, S.: 'Estimating Demand Response Load Impacts: Evaluation of Baseline Load Models for

Non-Residential Buildings in California Environmental Energy Technologies Division', 2008.

[11] Lake, C.: 'PJM Empirical Analysis of Demand Response Baseline Methods'. https://www.pjm.com/-/media/markets-ops/demand-response/pjm-analysis-of-dr-baseline-methods-full-report.ashx?la=en, accessed July 2021.

[12] California ISO: 'Baseline Accuracy Work Group Proposal', http://www.caiso.com/Documents/2017BaselineAccuracyWorkGroupFinalProposalNexant.pdf#search=customer baseline, accessed July 2021.

[13] Willoughby, L., Bode, J., St, M., Francisco, S.: '2012 San Diego Gas & Electric Peak Time Rebate Baseline Evaluation Prepared for: San Diego Gas & Electric Prepared by: The FSC Group Table of Contents', 2013.

[14] conEdison: 'Advanced Metering Infrastructure Business Plan', http://documents.dps.ny.gov/public/Common/ViewDoc.aspx?DocRefId=%7B0864BF10-42C8-40AF-8E57-833CD4FE6B07%7D, accessed July 2021.

[15] Australia Renewable Energy Agency (ARENA): 'Baselining the ARENA-AEMO Demand Response RERT Trial (September 2019)', 2019.

[16] Rossetto, N.: 'Measuring the Intangible: An overview of the methodologies for calculating customer baseline load in PJM'. *Florence Sch. Regul.*, 2018, (2018/5), pp. 1–10.

[17] NAESB: 'Business Practices for Measurement and Verification of Wholesale Electricity Demand Response', 2009.

[18] European Commission Smart Grids Task Force: 'Demand Side Flexibility: Perceived Barriers and Proposed Recommendations', 2019.

[19] Goldberg, M.L., Kennedy Agnew, G.: 'Measurement and Verification for Demand Response Prepared for the National Forum on the National Action Plan on Demand Response: Measurement and Verification Working Group', 2013.

[20] Mohajeryami, S., Doostan, M., Schwarz, P.: 'The impact of customer baseline load (CBL) calculation methods on peak time rebate program offered to residential customers'. *Electr. Power Syst. Res.*, 2016, **137**, pp. 59–65.

[21] Gabaldon, A.: 'REDYD 2050 Research Network on Distributed Energy Resources web page'. http://www.demandresponse.eu/, accessed January 2021.

[22] Shoreh, M.H., Siano, P., Shafie-khah, M., Loia, V., Catalão, J.P.S.: 'A survey of industrial applications of demand response'. *Electr. Power Syst. Res.*, 2016, **141**, pp. 31–49.

[23] Wijaya, T.K., Vasirani, M., Aberer, K.: 'When bias matters: an economic assessment of demand response baselines for residential customers'. *IEEE Trans. Smart Grid*, 2014, **5**(4), pp. 1755–1763.

[24] Lee, E., Jang, D., Kim, J.: 'A two-step methodology for free rider mitigation with an improved settlement algorithm: regression in CBL estimation and new incentive payment rule in residential demand response'. *Energies*, 2018, **11**(12), p. 3417.

[25] Pati, S., Ranade, S.J., Lavrova, O.: 'Methodologies for customer baseline load estimation and their implications'. *2020 IEEE Texas Power Energy Conference TPEC 2020*, 2020.

[26] Jiang, H., Zhang, Y., Muljadi, E., Zhang, J.J., Gao, D.W.: 'A short-term and high-resolution distribution system load forecasting approach using support vector regression with hybrid parameters optimization'. *IEEE Trans. Smart Grid*, 2018, **9**(4), pp. 3331–3350.

[27] Niu, D., Wang, Y., Wu, D.D.: 'Power load forecasting using support vector machine and ant colony optimization'. *Expert Syst. Appl.*, 2010, **37**(3), pp. 2531–2539.

[28] Pai, P.F., Hong, W.C.: 'Support vector machines with simulated annealing algorithms in electricity load forecasting'. *Energy Convers. Manag.*, 2005, **46**(17), pp. 2669–2688.

[29] Ruiz-Abellón, M. del C., Gabaldón, A., Guillamón, A.: 'Load forecasting for a campus university using ensemble methods based on regression trees'. *Energies*, 2018, **11**(8), p. 2038.

[30] Nie, H., Liu, G., Liu, X., Wang, Y.: 'Hybrid of ARIMA and SVMs for short-term load forecasting', *Energy Procedia,* Elsevier Ltd, 2012, pp. 1455–1460.

[31] Karthika, S., Margaret, V., Balaraman, K.: 'Hybrid short term load forecasting using ARIMA-SVM', *2017 Innovations in Power and Advanced Computing Technologies, i-PACT 2017,* Institute of Electrical and Electronics Engineers Inc., 2017, pp. 1–7.

[32] Ray, P., Sen, S., Barisal, A.K.: 'Hybrid methodology for short-Term load forecasting', *2014 IEEE International Conference on Power Electronics, Drives and Energy Systems, PEDES 2014*, Institute of Electrical and Electronics Engineers Inc., 2014.

[33] Li, K., Wang, F., Mi, Z., Fotuhi-Firuzabad, M., Duic, N., Wang, T.: 'Capacity and output power estimation approach of individual behind-the-meter distributed photovoltaic system for demand response baseline estimation'. *Appl. Energy*, 2019, **253**, p. 113595.

[34] Zarnikau, J., Thal, D.: 'The response of large industrial energy consumers to four coincident peak (4CP) transmission charges in the Texas (ERCOT) market'. *Util. Policy*, 2013, **26**, pp. 1–6.

[35] Lei, S., Mathieu, J., Jain, R.: 'Performance of existing baseline models in quantifying the effects of short-term load shifting of campus buildings'. *SLAC-R-11*, 2019.

[36] Gabaldón, A., García-Garre, A., Ruiz-Abellón, M.C., Guillamón, A., Álvarez-Bel, C., Fernandez-Jimenez, L.A.: 'Improvement of customer baselines for the evaluation of demand response through the use of physically-based load models'. *Util. Policy*, 2021, **70**, p. 101213.

[37] Wang, F., Li, K., Liu, C., Mi, Z., Shafie-Khah, M., Catalao, J.P.S.: 'Synchronous pattern matching principle-based residential demand response baseline estimation: mechanism analysis and approach description'. *IEEE Trans. Smart Grid*, 2018, **9**(6), pp. 6972–6985.

[38] Gabaldón, A., Álvarez, C., Ruiz-Abellón, M., *et al.*: 'Integration of methodologies for the evaluation of offer curves in energy and capacity markets through energy efficiency and demand response'. *Sustainability*, 2018, **10**(2), p. 483.

[39] Dinesh, C., Welikala, S., Liyanage, Y., Ekanayake, M.P.B., Godaliyadda, R.I., Ekanayake, J.: 'Non-intrusive load monitoring under residential solar power influx'. *Appl. Energy*, 2017, **205**, pp. 1068–1080.

[40] RTE (France): 'Règles pour la valorisation des effacements de consommation sur les marchés de l'énergie NEBEF 3.1', 2018.

[41] ISO/RTO Council: 'North American Wholesale Electricity Demand Response Program Comparison', https://isorto.org/reports-and-filings/, accessed January 2020.

[42] Yang, C., Xu, Q., Wang, X.: 'Strategy of constructing virtual peaking unit by public buildings' central air conditioning loads for day-ahead power dispatching'. *J. Mod. Power Syst. Clean Energy*, 2017, **5**(2), pp. 187–201.

[43] Chen, D., Irwin, D.: 'SunDance: black-box behind-the-meter solar disaggregation'. *Proc. Eighth Int. Conf. Futur. Energy Syst.*, vol. 11, 2017.

[44] Xuan, Z., Gao, X., Li, K., Wang, F., Ge, X., Hou, Y.: 'PV-load decoupling based demand response baseline load estimation approach for residential customer with distributed PV system'. *IEEE Trans. Ind. Appl.* 2020, **56**(6), pp. 6128–6137.

[45] NYISO: 'Emergency Demand Response Program Manual', https://www.nyiso.com/demand-response, accessed January 2020.

[46] Jiang, B., Farid, A.M., Youcef-Toumi, K.: 'Demand side management in a day-ahead wholesale market: a comparison of industrial & social welfare approaches'. *Appl. Energy*, 2015, **156**, pp. 642–654.

[47] Chen, X., Kleit, A.N.: 'Money for nothing? Why FERC Order 745 should have died'. *Energy J.*, 2016, **37**(2), pp. 201–222.

[48] Schittekatte, T., Deschamps, V., Meeus, L.: 'The regulatory framework for independent aggregators'. *The Electricity Journal*, 2021, **34**(6), 106971.

Chapter 15

Modeling and optimizing the value of flexible industrial processes in the UK electricity market

Dimitrios Papadaskalopoulos[1,2], *Makedon Karasavvidis*[2], *Gerasimos Takis*[1], *Athanasios Botsis*[1] *and Anastasios Oulis Rousis*[2]

Despite its comparative advantages with respect to residential and commercial demand response (DR), industrial DR (IDR) in general, and modeling of different types of flexible industrial processes in particular, has received relatively limited research attention, with previous work having only explored limited and industry-sector-specific subsets of such processes. This chapter adopts an alternative, sector-agnostic modeling approach and develops generic models of all conceivable types of flexible industrial processes, with the aim to shed light on their key operating differences and assist industrial consumers interested in IDR schemes to identify and assess the types that are more relevant to their systems. In this context, this chapter identifies and discusses seven different types: (1) uninterruptible processes with fixed power, (2) interruptible processes with fixed power, (3) uninterruptible processes with discretely adjustable power, (4) interruptible processes with discretely adjustable power, (5) uninterruptible processes with continuously adjustable power, (6) interruptible processes with continuously adjustable power, and (7) material storage buffers.

For each of these types, a mathematical model capturing its techno-economic operating characteristics is developed, including: (a) input parameters, (b) decision variables, (c) operating constraints, and (d) inconvenience cost function. These models are integrated into the electricity costs' minimization problem of an industrial consumer participating in the energy market through a real-time pricing (RTP) scheme. A case study concerning an actual industrial consumer in the UK is investigated, employing its real demand data and a target industrial process as well as real price data from the UK wholesale market. Different scenarios are examined regarding the flexibility type of the target process, and the results demonstrate, quantify, and compare the economic benefits generated by different flexibility types. Specifically, the

[1]Decentralised Energy Solutions Ltd, London, UK
[2]Imperial College London, London, UK

yearly electricity cost savings are in the range between 10.71% and 23.13% with respect to the baseline electricity costs of the target process alone and between 1.58% and 3.41% with respect to the baseline electricity costs of the industrial consumer as a whole.

15.1 Introduction

15.1.1 Decarbonization challenges and value of demand response

Energy systems across the world are currently undergoing fundamental changes, mainly driven by the continuously increasing levels of greenhouse gases emission in the atmosphere and the associated climate change concerns. In response to such concerns, numerous governments have taken significant initiatives towards the decarbonization of energy systems. In the United Kingdom (UK), the most recent advice by the Committee on Climate Change (CCC) to the UK Government has recommended an objective to reduce the carbon intensity of power generation from the current levels of around 350 gCO_2/kWh to around 100 gCO_2/kWh in 2030 and potentially 25 gCO_2/kWh in 2050 [1].

Such ambitious decarbonization objectives are to be achieved through two key avenues, both of which introduce critical techno-economic challenges to electricity systems. The first one is associated with the generation side of electricity systems and involves the large-scale integration of renewable energy sources (RES) and the partial phase-out of conventional, CO_2-intensive generation technologies. Nevertheless, RES are inherently characterized by high output variability, limited controllability and lack of inertia. As a result, the balancing burden and costs for system operators are significantly aggravated, while conventional electricity producers face increased price volatility risks and the need to operate part-loaded in order to provide the required balancing services [2].

The second avenue is associated with the demand side of electricity systems and involves the large-scale electrification of transport, heating and cooling sectors, since conventional gas/oil fired technologies are responsible for the emission of a significant portion of the total CO_2 emissions. Nevertheless, due to the natural energy intensity of transport, heating and cooling sectors, this decarbonization potential is accompanied by the introduction of a considerable amount of new electricity demand. Furthermore, electricity demand peaks are expected to get disproportionally higher (than the increase in the total electricity consumption), due to the temporal patterns of end users' travelling, heating and cooling requirements, driving capital intensive investments in new generation and network capacity [3].

In this setting, demand response (DR) technologies have attracted great interest by governments, industry and academia due to their significant potential in addressing all the above decarbonization challenges. DR is defined as the flexible modification of electricity consumption patterns with respect to its baseline patterns and entails either reducing the electrical demand during emergency conditions, or, more importantly, flexibly redistributing electrical demand in time without compromising the service quality delivered to end users [4,5].

Specifically, DR has been demonstrated to yield numerous system benefits, including: (a) reduction of energy costs, by avoiding consumption during periods with low RES output and rescheduling consumption during periods with abundant RES output, (b) supporting system balancing through the provision of frequency response and reserve services, and (c) avoiding/deferring investments in new generation and network capacity by reducing demand peaks. According to the studies undertaken in [1], the potential size of these benefits for the UK electricity system cannot be neglected: it is between £3.2bn and £4.7bn per year in a system meeting the benchmark carbon emissions target of 100 gCO_2/kWh in 2030, and up to £7.8bn per year in a system meeting a more ambitious carbon emissions target of 50 gCO_2/kWh.

15.1.2 Industrial DR: significance and relevant work

Most of the previous work on DR has focused on modeling, optimizing and analyzing the value of residential and commercial DR by exploring different flexible demand technologies, such as electric vehicles with smart charging and vehicle-to-grid capabilities, thermal appliances (e.g., heating, ventilation and air conditioning systems, water heaters, refrigerators) and wet appliances (e.g., washing machines, tumble dryers).

Contrastingly, the potential of industrial DR (IDR) has received relatively limited attention. We believe that this constitutes a major knowledge gap that prevents the exploitation of a hidden source of substantial flexibility, for the following reasons [6,7]:

(a) Industrial demand represents a significant share of the total electricity consumption (e.g., 25% in the UK in 2020 [8]).
(b) Considering their high energy costs and competitiveness concerns, industrial consumers have strong economic incentives to actively participate in the electricity market.
(c) In contrast with small-scale residential consumers, the majority of industrial consumers already possess the required monitoring, control and communication infrastructure to enable participation in the electricity market, and their large size can avoid the technical and economic challenges of aggregation.
(d) Due to COVID-19 lockdown and work-from-home measures adopted worldwide, the criticality of residential demand has been enhanced, while research activities involving interactions with residential consumers (e.g., interviewing residential consumers, accessing residential properties to install monitoring and control equipment, obtaining residential demand data) have been seriously affected.

Previous research works on IDR can be firstly classified into two broad categories according to their scope. The first one focuses on the electricity system's perspective, by analyzing the value of IDR for the electricity system as a whole. Although our work does not belong to this category, it is worth mentioning two representative references for the interested reader [6,9]. In [6], the authors focus on the German electricity generation system and analyses the value of IDR in providing tertiary reserve; the results

Table 15.1 Summary of previous work on optimizing industrial consumers' electricity costs through IDR (FP: flexible processes, EG: electricity generation, EES: electrical energy storage, ETV: electricity and thermal energy vectors, ToU: time-of-use, RTP: real-time pricing, CPP: critical peak pricing, IBR: inclining block rates)

Reference	Industry sector	Flexibility source	Market segment
[10]	Steel	FP (2 types)	Energy market (ToU)
[11]	Oxygen generation	FP (2 types), EG, EES	Energy market (RTP)
[12]	Agnostic	ETV	Energy market (ToU)
[13]	Refinery	FP (2 types)	Energy market (RTP)
[14]	Steel	FP (2 types), EG, EES	Energy market (RTP, ToU, CPP, IBR)
[15]	Agnostic	FP (2 types), EG	Matching demand with on-site RES
[16]	Cement	FP (2 types), EES	Ancillary services
[17]	Cement	FP (2 types)	Energy market (RTP), Carbon market
[18]	Steel	FP (3 types), EES	Energy market (RTP)
[19]	Agnostic	FP (1 type), EG, EES	Energy market (RTP, bilateral contracts)
[20]	Tire manufacturing	FP (2 types), EG, EES, ETV	Energy market (ToU)
[21]	Flour mill	FP (2 types)	Energy market (RTP)
[22]	Agnostic	FP (2 types), EG, EES	Energy market (ToU)
This work	Agnostic	FP (7 types)	Energy market (RTP)

demonstrate that IDR yields significant capital cost savings by avoiding investments in peaking (gas) generation capacity. The authors of [9] pursue a wider objective by analyzing both long-term and short-term economic benefits of IDR in the European electricity system (including generation, transmission and distribution sectors) through a novel whole-system modeling framework; these benefits are demonstrated to be in the order of billion Euros per year, highlighting the motivation for further exploring IDR initiatives. Due to the system focus, these works employ generic, high-level models for the representation of industrial demand flexibility, without delving deeper into the flexibility characteristics of different industrial loads and processes.

The second category, to which our work belongs, focuses on the industrial energy consumers' perspective, by developing optimization models for minimizing individual consumers' electricity costs through exploitation of IDR [10–22]. In the context of summarizing this stream of previous work and manifesting where our work stands, Table 15.1 classifies references belonging to this category based on three criteria. The first criterion involves the industry sectors under study, with significant diversity observed among the different relevant references.

The second criterion involves the sources of underlying flexibility that enable the realization of IDR. At this point, it should be stressed that industrial demand flexibility includes a very large number of diverse energy technologies, systems and assets, which vary greatly according to the specific sector under study, the production processes, as well as the perceptions, preferences and requirements of the industrial consumers. Nevertheless, we believe that such diversity can be encapsulated by four different classes of flexibility sources. The first one includes flexible industrial processes, the electricity demand requirements of which can be redistributed in time; such processes have been explored in [10–22]. The second class includes on-site (i.e., installed at the industrial consumers' premises) electricity generation assets, which have been explored in [11,14,15,19,20,22]. The third class includes on-site electrical energy storage assets, which exhibit the flexibility to act as both electricity demand (when charging) and generation (when discharging) which have been explored in [11,14,16,18–20,22]. Finally, the fourth class includes flexibility sources associated with the interaction between electricity and thermal (heating/cooling) energy vectors, such as electric heat pumps and thermal energy storage [12,20].

The last criterion involves the market segments and the pricing schemes through which the exploitation of IDR yields economic benefits for the industrial consumers. It can be observed that the majority of relevant references have focused on the energy market, where industrial consumers naturally aim at buying energy at low prices and selling energy (in cases of on-site generation and storage) at high prices. This stream of previous work has explored different energy pricing schemes mobilizing IDR, such as time-of-use (ToU) pricing, and real-time pricing (RTP).

15.1.3 Chapter motivation and contributions

This work belongs to the second category discussed in Section 15.1.2 (i.e., industrial energy consumers' perspective), with a particular focus on modeling different types of flexible industrial processes and integrating these models into the electricity costs' minimization problem of an industrial consumer. This focus is driven by two motivations. First of all, models of the other three classes of industrial demand flexibility (i.e., on-site electricity generation, on-site electrical energy storage, and flexibility sources associated with the interaction between electricity and thermal energy vectors) have been also explored in previous works focusing on residential and commercial DR and are well established. On the other hand, flexible industrial processes constitute a differentiating flexibility source with respect to residential and commercial DR applications that has received limited attention. Second, previous work modeling this flexibility source has only explored a limited subset of particular types of flexible industrial processes (up to three different types in [18]), and the developed models are mostly driven by the particularities of the industry sector under study (Table 15.1).

This chapter adopts an alternative, sector-agnostic modeling approach and develops generic models of all conceivable types of flexible industrial processes, with the aim to shed light on their key operating differences and assist industrial consumers interested in IDR schemes to identify and assess the types that are more relevant to their systems. In this context, this chapter identifies and discusses seven different types: (1) uninterruptible processes with fixed power, (2) interruptible processes with fixed power, (3) uninterruptible processes with discretely adjustable power, (4) interruptible processes with discretely adjustable power, (5) uninterruptible processes with continuously adjustable power, (6) interruptible processes with continuously adjustable power, and (7) material storage buffers. For each of these types, a mathematical model capturing its techno-economic operating characteristics is developed, including: (a) input parameters, (b) decision variables, (c) operating constraints, and (d) inconvenience cost function; the latter encapsulates operating costs driven by rescheduling the timing of the processes with respect to the baseline operating patterns.

These models are integrated into the electricity costs' minimization problem of an industrial consumer; in line with the majority of previous works (Table 15.1), our focus lies in the participation of the industrial consumer in the energy market through an RTP scheme, since we believe that RTP closes the existing gap between wholesale and retail electricity market segments and promises a more active market participation of IDR and higher subsequent benefits for the whole electricity system. Nevertheless, we believe that our developed models of flexible industrial processes are generic and are applicable to problems with different pricing schemes and participation in ancillary services markets (with the latter constituting an area of our future work).

Finally, a case study concerning an actual industrial consumer in the UK is investigated, employing its real demand data and a target industrial process (i.e., a process which the consumer is interested in operating flexibly) as well as real price data from the UK wholesale market. Different scenarios are examined regarding the flexibility type of the target process, and the results demonstrate, quantify, and compare the economic benefits generated by different flexibility types. Specifically, the yearly electricity cost savings are in the range between 10.71% and 23.13% with respect to the baseline electricity costs of the target process alone and between 1.58% and 3.41% with respect to the baseline electricity costs of the industrial consumer as a whole; we believe that these benefits are very significant, considering that the target process corresponds to approximately 17% of the consumer's total electricity consumption.

15.1.4 Chapter outline

The rest of this chapter is organized as follows. Section 15.2 details the developed models of different types of flexible industrial processes and integrates them into the electricity costs' minimization problem of an industrial consumer. Section 15.3 presents the examined case study and the obtained results. Finally, Section 15.4 discusses conclusions and future extensions of this work.

15.2 Modeling framework

15.2.1 Assumptions and generic formulation of industrial consumer's optimization problem

As discussed in Section 15.1.3, our focus lies in modeling different types of flexible industrial processes. The first key assumption of our work is that such flexibility does not entail reduction of the overall electrical energy consumption of the considered processes, but only redistribution of their electricity requirements in time. This assumption is also adopted in all previous relevant works (Table 15.1), since it is widely recognized that industrial processes require certain fixed levels of energy to ensure that their production performance is not compromised. Therefore, IDR schemes entailing reduction of the overall electricity consumption are generally limited to emergency schemes deployed when the electricity grid is under massive pressure (e.g., during generation or network outages). On the other hand, IDR schemes entailing redistribution of such fixed energy requirements in time exhibit much higher applicability and acceptability by industrial consumers. For these reasons, the developed models of all seven types of flexible industrial processes presented in the following sections respect this fixed energy constraint.

Despite this higher acceptability of temporal redistribution of industrial processes, it is also widely recognized that this entails certain inconvenience costs, i.e., operating costs incurred by the industrial users due to changes in the schedule of processes with respect to the baseline or preferred operating patterns. For example, these may include additional labor costs for operating processes outside the normal temporal horizons. For this reason, the developed models of all seven types of flexible industrial processes include an inconvenience cost function which is accounted for (along with the electricity cost function) in the industrial consumer's optimization problem.

Finally, since our focus lies in the mobilization of IDR through an RTP scheme, we assume that the considered industrial consumer possesses the required communication and control infrastructure to receive the RTP signals from its electricity supplier and accordingly optimize the schedules of its flexible processes in an automated fashion. It should be noted that RTP schemes generally include -on top of the wholesale electricity prices – a number of non-energy cost components (e.g., associated with network charges, decarbonization subsidies, etc.); such cost components are out of the scope of this work, and we thus assume that the RTP signals correspond to the wholesale electricity prices.

Considering the above assumptions, we provide below a generic formulation of the considered industrial consumer's optimization problem. The objective function (15.1) lies in minimizing the total operating cost of the industrial consumer across the considered scheduling horizon (e.g., daily in the context of the case study presented in Section 15.3) which constitutes the sum of the inconvenience cost associated with flexible processes (first term) and their electricity cost (second term). The latter depends on their power demand and the RTP signal at each time period, considering the resolution of the scheduling horizon (e.g., hourly in the context of the case study presented in Section 15.3). The input parameters of the problem include the RTP signals along with an additional set of input parameters that apply to each specific type of flexible industrial processes. The decision variables of the problem include the power demands of the flexible processes along with an additional set of type-specific decision variables. The inconvenience cost function and constraints of the problem are also type-specific. All type-specific elements of the problem (input parameters, decision variables, inconvenience cost function, constraints) are provided in dedicated Sections 15.2.2–15.2.6.

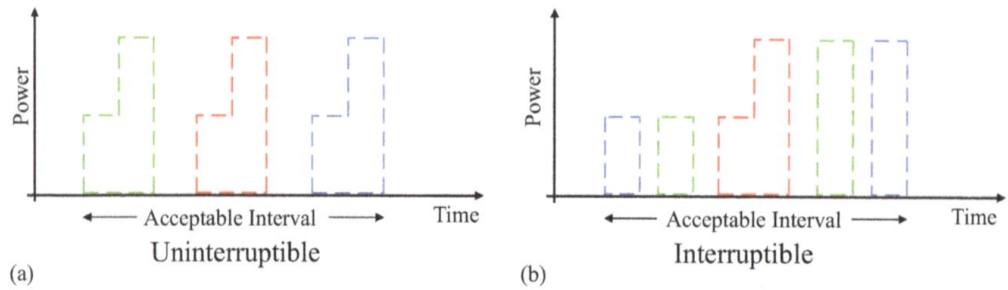

Figure 15.1 *Illustration of flexibility of (a) uninterruptible and (b) interruptible process with fixed power*

Objective function:

$$\min \sum_{j} \left[C_j^{INC} + \sum_{t} d_{j,t} \cdot \lambda_t \cdot \tau \right] \tag{15.1}$$

Input parameters: $\{\lambda_t\}\forall t$ and type-specific input parameters (Sections 15.2.2–15.2.6)
Decision variables: $\{d_{j,t}\}\forall j, t$ and type-specific decision variables (Sections 15.2.2–15.2.6)
Constraints: type-specific operating constraints (Sections 15.2.2–15.2.6)
where:

$j \in J$	index and set of flexible processes
$t \in T$	index and set of time periods belonging to the scheduling horizon
τ	time resolution of scheduling horizon (h)
λ_t	RTP signal at period t (£/MWh)
$d_{j,t}$	power demand of process j at period t (kW)
C_j^{INC}	inconvenience cost function of process j (£) (type-specific, Sections 15.2.2–15.2.6)

15.2.2 *Uninterruptible processes with fixed power*

The first type includes processes that can be temporally shifted within an acceptable (by the industrial user) temporal interval, but (a) their power profile, generally consisting of a number of steps (which may also represent interdependent, consecutive sub-processes of the same process) is fixed and cannot be modified; and (b) they cannot be interrupted after their initiation. Therefore, their flexibility lies only in the ability to shift their whole power profile earlier or later in time, as illustrated in Figure 15.1(a) for an example of a process with 2 steps. Curves with different colors indicate different feasible options for the execution of the process: (i) the red curve indicates that the process is executed at its baseline execution time, (ii) the green curve indicates that the process is executed at an earlier (than its baseline) time and

(iii) the blue curve indicates that the process is executed at a later (than its baseline) time.

The mathematical model of this type includes the following elements:
Input parameters: $\{T_j^{DUR}, t_j^{BAS}, t_j^{IN}, t_j^{TER}, C_j^-, C_j^+\}\forall j, \quad \{D_{j,s}^{ST}\}\forall j, s \in S_j$
Decision variables: $\{u_{j,t}^{IN}, d_{j,t}\}\forall j, t, \quad \{\delta_j^-, \delta_j^+\}\forall j$
Inconvenience cost function:

$$C_j^{INC} = (C_j^- \cdot \delta_j^- + C_j^+ \cdot \delta_j^+) \cdot \tau, \forall j \tag{15.2}$$

Constraints:

$$\delta_j^- = \sum_{y=0}^{y=t_j^{BAS}-t_j^{IN}} \Bigg[1 - \overbrace{\sum_{x=0}^{x=y} u_{j,t_j^{BAS}-x}^{IN}}^{\textbf{1 for each period of earlier initiation}} - \overbrace{\sum_{o=t_j^{BAS}+1}^{o=t_j^{TER}-T_j^{\mathrm{DUR}}+1} u_{j,o}^{IN}}^{\textbf{1 if initiated later}} \Bigg], \quad \forall j, \tag{15.3}$$

$$\delta_j^+ = \sum_{y=t_j^{BAS}}^{y=t_j^{TER}-T_j^{\mathrm{DUR}}+1} \Bigg[1 - \overbrace{\sum_{x=t_j^{BAS}}^{x=y} u_{j,x}^{IN}}^{\textbf{1 for each period of later initiation}} - \overbrace{\sum_{o=t^{\mathrm{in}}}^{o=t_j^{BAS}-1} u_{j,o}^{IN}}^{\textbf{1 if initiated earlier}} \Bigg], \quad \forall j, \tag{15.4}$$

$$d_{j,t} = \sum_{s=1}^{s=T_j^{\mathrm{DUR}}} z_{j,t+1-s} \cdot D_{j,s}^{\mathrm{ST}}, \qquad \forall j, t, \tag{15.5}$$

$$\sum_{t<t_j^{IN}} u_{j,t}^{IN} = 0, \qquad \forall j, \tag{15.6}$$

$$\sum_{t>t_j^{TER}-T_j^{\mathrm{DUR}}+1} u_{j,t}^{IN} = 0, \qquad \forall j, \tag{15.7}$$

$$\sum_{t_j^{IN}}^{t_j^{TER}-T_j^{\mathrm{DUR}}+1} u_{j,t}^{IN} = 1, \qquad \forall j, \tag{15.8}$$

where

$s \in S_j$	index and set of steps of process j
T_j^{DUR}	duration of process j
t_j^{BAS}	baseline initiation time of process j
t_j^{IN}	earliest acceptable initiation time of process j

t_j^{TER}	latest acceptable termination time of process *j*
C_j^-	marginal inconvenience cost of earlier initiation of process *j* (£/h)
C_j^+	marginal inconvenience cost of later initiation of process *j* (£/h)
$D_{j,s}^{ST}$	power demand of step *s* of process *j* (kW)
$u_{j,t}^{IN}$	binary variable expressing whether process *j* is initiated at period *t* (1) or not (0)
δ_j^-	extent of earlier initiation of process *j*
δ_j^+	extent of later initiation of process *j*

Based on this model, the industrial user determines the baseline initiation, earliest acceptable initiation, and latest acceptable termination times (which generally reflect production performance objectives and operating practices such as labor shifts), and the marginal inconvenience costs of earlier and later initiation with respect to the baseline initiation time. The primary decision of the model is the period at which the process in initiated (i.e., the period for which $u_{j,t}^{IN} = 1$). As reflected in (15.2), the inconvenience cost is proportional to the temporal extents of earlier or later initiation (only one of its two terms can be non-zero); these temporal extents constitute dependent decision variables, depending on the primary decision variables $u_{j,t}^{IN}$, as determined by constraints (15.3)–(15.4). The same holds for the power demands of the process, which are determined by constraint (15.5). Constraints (15.6)–(15.7) ensure that the process cannot be executed outside the acceptable temporal interval (defined as the interval between the earliest acceptable initiation and latest acceptable termination times). Constraint (15.8) ensures that the process should be executed once during the acceptable temporal interval.

15.2.3 Interruptible processes with fixed power

The processes of the second type differ from those of the first type (Section 15.2.2) in that they exhibit the additional flexibility of being interruptible after the execution of each of their steps; this may represent sub-processes which do not necessarily need to be consecutively executed. Their flexibility is illustrated in Figure 15.1(b) for an example of a process with 2 steps. Curves with different colors indicate different feasible options for the execution of the process: (i) the red curve indicates that both steps of the process are executed at their baseline execution times, (ii) the green curve indicates that the first step is executed at an earlier (than its baseline) time, the process is then interrupted, and the second step is executed at a later (than its baseline) time, and (iii) the blue curve is similar to the green one, with the difference that the interruption interval between the two steps is longer.

The mathematical model of this type includes the following elements:
Input parameters: $\{T_j^{DUR}\}\forall j, \quad \{t_{j,s}^{BAS}, t_{j,s}^{IN}, t_{j,s}^{TER}, C_{j,s}^-, C_{j,s}^+, D_{j,s}^{ST}\}\forall j, s \in S_j$
Decision variables: $\{u_{j,t,s}^{IN}\}\forall j, t, s \in S_j, \quad \{d_{j,t}\}\forall j, t, \quad \{\delta_{j,s}^-, \delta_{j,s}^+\}\forall j, s \in S_j$
Inconvenience cost function:

$$C_j^{INC} = \sum_s \left[(C_{j,s}^- \cdot \delta_{j,s}^- + C_{j,s}^+ \cdot \delta_{j,s}^+) \cdot \tau \right], \forall j \tag{15.9}$$

Constraints:

$$\delta^{-}_{j,s} = \sum_{y=0}^{y=t^{BAS}_{j,s}-t^{IN}_{j,s}} \Bigg[\overbrace{1 - \sum_{x=0}^{x=y} u^{IN}_{j,t^{BAS}_{j,s}-x,s}}^{\textbf{1 for each period of earlier execution}} - \overbrace{\sum_{o=t^{BAS}_{j,s}+1}^{o=t^{TER}_{j,s}} u^{IN}_{j,o,s}}^{\textbf{1 if executed later}} \Bigg], \quad \forall j,s, \tag{15.10}$$

$$\delta^{+}_{j,s} = \sum_{y=t^{BAS}_{j,s}}^{y=t^{TER}_{j,s}} \Bigg[\overbrace{1 - \sum_{x=t^{BAS}_{j,s}}^{x=y} u^{IN}_{j,x,s}}^{\textbf{1 for each period of later execution}} - \overbrace{\sum_{o=t^{IN}_{j,s}}^{o=t^{BAS}_{j,s}-1} u^{IN}_{j,o,s}}^{\textbf{1 if executed earlier}} \Bigg], \quad \forall j,s, \tag{15.11}$$

$$d_{j,t} = \sum_{s} u^{IN}_{j,t,s} \cdot D^{ST}_{j,s}, \qquad \forall j,t, \tag{15.12}$$

$$\sum_{t<t^{IN}_{j,s}} u^{IN}_{j,t,s} = 0, \qquad \forall j,s, \tag{15.13}$$

$$\sum_{t>t^{TER}_{j,s}} u^{IN}_{j,t,s} = 0, \qquad \forall j,s, \tag{15.14}$$

$$\sum_{t^{IN}_{j,s}}^{t^{TER}_{j,s}} u^{IN}_{j,t,s} = 1, \qquad \forall j,s, \tag{15.15}$$

$$u^{IN}_{j,t,s+1} \leq \sum_{y=1}^{y=t-1} u^{IN}_{j,y,s}, \qquad \forall j, s \in \left[1, |S_j| - 1\right], t, \tag{15.16}$$

where

$t^{BAS}_{j,s}$ baseline execution time of step s of process j
$t^{IN}_{j,s}$ earliest acceptable execution time of step s of process j
$t^{TER}_{j,s}$ latest acceptable execution time of step s of process j
$C^{-}_{j,s}$ marginal inconvenience cost of earlier execution of step s of process j (£/h)
$C^{+}_{j,s}$ marginal inconvenience cost of later execution of step s of process j (£/h)
$u^{IN}_{j,t,s}$ binary variable expressing whether step s of process j is executed at period t (1) or not (0)
$\delta^{-}_{j,s}$ extent of earlier execution of step s of process j
$\delta^{+}_{j,s}$ extent of later execution of step s of process j

This model is similar to the model of Section 15.2.2, with the difference that its elements are associated with each step of the process, and not just with the process as a whole. Specifically, the industrial user determines baseline, earliest and latest execution times as well as marginal inconvenience costs of earlier and later execution, for each step of the process. The primary decisions of the model are the periods at which each step of the process is executed. As reflected in (15.9), the inconvenience cost is equal to the sum of the earlier/later execution costs of each step. The temporal extents of earlier/later execution of each step as well as the power demands of the process constitute dependent decision variables, determined by constraints (15.10)–(15.11) and (15.12), respectively. Constraints (15.13)–(15.14) ensure that each step of the process cannot be executed outside its acceptable temporal interval. Constraint (15.15) ensures that each step of the process should be executed once during its acceptable temporal interval. Constraint (15.16) ensures that each step of the process can be executed only if the preceding step has already been executed.

15.2.4 Uninterruptible and interruptible processes with discretely adjustable power

The processes of the third and fourth types differ from those of the previous two types (Sections 15.2.2 and 15.2.3) in that they exhibit the additional flexibility of adjustable power profiles, as long as their total electricity consumption within an interval specified by their users is not affected (i.e., they respect the fixed energy constraint, as explained in Section 15.2.1). Such processes may represent manufacturing machines that can run with alternative power levels. Nevertheless, for these two types examined in this section, this power adjustability involves a certain set of discrete power levels (their power cannot be continuously adjusted). The flexibility of uninterruptible and interruptible processes with discretely adjustable power is illustrated in Figure 15.2(a) and 15.2(b), respectively, for an example of a process with 2 steps and 3 alternative power levels. Curves with different colors indicate different feasible options for the execution of the process: (i) the red curve indicates that both steps of the process are executed with their baseline power profile, involving the lowest power level P1 for the first step and the highest power level P3 for the second step, (ii) the green curve indicates that both steps are executed with the intermediate power level P2, and (iii) the blue curve indicates that the first step is executed with the highest power level P3 and the second step with the lowest power level P1.

The mathematical model of uninterruptible processes with discretely adjustable power levels includes the following elements:
Input parameters: $\{t_j^{IN}, t_j^{TER}, t_j^{B1}, t_j^{B2}, K_j^-, K_j^+, E_j\}\forall j, \quad \{D_{j,l}^{LVL}\}\forall j, l \in L_j$
Decision variables: $\{u_{j,t,l}^{LVL}\}\forall j, t, l \in L_j, \quad \{d_{j,t}\}\forall j, t$
Inconvenience cost function:

$$C_j^{INC} = K_j^- \cdot \sum_{t=t_j^{IN}}^{t<t_j^{B1}} d_{j,t} \cdot (t_j^{B1} - t) \cdot \tau^2 + K_j^+ \cdot \sum_{t>t_j^{B2}}^{t=t_j^{TER}} d_{j,t} \cdot (t - t_j^{B2}) \cdot \tau^2, \forall j \quad (15.17)$$

Constraints:

$$d_{j,t} = \sum_l u_{j,t,l}^{LVL} \cdot D_{j,l}^{LVL}, \forall j, t, \quad (15.18)$$

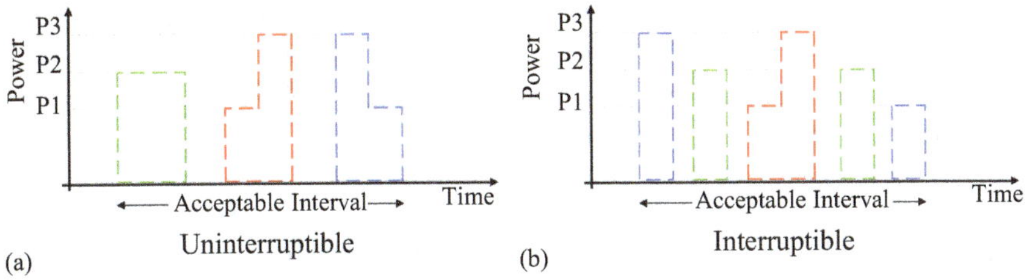

Figure 15.2 Illustration of flexibility of (a) uninterruptible and (b) interruptible process with discretely adjustable power

$$\sum_{l} u_{j,t,l}^{LVL} \leq 1, \forall j, t, \tag{15.19}$$

$$\sum_{t<t_j^{IN}} u_{j,t,l}^{LVL} = 0, \forall j, l, \tag{15.20}$$

$$\sum_{t>t_j^{TER}} u_{j,t,l}^{LVL} = 0, \forall j, l, \tag{15.21}$$

$$\sum_{t_j^{IN}}^{t_j^{TER}} d_{j,t} \cdot \tau = E_j, \forall j, \tag{15.22}$$

$$\sum_{l} u_{j,t+1,l}^{LVL} - \sum_{l} u_{j,t,l}^{LVL} \leq 1 - \frac{\sum_{m=1}^{m=t} \sum_{l} u_{j,m,l}^{LVL}}{|T|}, \forall j, t \in [1, |T| - 1], \tag{15.23}$$

Where:

$l \in L$	index and set of discrete (non-zero) power levels of process j
t_j^{B1}	baseline initiation time of process j
t_j^{B2}	baseline termination time of process j
K_j^-	marginal inconvenience cost of earlier consumption of energy by process j (£/kWh/h)
K_j^+	marginal inconvenience cost of later consumption of energy by process j (£/kWh/h)
E_j	total energy required by process j (kWh)
$D_{j,l}^{LVL}$	discrete (non-zero) power level l of process j (kW)
$u_{j,t,l}^{LVL}$	binary variable expressing whether power level l of process j is selected at period t (1) or not (0)

Based on this model, the industrial user determines the acceptable execution interval $\left[t_j^{IN}, t_j^{TER}\right]$, the baseline execution interval $\left[t_j^{B1}, t_j^{B2}\right]$ (implying that this constitutes a subset of the acceptable execution interval, i.e., $t_j^{IN} \leq t_j^{B1} \leq t_j^{B2} \leq t_j^{TER}$) and the marginal inconvenience costs of consuming energy earlier/later than the baseline interval. The primary decision of the model is the selected power level at each period t (which is either equal to one of the alternative discrete power levels l if $u_{j,t,l}^{LVL} = 1$, or equal to zero if $\sum_l u_{j,t,l}^{LVL} = 0$). As reflected in (15.17), the inconvenience cost is proportional to the amount of energy consumed earlier/later than the baseline interval as well as to the temporal extent of such earlier/later energy consumption (i.e., it increases as energy is obtained further before or further after the baseline interval). The power demands of the process constitute dependent decision variables, determined by constraint (15.18). Constraint (15.19) ensures that at maximum one of the alternative discrete power levels can be selected at each period. Constraints (15.20) and (15.21) ensure that the process cannot be executed outside the acceptable temporal interval. Constraint (15.22) ensures that the fixed energy requirements are consumed within the acceptable interval. Constraint (15.23) ensures that the process cannot be interrupted after its initiation.

Therefore, the only difference of the mathematical model of interruptible processes with discretely adjustable power levels is that constraint (15.23) is omitted.

15.2.5 Uninterruptible and interruptible processes with continuously adjustable power

The processes of the fifth and sixth types differ from those of the previous two types (Section 15.2.4) in that they exhibit the additional flexibility of continuous (instead of discrete) power adjustability, i.e., their power can take any value between a minimum and a maximum limit.

The mathematical model of this type includes the following elements:
Input parameters: $\{t_j^{IN}, t_j^{TER}, t_j^{B1}, t_j^{B2}, K_j^-, K_j^+, E_j, \underline{D}_j, \overline{D}_j\} \forall j$
Decision variables: $\{u_{j,t}^{ON}, d_{j,t}\} \forall j, t$
Inconvenience cost function:

$$C_j^{INC} = K_j^- \cdot \sum_{t=t_j^{IN}}^{t<t_j^{B1}} d_{j,t} \cdot (t_j^{B1} - t) \cdot \tau^2 + K_j^+ \cdot \sum_{t>t_j^{B2}}^{t=t_j^{TER}} d_{j,t} \cdot (t - t_j^{B2}) \cdot \tau^2, \forall j \quad (15.24)$$

Constraints:

$$\sum_{t<t_j^{IN}} u_{j,t}^{ON} = 0, \forall j, \quad (15.25)$$

$$\sum_{t>t_j^{TER}} u_{j,t}^{ON} = 0, \forall j, \quad (15.26)$$

$$u_{j,t}^{ON} \cdot \underline{D}_j \leq d_{j,t} \leq u_{j,t}^{ON} \cdot \overline{D}_j, \forall j, \quad (15.27)$$

$$\sum_{t_j^{IN}}^{t_j^{TER}} d_{j,t} \cdot \tau = E_j, \forall j, \tag{15.28}$$

$$u_{j,t+1}^{ON} - u_{j,t}^{ON} \leq 1 - \frac{\sum_{m=1}^{m=t} \sum_{l} u_{j,m}^{ON}}{\mid T \mid}, \forall j, t \in [1, \mid T \mid -1], \tag{15.29}$$

where

$\underline{D}_j$	minimum (non-zero) power limit of process j (kW)
$\overline{D}_j$	maximum power limit of process j (kW)
$u_{j,t}^{ON}$	binary variable expressing whether process j is active (consuming non-zero power) at period t (1) or not (0)

The inputs determined by the industrial user and the inconvenience cost function (15.24) are identical to the ones of the model of Section 15.2.4. The primary decisions of the model include both the binary states $u_{j,t}^{ON}$ and the power demands $d_{j,t}$ of the process, i.e., in contrast to all the previous models, these power demands constitute independent rather than dependent decision variables. Constraints (15.25) and (15.26) ensure that the process cannot be executed outside the acceptable temporal interval. Constraint (15.27) ensures that the power demands respect the minimum and maximum power limits of the process. Constraint (15.28) ensures that the fixed energy requirements are consumed within the acceptable interval. Constraint (15.29) ensures that the process cannot be interrupted after its initiation.

Therefore, the only difference of the mathematical model of interruptible processes with continuously adjustable power levels is that constraint (15.29) is omitted.

15.2.6 Material storage buffers

Many realistic industrial plants involve interdependent processes, in the sense that one process produces a particular material which is then consumed by another process. In case that only a subset of these processes exhibit flexibility (i.e., they belong to one of the flexible types presented in Sections 15.2.2–15.2.5), while the rest of the processes are completely inflexible (i.e., they cannot be shifted in time nor adjust their power profile), such interdependencies imply that the flexibility of the subset of flexible processes cannot be actually exploited due to the criticality of the subset of inflexible processes.

The deployment of material storage buffers, which can store materials produced/consumed by different processes, can address this challenge and enable exploitation of the available flexibility in the industrial plant. This is because such buffers effectively decouple the operation of the interdependent processes, by storing materials produced by preceding processes and providing these materials to succeeding processes, at decoupled timeframes. Therefore, the power demand of flexible processes can be redistributed in time without compromising the operation of dependent inflexible processes. In this context, it should be noted that the deployment of such buffers does not constitute a primary flexibility potential, as an industrial plant involving only inflexible processes cannot exhibit flexibility irrespectively of the inclusion of such buffers; it rather constitutes an enabler for the exploitation of available flexibility in a subset of the processes.

The mathematical model we present below is generic, in that it allows the representation of any number of interdependent flexible and inflexible processes, any number of material storage buffers and any possible configuration of the links between processes and buffers. Furthermore, our modeling approach can accommodate any of the flexibility types presented in Sections 15.2.2–15.2.5 for the subset of flexible processes; however, for simplicity and brevity reasons, the model below assumes that all flexible processes belong to the first type i.e., they are uninterruptible processes with fixed power (Section 15.2.2).

Input parameters: $\{T_j^{DUR}, t_j^{BAS}, t_j^{IN}, t_j^{TER}, C_j^-, C_j^+\}\forall j$, $\{D_{j,s}^{ST}\}\forall j, s \in S_j$, $\{PM_{j,m,s}^F\}\forall j \in J \cap G_m, m, s \in S_j$, $\{PM_{i,m,t}^{IF}\}\forall i \in I \cap G_m, m, t$, $\{CM_{j,m,s}^F\}\forall j \in J \cap F_m, m$, $s \in S_j$, $\{CM_{i,m,t}^{IF}\}\forall i \in I \cap F_m, m, t$, $\{\underline{Y}_m, \overline{Y}_m, y_{m,0}\}\forall m$
Decision variables: $\{u_{j,t}^{IN}, d_{j,t}\}\forall j, t$, $\{\delta_j^-, \delta_j^+\}\forall j$, $\{pm_{j,m,t}^F\}\forall j \in J \cap G_m, m, t$, $\{cm_{j,m,t}^F\}\forall j \in J \cap F_m, m, t$, $\{y_{m,t}\}\forall m, t$

Inconvenience cost function:

$$C_j^{INC} = (C_j^- \cdot \delta_j^- + C_j^+ \cdot \delta_j^+) \cdot \tau, \forall j \tag{15.30}$$

Constraints:

$$(15.3) - (15.8), \tag{15.31}$$

$$pm_{j,m,t}^F = \sum_{s=1}^{s=T_j^{DUR}} u_{j,t+1-s}^{IN} \cdot PM_{j,m,s}^F, \forall j \in J \cap G_m, m, t, \tag{15.32}$$

$$cm_{j,m,t}^F = \sum_{s=1}^{s=T_j^{DUR}} u_{j,t+1-s}^{IN} \cdot CM_{j,m,s}^F, \forall j \in J \cap F_m, m, t, \tag{15.33}$$

$$\begin{aligned} y_{m,t} = y_{m,t-1} + \sum_{j \in J \cap G_m} pm_{j,m,t}^F + \sum_{i \in I \cap G_m} PM_{i,m,t}^{IF} - \sum_{j \in J \cap F_m} cm_{j,m,t}^F \\ - \sum_{i \in I \cap F_m} CM_{i,m,t}^{IF}, \forall m, t, \end{aligned} \tag{15.34}$$

$$\underline{Y}_m \le y_{m,t} \le \overline{Y}_m, \forall m, t, \tag{15.35}$$

$$y_{m,0} = y_{m,|T|}, \forall m, \tag{15.36}$$

where

$m \in M$	index and set of material storage buffers
$i \in I$	index and set of inflexible processes
G_m	set of processes that input material to buffer m
F_m	set of processes that output material from buffer m
$PM_{j,m,s}^{F}$	amount of material produced by step s of flexible process j and inputted to buffer m (kg)
$PM_{i,m,t}^{IF}$	amount of material produced by inflexible process i and inputted to buffer m at period t (kg)
$CM_{j,m,s}^{F}$	amount of material consumed by step s of flexible process j and outputted from buffer m (kg)
$CM_{i,m,t}^{IF}$	amount of material consumed by inflexible process i and outputted from buffer m at period t (kg)
$\underline{Y}_m, \overline{Y}_m, y_{m,0}$	minimum, maximum, and initial material content of buffer m (kg)
$pm_{j,m,t}^{F}$	amount of material produced by flexible process j and inputted to buffer m at period t (kg)
$cm_{j,m,t}^{F}$	amount of material produced by flexible process j and outputted from buffer m at period t (kg)
$y_{m,t}$	material content of buffer m at period t (kg)

Beyond the inputs determined by the industrial user (which are identical to the ones of the model of Section 15.2.2), the model assumes knowledge of the amounts of material produced/consumed by each flexible and inflexible process, and inputted to/outputted from each buffer, as well as the minimum, maximum, and initial (at the start of the scheduling horizon) contents of each buffer. The primary decisions $u_{j,t}^{IN}$, inconvenience cost function (15.30) and constraints (15.31) are the same as in the model of Section 15.2.2; this implies our assumption that the utilization of buffers does not entail any inconvenience cost. The time-specific amounts of material produced/consumed by each flexible process, and inputted to/outputted from each buffer, constitute dependent decision variables, determined by constraints (15.32) and (15.33). Constraint (15.34) expresses the material content balance equation of each buffer, implying that its material content at each period depends on its material content at the previous period plus/minus the material inputted to/outputted from it. Constraint (15.35) ensures that the material content of each buffer respects its minimum and maximum content limits (reflecting its material storage capacity). Constraint (15.36) expresses the material content neutrality assumption for each buffer i.e., that the material content at the start and the end of the examined scheduling horizon are assumed equal.

15.3 Case study

15.3.1 Description and input data

The examined case study involves an actual industrial consumer in the UK, who has participated in the INFINITE project. This consumer has provided us with their actual (metered) demand data over a full year (January–December 2020) and has identified a target industrial process which they currently operate in an inflexible fashion, but they are interested in operating flexibly, since it corresponds to approximately 17% of their total electricity consumption over this year. This process needs to be executed once every day and its current (baseline) temporal and power demand parameters are generally identical every day. In this context, a daily scheduling horizon with an hourly resolution is employed in the case study. As discussed in Section 15.2.1, we have assumed that the RTP signals communicated to the industrial user correspond to the UK wholesale electricity prices, which have been derived (for the same year January–December 2020) from [23].

In the baseline scenario i.e., the current operating pattern, the process involves 4 hourly steps which are consecutively (without interruptions) executed at hours 11:00–12:00, 12:00–13:00, 13:00–14:00 and 14:00–15:00 (i.e., at periods $t = 12$, $t = 13$, $t = 14$ and $t = 15$ of our model), with their respective power demands being 250 kW, 550 kW, 700 kW and 350 kW. Since the industrial consumer has been uncertain around the specific types of flexibility that can be deployed for the target process, the case study examines different scenarios, considering all seven types investigated in Section 15.2. The assumed values of the input parameters applying to each type are presented in Table 15.2. Considering that the industrial consumer currently operates the process in an inflexible fashion and does not have prior experience on flexible operation, these values have been determined in collaboration with the authors, with the high-level aim to respect the key operating characteristics of the target process and allow a consistent comparison of the different flexibility types. It should be noted that all inconvenience cost parameters ($C_j^-, C_j^+, C_{j,s}^-, C_{j,s}^+, K_j^-, K_j^+$) have been assumed equal to zero, because of uncertainties and ambiguities faced by the industrial consumer in evaluating the economic implications of different flexibility types; therefore, the objective function of the industrial consumer's optimization problem includes only the electricity cost (despite the generality of the models presented in Section 15.2).

15.3.2 Benefits of flexibility types with fixed power

We start our analysis by presenting the benefits generated by the two flexibility types with fixed power (uninterruptible and interruptible). In both of these scenarios, the duration and power profile of the target process are not altered with respect to the baseline scenario (Table 15.2), but the execution of the process can be shifted earlier or later in time, with an allowable execution interval between $t = 8$ and $t = 19$; in other words, the target process can be shifted by 4 hours earlier or later with respect to the baseline scenario. Figure 15.3 illustrates the schedule of the target process under the three compared scenarios (baseline, uninterruptible with fixed power, interruptible with fixed power) for one of the days examined in the case study (23 February 2020) along with the RTP signals for the same day.

In the scenario where the target process becomes a flexible uninterruptible process with fixed power (indicated in yellow), its initiation time is shifted 3 hours later with respect to the baseline scenario (indicated in red), i.e., from $t = 12$ to $t = 15$, in order to exploit the lower RTP later in the day, and particularly execute the third step of the process – which is characterized by the highest electricity requirements – at $t = 17$,

Table 15.2 Assumed values of input parameters in the case study

Input parameter	Uninterruptible fixed power	Interruptible fixed power	Discretely adjustable power	Continuously adjustable power	Material storage buffer
T_j^{DUR}	4	4	–	–	2
t_j^{BAS}	12	–	–	–	12
$\{t_j^{IN}, t_j^{TER}\}$	{8,19}	–	{8,19}	{8,19}	{8,19}
$\{C_j^-, C_j^+\}$	{0,0}	–	–	–	{0,0}
$\{D_{j,s}^{ST}\}$	{250,550, 700,350}	{250,550, 700,350}	–	–	{250,550}
$t_{j,s}^{BAS}$	–	{12,13,14,15}	–	–	–
$t_{j,s}^{IN}$	–	{8,9,10,11}	–	–	–
$t_{j,s}^{TER}$	–	{16,17,18,19}	–	–	–
$\{C_{j,s}^-, C_{j,s}^+\}$	–	$\{0,0\}\forall s$	–	–	
$\{t_j^{B1}, t_j^{B2}\}$	–	–	{12,15}	{12,15}	–
$\{K_j^-, K_j^+\}$	–	–	{0,0}	{0,0}	–
E_j	–	–	1,850	1,850	–
$\{D_{j,l}^{LVL}\}$	–	–	{250,550, 700,350}	–	–
$\{\underline{D}_j, \overline{D}_j\}$	–	–	–	{250,700}	–
$PM_{j,m,s}^F$	–	–	–	–	{0,50}
$PM_{i,m,t}^{IF}$	–	–	–	–	$\{0\}\forall t$
$CM_{j,m,s}^F$	–	–	–	–	{0,0}
$CM_{i,m,t}^{IF}$	–	–	–	–	$\{0\}\forall t \neq 14$ $\{50\}\forall t = 14$
$\underline{Y}_m$	–	–	–	–	10
$\overline{Y}_m$	–	–	–	–	110
$y_{m,0}$	–	–	–	–	60

since this hour exhibits a very low RTP. As a result, the daily electricity cost of the process is reduced by 16.41% with respect to the baseline scenario.

In the scenario where the target process exhibits the additional flexibility of being interruptible (indicated in green), its first two steps are executed at the earliest acceptable time ($t = 8$ and $t = 9$) in order to exploit the lower RTP, and particularly execute the second step of the process at $t = 9$ which exhibits the lowest RTP of the day. The process is then interrupted for 7 hours, and its last two steps are executed at $t = 17$ and $t = 18$ (as in the uninterruptible scenario) in order to exploit the very low RTP of $t = 17$ for the third step of the process. As a result of this additional flexibility, the daily electricity cost of the process is now reduced by 28.53% with respect to the baseline scenario.

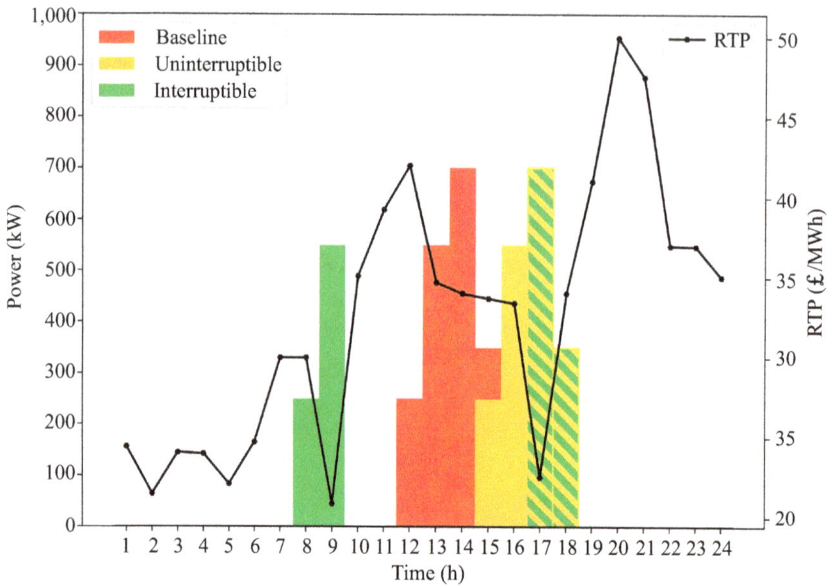

Figure 15.3 *Schedule of target process under baseline, uninterruptible with fixed power and interruptible with fixed power scenarios (23 February 2020)*

15.3.3 Benefits of flexibility types with adjustable power

We continue our analysis by presenting the benefits generated by the flexibility types with adjustable power (discretely adjustable and continuously adjustable). In these scenarios, the total energy requirement of the target process is not altered with respect to the baseline scenario (i.e., it is equal to $250\,\text{kW} \cdot 1h + 550\,\text{kW} \cdot 1h + 700\,\text{kW} \cdot 1h + 350\,\text{kW} \cdot 1h = 1{,}850\,\text{kWh}$, see Table 15.2) and the allowable execution interval remains the same with the two scenarios with fixed power examined in Section 15.3.2 (between $t = 8$ and $t = 19$) for comparison consistency purposes. In the same logic, the assumed power adjustability is consistent with the power levels of the baseline scenario: (a) in the discretely adjustable power scenario, the 4 alternative discrete power levels are identical to the power levels of the 4 steps of the process in the baseline scenario, and (b) in the continuously adjustable power scenario, the minimum and maximum power limits correspond to the minimum and maximum power of the process in the baseline scenario.

Figure 15.4 illustrates the schedule of the target process under four compared scenarios (baseline, uninterruptible with fixed power, uninterruptible with discretely adjustable power, uninterruptible with continuously adjustable power) for one of the days examined in the case study (23 February 2020) along with the RTP signals for the same day.

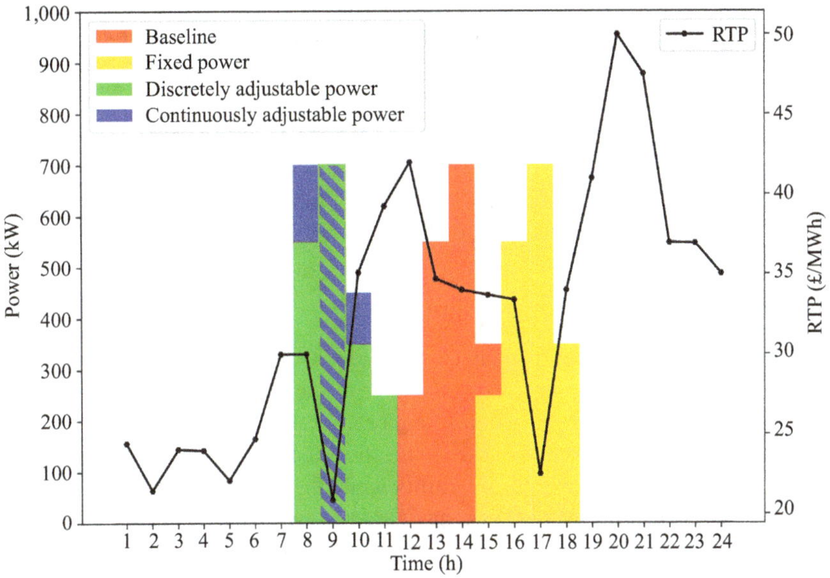

Figure 15.4 Schedule of target process under baseline, uninterruptible with fixed power, uninterruptible with discretely adjustable power and uninterruptible with continuously adjustable power scenarios (23 February 2020)

In the scenario where the target process becomes a flexible uninterruptible process with discretely adjustable power (indicated in green), it is executed at the earliest acceptable window (from $t = 8$ to $t = 11$), in contrast to the respective scenario with fixed power (indicated in yellow) where it is executed from $t = 15$ to $t = 18$, and despite the fact that the two scenarios exhibit the same allowable execution interval. The reason behind this effect is that the power adjustability allows the process to alter the power sequence of its 4 steps and exploit the lowest RTP of the day (at $t = 9$) by consuming the maximum possible power (700 kW) at this hour. As a result, the daily electricity cost of the process is now reduced by 18.40% with respect to the baseline scenario (while the respective benefit in the scenario with fixed power is 16.41%).

In the scenario where the target process exhibits the additional flexibility of continuous power adjustability (indicated in blue), its duration of execution is reduced to 3 hours (from $t = 8$ to $t = 10$), despite the fact that its minimum and maximum power limits are identical to the respective scenario with discrete power adjustability. This is because the continuous power adjustability allows it to consume the total energy requirement (1,850 kWh) through the 3-hour combination of 700 kW/700 kW/450 kW, while such a 3-hour combination is not feasible in the scenario with discrete power adjustability since the power level of 450 kW does not belong to the set of alternative discrete power levels; the closest feasible combinations are 700 kW/700 kW/550 kW (yielding a consumed energy of 1,950 kWh) and 700 kW/550 kW/550 kW (yielding a consumed energy of 1,800 kWh). Based on this feasible 3-hour combination, the target process consumes the maximum possible power (700 kW) at the first two hours ($t = 8$ to $t = 9$) which constitute the most favorable consecutive-2-hour interval within the allowable execution interval. As a result of this additional flexibility, the daily electricity cost of the process is now reduced by 21.17% with respect to the baseline scenario.

15.3.4 Benefits of material storage buffers

We conclude our analysis by presenting the benefits generated by material storage buffers. As discussed in Section 15.2.6, the value of such buffers lies in cases where the industrial consumer operates interdependent processes (in the sense that one process produces a particular material which is then consumed by another process), and only a subset of these processes exhibit flexibility while the rest are completely inflexible. In order to simulate such a case, we have assumed that the examined consumer's 4-step target process (Section 15.3.1) is broken down to two consecutive 2-step sub-processes: (a) a flexible sub-process, consisting of the first 2 steps of the original target process and producing 50 kg of a particular material at the end of its second step (Table 15.2) and (b) an inflexible sub-process, consisting of the last 2 steps of the original target process and consuming 50 kg of the same material at the beginning of its first step. Furthermore, following the convention of Section 15.2.6, we assume that the flexible sub-process is an uninterruptible process with fixed power (although our modeling approach can accommodate any of the examined flexibility types, see Section 15.2.6). Considering this case, we compare the baseline scenario (without a material storage buffer) against a scenario where a material storage buffer sits between the two sub-processes and can store the same material.

Figures 15.5 and 15.6 illustrate the schedule of the target process under the two compared scenarios for two different days examined in the case study (23 February 2020 and 28 February 2020) along with the respective RTP signals. In the baseline scenario, the flexibility of the first sub-process cannot be exploited, since shifting its execution earlier or later in time would imply that the second (inflexible) sub-process would not have the material required for its execution. Therefore, the whole target process is operated in an inflexible fashion, i.e., executed consecutively from $t = 12$ to $t = 15$.

The introduction of the material storage buffer (simplistically represented as a battery in Figures 15.5 and 15.6) enables exploitation of the flexibility of the first sub-process, by decoupling the scheduling of the latter from the scheduling of the second (inflexible) sub-process. In the first of the two examined days (Figure 15.5), the execution of the flexible sub-process is shifted 4 hours earlier with respect to the baseline scenario, i.e., from $t = 12$ to $t = 8$, in order to exploit the most favorable consecutive-2-hour interval within the allowable execution interval. The material produced by this sub-process is then stored at the buffer, increasing its content from 60 kg (at the beginning of the day, see Table 15.2) to 110 kg (after the termination

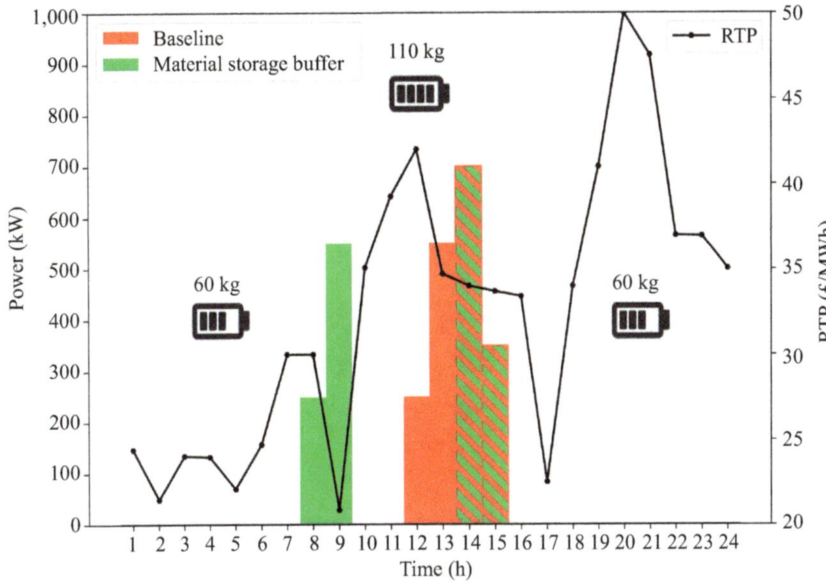

Figure 15.5 Schedule of target process under baseline and material storage buffer scenarios (23 February 2020)

of the flexible sub-process, see Figure 15.5); then, the same amount of material is fed to the second (inflexible) sub-process at its required initiation time ($t = 14$). In the second examined day (Figure 15.6), the execution of the flexible sub-process is now shifted 4 hours later with respect to the baseline scenario, i.e., from $t = 12$ to $t = 16$, as the RTP pattern is now different, and the most favorable consecutive-2-hour interval (within the allowable execution interval) comprises of $t = 16$ and $t = 17$. The amount of material required by the inflexible sub-process is now provided by the buffer, reducing its content from 60 kg (at the beginning of the day) to 10 kg (after the inflexible sub-process, see Figure 15.6); then, the buffer is replenished with the same amount of material after the execution of the flexible sub-process.

15.3.5 Summary of benefits of different flexibility types

Table 15.3 summarizes the results of the case study by presenting the benefits generated in each flexibility type scenario with respect to the baseline scenario. Specifically, the table presents the generated benefits in terms of the yearly electricity cost savings (a) of the target process alone and (b) of the examined industrial consumer as a whole, considering also its electricity demand that is not associated with the target process. The range of the former lies between 10.71% and 23.13%, while the range of the latter lies between 1.58% and 3.41% (since the target process constitutes a part

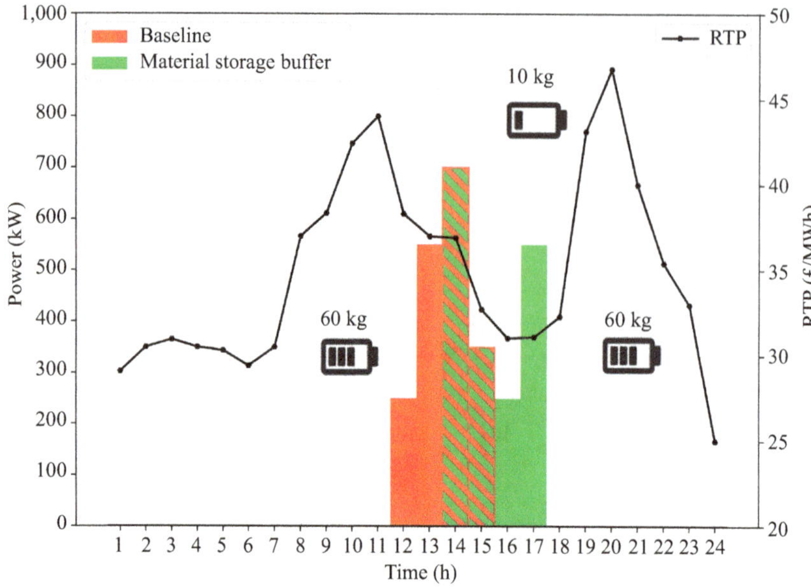

Figure 15.6 Schedule of target process under baseline and material storage buffer scenarios (28 February 2020)

of the examined consumer's total demand). Following the trends discussed in Sections 15.3.2–15.3.4, flexibility types allowing interruptibility of the target process yield higher benefits compared to the respective uninterruptible types. Furthermore, flexibility types allowing power adjustability yield higher benefits compared to the respective types with fixed power, with continuous power adjustability yielding higher benefits compared to discrete power adjustability. The scenario with a material storage buffer cannot be consistently compared with the other (apart from the baseline) scenarios since (a) the deployment of material storage buffers does not constitute a primary flexibility potential but rather an enabler for the exploitation of other flexibility types (Section 15.2.6), (b) only one flexibility type has been examined in conjunction with a material storage buffer (i.e., interruptibility with fixed power, see Section 15.3.4), and (c) only a part of the target process's demand can be flexibly redistributed in time (i.e., its first two steps, see Section 15.3.4).

15.4 Conclusions and future work

In the context of addressing the relatively limited and industry-sector-specific research attention to IDR, this chapter has developed generic, sector-agnostic models of all conceivable types of flexible industrial processes, with the aim to shed light on their

Table 15.3 Summary of yearly electricity cost savings quantified in the case study

Flexibility type	Process electricity cost	Total electricity cost
Uninterruptible with fixed power	17.02%	2.51%
Interruptible with fixed power	18.65%	2.75%
Uninterruptible with discretely adjustable power	19.28%	2.84%
Interruptible with discretely adjustable power	21.02%	3.10%
Uninterruptible with continuously adjustable power	21.68%	3.20%
Interruptible with continuously adjustable power	23.13%	3.41%
Material storage buffer	10.71%	1.58%

key operating differences and assist industrial consumers interested in IDR schemes to identify and assess the types that are more relevant to their systems. In this context, seven different types have been identified and discussed: (1) uninterruptible processes with fixed power, (2) interruptible processes with fixed power, (3) uninterruptible processes with discretely adjustable power, (4) interruptible processes with discretely adjustable power, (5) uninterruptible processes with continuously adjustable power, (6) interruptible processes with continuously adjustable power, and (7) material storage buffers. For each of these types, a mathematical model capturing its techno-economic operating characteristics has been developed, including: (a) input parameters, (b) decision variables, (c) operating constraints, and (d) inconvenience cost function. These models have been integrated into the electricity costs' minimization problem of an industrial consumer participating in the energy market through an RTP scheme.

A case study concerning an actual industrial consumer in the UK has been investigated, employing its real demand data and a target industrial process as well as real price data from the UK wholesale market. Different scenarios have been examined regarding the flexibility type of the target process, and the results have demonstrated, quantified and compared the economic benefits generated by different flexibility types. Specifically, the results demonstrate the additional value of the target process's interruptibility and power adjustability (discrete and continuous) as well as the deployment of material storage buffers. Overall, the yearly electricity cost savings are in the range between 10.71% and 23.13% with respect to the baseline electricity costs of the target process alone and between 1.58% and 3.41% with respect to the baseline electricity costs of the industrial consumer as a whole; we believe that these benefits are very significant, considering that the target process corresponds to approximately 17% of the consumer's total electricity consumption.

Although this chapter has focused on the exploitation of the flexibility of the seven identified types of flexible industrial processes in the energy market only, we believe that the developed models can be extended to exploitation of such flexibility in ancillary services markets, which receive continuously increasing interest by both industry and academia and can generate significant revenues for industrial consumers. This extension constitutes an area of future work and requires a comprehensive review and modeling incorporation of the complex technical requirements and market mechanisms employed by system operators for the procurement of different types of ancillary services.

Acknowledgement

This work has been partially funded by Innovate UK (as part of UK Research and Innovation) under grant number 78752 – for more information: https://gtr.ukri.org/projects?ref=78752.

References

[1] Shakoor A, Davies G, Strbac G, *et al.* Roadmap for Flexibility Services to 2030: A report to the Committee on Climate Change; 2017.
[2] Strbac G, Papadaskalopoulos D, Chrysanthopoulos N, *et al.* Decarbonization of electricity systems in Europe: market design challenges. IEEE Power and Energy Magazine. 2021;19(1):53–63.
[3] Papadaskalopoulos D, Strbac G, Mancarella P, *et al.* Decentralized participation of flexible demand in electricity markets—Part II: application with electric vehicles and heat pump systems. IEEE Transactions on Power Systems. 2013;28(4):3667–3674.
[4] Albadi M, El-Saadany E. A summary of demand response in electricity markets. Electric Power Systems Research. 2008;78(11):1989–1996.
[5] Strbac G. Demand side management: benefits and challenges. Energy Policy. 2008;36(12):4419–4426.
[6] Paulus M, Borggrefe F. The potential of demand-side management in energy-intensive industries for electricity markets in Germany. Applied Energy. 2011;88(2):432–441.
[7] Shoreh MH, Siano P, Shafie-khah M, *et al.* A survey of industrial applications of demand response. Electric Power Systems Research. 2016;141:31–49.
[8] Digest of UK Energy Statistics (DUKES) 2021, Chapter 5: Electricity. Department for Business, Energy & Industrial Strategy; 2021. Available from: https://www.gov.uk/government/collections/digest-of-uk-energy-statistics-dukes.
[9] Papadaskalopoulos D, Moreira R, Strbac G, *et al.* Quantifying the potential economic benefits of flexible industrial demand in the European power system. IEEE Transactions on Industrial Informatics. 2018;14(11):5123–5132.
[10] Ashok S. Peak-load management in steel plants. Applied Energy. 2006;83(5):413–424.
[11] Ding Y, Hong S, Li X. A demand response energy management scheme for industrial facilities in smart grid. IEEE Transactions on Industrial Informatics. 2014;10(4):2257–2269.

[12] Molavi H, Ardehali MM. Utility demand response operation considering day-of-use tariff and optimal operation of thermal energy storage system for an industrial building based on particle swarm optimization algorithm. Energy and Buildings. 2016;127:920–929.

[13] Reka SS, Ramesh V. Industrial demand side response modelling in smart grid using stochastic optimisation considering refinery process. Energy and Buildings. 2016;127:84–94.

[14] Gholian A, Mohsenian-Rad H, Hua Y. Optimal industrial load control in smart grid. IEEE Transactions on Smart Grid. 2016;7(5):2305–2316.

[15] Beier J, Thiede S, Herrmann C. Energy flexibility of manufacturing systems for variable renewable energy supply integration: real-time control method and simulation. Journal of Cleaner Production. 2017;141:648–661.

[16] Zhang X, Hug G, Kolter JZ, *et al.* Demand response of ancillary service from industrial loads coordinated with energy storage. IEEE Transactions on Power Systems. 2018;33(1):951–961.

[17] Summerbell DL, Khripko D, Barlow C, *et al.* Cost and carbon reductions from industrial demand-side management: study of potential savings at a cement plant. Applied Energy. 2017;197:100–113.

[18] Huang X, Hong SH, Li Y. Hour-ahead price based energy management scheme for industrial facilities. IEEE Transactions on Industrial Informatics. 2017;13(6):2886–2898.

[19] Angizeh F, Parvania M, Fotuhi-Firuzabad M, *et al.* Flexibility scheduling for large customers. IEEE Transactions on Smart Grid. 2019;10(1):371–379.

[20] Wang J, Shi Y, Zhou Y. Intelligent demand response for industrial energy management considering thermostatically controlled loads and EVs. IEEE Transactions on Industrial Informatics. 2019;15(6):3432–3442.

[21] Zaidi BH, Khan SS, Farooqui FN, *et al.* Demand response for industrial facilities. In: 2020 IEEE Transportation Electrification Conference Expo (ITEC); 2020. p. 760–765.

[22] Huang C, Zhang H, Song Y, *et al.* Demand response for industrial micro-grid considering photovoltaic power uncertainty and battery operational cost. IEEE Transactions on Smart Grid. 2021;12(4):3043–3055.

[23] ENTSO-E Transparency Platform [homepage on the Internet]. ENTSO-E; 2021. Available from: https://transparency.entsoe.eu/.

Chapter 16

Case study of Aran Islands: optimal demand response control of heat pumps and appliances

Marko Jelić[1], Dea Pujić[1], Nikola Tomašević[1], Paulo Lissa[2,3,4], Dayanne Peretti Correa[2,3,4] and Marcus Keane[2,3,4]

Demand response has proven to be a crucial mechanism in the process of flexibility exploitation on the demand side. Throughout the years, demand response has evolved exploiting more and more previously untapped potential energy sources. In that process, residential users have provided a significant buffering capacity for balancing energy production and demand, but this came with a few challenges. With more and more households transitioning from being purely energy users to smart homes and energy prosumers with distributed renewable energy generation, new possibilities have opened up for integrated optimisation approaches that make the best use of both locally generated and grid-supplied energy as well as energy storage systems.

16.1 Origins of demand response programmes

The global share of energy consumption, as analysed by the United Nations Environment Programme [1], can be disaggregated into different sectors. According to this report, 30% of final energy consumption and 28% of CO_2 emissions can be attributed to buildings. Interestingly, electricity consumption in building operation is said to represent around 55% of global electricity consumption. When specifically looking at residential buildings, they are reported to contribute 22% of final energy consumption and 17% of CO_2 emissions. Therefore, any reduction in energy consumption or increase in energy efficiency in the residential sector goes a long way towards fighting the ongoing climate battle.

Several elements have been noted in literature as key enablers of a transition to a greener future by analysing the problem of decarbonisation of the energy system at

[1]School of Electrical Engineering, Institute Mihajlo Pupin, University of Belgrade, Belgrade, Serbia
[2]College of Science and Engineering, National University of Ireland, Galway, Ireland
[3]Informatics Research Unit for Sustainable Engineering (IRUSE), Galway, Ireland
[4]Ryan Institute, National University of Ireland, Galway, Ireland

different scales, from the entire energy sector [2] to small and isolated grids that can be found on geographical islands [3]. Of these, two are crucial for residential energy use optimisation. First is the increasing prominence of distributed renewable energy generation (photovoltaic panels, wind turbines, etc.) and implementation of highly efficient sustainable technologies (like heat pumps) to replace legacy energy devices with large carbon footprints. However, with intermittent renewable sources becoming more commonplace, the delicate balance between the supply and demand has been jeopardised. Therefore, the second enabler is the utilisation of demand-side flexibility to aid in sustaining this equilibrium. Although a comprehensive classification of different mechanisms by which demand-side flexibility can be exploited is not yet well established within the related literature, it is generally well understood that demand-side management (DSM) and demand response (DR) are the crucial instruments in this domain. Although often used interchangeably by mistake, these two terms signify different approaches to load modification. DSM generally depicts long-term efficiency improvements that, overall, result in load reduction over time and aid the process of achieving full energy autonomy. On the other hand, DR refers to short-term load modifications that are made in reference to external impulses or incentives, thus helping maintain the stability of the wider power supply system.

As estimated by [4] at the time, 20% of power generation capacity was utilised only to fulfil peak demand levels which were present only about 5% of the time. This discrepancy has resulted in high operational costs of the power supply network as well as negative implications on emissions. However, by employing mechanisms such as DR, load levels should be able to exhibit more flexibility and, as a result, allow for easier load balancing between the supply and demand. Load modification techniques such as "peak curtailment/shaving" and "valley filling", as review by [5], can produce more balanced load curves that are less challenging to match with appropriate generation facilities. DR is an especially important tool in this regard as it can guide load modifications using its several different variants, as will be discussed in the following sections.

16.1.1 Traditional (industrial) DR applications

The first implementations of DR programmes can be traced back to the latter half of the 20th century. At that time, attempts at DR integration were primarily focused on large commercial and industrial customers. They were selected mainly due to an already present high level of automation as this ensures easier control of assets without additional devices and retrofits. Furthermore, industrial customers were also able to provide a significant amount of flexibility which facilitated contracting much more efficient as opposed to individual residential consumers which require some form of aggregation to provide a noteworthy impact.

Initial approaches were based on direct load controls through the so-called "explicit DR" which entails that the flexibility provider offers direct control over some of their assets at predefined time intervals and frequency while being offered monetary reimbursements in line with the provided capacity. This system has allowed utilities to make use of a portion of industrial demand levels as a buffer. In time, this process has evolved into a so-called "implicit DR" where the exchange between demand flexibility and monetary reimbursement is conducted via a variable energy price tariff. By alternating between low and high price intervals, the periods during which demand should be increased or decreased are implicitly encoded. When optimising their demand, users essentially attempt to reduce their operating costs by aligning the demand with the tariff profile, with resulting savings representing the previously directly agreed upon monetary reimbursement.

16.1.2 Transition towards the residential sector

Understandably, the concept of direct control over household appliances in the residential sector, even if the technical challenges of deploying the necessary equipment are ignored, is met with resistance by dwellers. Due to specific aspects of how human behaviour influences energy consumption habits, different attempts to utilise specifically price-based DR approaches have shown positive results in this domain [6]. Arguably, the most prominent implementation of implicit DR which has been in use for some time is time-of-use tariffs (also commonly implemented as night/day or peak/off-peak tariffs). However, the inclusion of distributed renewable generation as well as various controllable devices, especially as more and more households embrace the concept of smart homes, calls for an integrated approach to load modifications based on current conditions in order to maintain effectiveness in providing a balance between the supply and demand. One such solution, along with a set of results from a real-world use case, will be presented in this chapter.

16.2 RESPOND control loop and methodology

In order to provide a holistic solution to the problem of energy management for residential smart homes, the answer provided by the consortium of H2020 RESPOND project*, depicted in this chapter, utilises a set of smart services in conjunction with edge sensors and actuators to facilitate efficient day-to-day operation. This section presents different components of the proposed platform which, through synergistic operation, aim to integrate DR-supported optimisation into the operation of appliances, storage systems and heat pumps.

16.2.1 IoT backend platform

Considering the various types of data that need to be processed and stored in order to facilitate the operation of an Internet of things (IoT)-based system for home energy management, a complex heterogeneous platform for data handling and management was deployed as one of the primary components for this system. The RESPOND IoT platform was composed of a set of various data repositories:

- Semantic repository which contained metadata regarding users, sensors, equipment such as: characteristics of the photovoltaic panels (total capacity, slope, etc.), energy storage (battery capacity, maximum charge and discharge rates, etc.).

*www.project-respond.eu

- Influx database (DB) which contained time series measurements from the field devices such as electrical demand, renewable production measurements, temperature measurements, etc.
- Relational database which was used as an intermediary log for interaction between different services, as a repository for data to be shown in the accompanying mobile phone app as well as for user management for platform access.

All of these data stores were connected to the field level equipment via a Message Queuing Telemetry Transport (MQTT) broker. Since the integration point between different components of the system is located within the cloud platform, each service is envisioned to be able to obtain the required input parameters from the platform, as well as to store the outputs into the corresponding data repositories. Even external services like weather forecasting which are utilised within the system are integrated with the aforementioned data repositories.

16.2.2 Forecasting services

In order to be able to shift the demand and adapt it accordingly, as explained in the first section of this chapter, it is of utmost importance to have an estimation of the expected, baseline, renewable energy on-site production and demand in the first place. Hence, within the RESPOND project, both production and demand forecasting models have been developed independently with the goal of providing predictions for the same horizon and resolution which will be exploited by subsequent services like the optimisation. Since, at the time of development, historical production data was lacking while there were sufficient logs of previous demand measurements, the models were developed to provide 24 hour-ahead hourly forecasts with the demand forecasting model being data-driven and the production forecasting service realised using physical models.

The production forecasting service is envisioned to map predicted meteorological parameters to the expected production of available renewable energy sources (RES). In the particular use case that is presented in this chapter, the pilot site was equipped with solar photovoltaic (PV) panels. The inputs for the corresponding physical model, presented in [7], can be classified into one of the following two categories:

- **Dynamic parameters:** in order to be able to provide the expected production, external meteorological conditions are required. Therefore, as inputs for this model, global solar radiation and cloud coverage were necessary. Additionally, apart from the weather, temporal parameters are correlated with the production, as it is highly dependent on the instantaneous solar position. Therefore, current time and date are included as the inputs, as well.
- **Static parameters:** as usual when utilising physical models, apart from the dynamic parameters, which are also common for data-driven models, the following physical and geographical parameters were necessary: slope and azimuth of the PV cell surface, temperature coefficient and surface area of the PV cells, rated capacity of the PV array, nominal operating cell temperature, longitude, latitude and time zone offset.

When performance of the utilised model is considered, it achieved a mean absolute error of 21% and, as was expected, was not as accurate as machine learning (ML) models would be if historical data had been available. Nevertheless, it was precise enough to be utilised as an indicative input for optimisation purposes.

Regarding demand forecasting, a couple of different ML models were tested and the k-nearest neighbour (kNN) algorithm was chosen for utilisation since it had the most accurate predictions. Similarly regarding the production forecast, the demand forecasting service was designed to provide 24 hour-ahead forecast with an hourly resolution. As explained in more details in [8], the kNN model predicts expected demand depending on the previous demand and a set of the time variables extracted from the date and time such as day of the month, the season, the day of the week and a Boolean variable indicating whether it is a working day or not. The previous electrical load is obtained from the Influx DB, whilst the outputs of the service are stored in MySQL DB, as was the case with the production forecaster. Since the COVID pandemic has impacted the validation period of the system, the models have been adapted accordingly [9]. This improvement was necessary since household electricity demand has changed as a result of the fact that most of the users started working from home.

16.2.3 Optimisation services

Constantly analysing the expected renewable generation and attempting to align the demand in accordance to its profile, as well as various DR requests, is a cumbersome task that very few residential users want to constantly take upon themselves to resolve. In order to ensure cost-effective and energy-efficient operation, significant efforts have to be invested in order to make best use of, for example, varying energy prices, or local energy generation and storage. Therefore, the optimisation services within the RESPOND control platform was envisioned as an integrative component that would be capable of assessing multiple aspects of the energy management problem and automatically providing the best course of action by analysing the:

- arrangement of the underlying energy infrastructure components including all relevant energy carriers, converters and storage systems of individual energy prosumers.
- forecasted production profile from all locally available renewable energy sources obtained by the corresponding forecasting service.
- forecasted energy demand as well as corresponding load flexibility constraints obtained by the corresponding forecasting service.
- limitations of the grid connection and other components of the system.

In order to make the best use of contemporary computational power while simultaneously guaranteeing that all available resources are being used in the most efficient manner, adequate models of the energy systems are built and optimised using an appropriate solver engine. The methodology that was utilised within the control platform is based on the Energy Hub modelling approach. Originally presented by [10] with various subsequent implementations in literature, as revised in [11], owing to its flexible nature, the Energy Hub can be utilised in a variety of problems, from single carrier electric energy systems [12] to complex multi-stage hybrid systems that involve, for example, both electrical and thermal domains with adequate converters [13].

Following this concept, a corresponding model was developed for the use cases that will be discussed in the following sections of this chapter. Since the optimisation in this case focuses on the electrical domain as it is supplemented with smart sensors and actuators that are integrated in the platform, the model depicts different electric energy sources, a battery storage system (where applicable), converters and loads, as illustrated in Figure 16.1. In accordance with this structure, a corresponding set of variables, bounds and constraints can be derived and implemented as a mixed-integer linear programming (MILP) problem. This choice allows for the energy flow for each model to be efficiently evaluated in sub-second times facilitated by the simplicity of a MILP model that does not require the implementation of numerical solvers as this type of precision is generally regarded as unnecessary in similar modelling problems.

The forecasting services, which are to be evaluated before the optimisation, define a portion of the variables of the model. For example, the RES forecast depicted previously defines the available energy from the PV array while the demand forecast provides a reference based on which some form of demand flexibility can be implemented. In line with general findings from [14], the overall flexibility in a community-oriented project is estimated at around 20% while a use case study depicted in [15] hits at the possibility of peak load reduction of 30% with extreme energy tariff manipulations. These findings provide a range of values that can provide context for the flexibility margin around the forecasted load profile in which upwards and downwards load modifications are made.

An important feature in the optimisations service is the integration of both explicit (direct) and implicit (price-based) DR. Namely, since the operation of each mode is optimised with operational costs set as the main criterion, the output reflected the most cost-effective solution with varying energy prices in mind. With implicit DR natively supported using this setup, an additional term in the criterion is added such that, if an external grid-side entity or aggregator wishes to request a certain demand level at a predefined time interval, the deviation of the output demand curve is separately

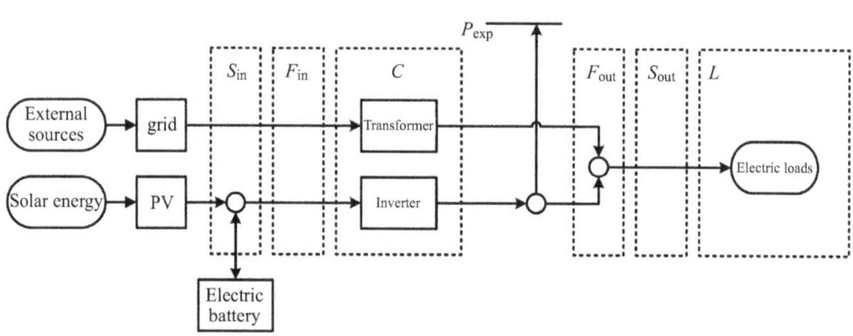

Figure 16.1 Illustration of the energy hub layout of a house in Aran Islands

and highly penalised. This is done such that the output optimal demand reflects the required profile as closely as possible, thus also facilitating explicit DR requests.

Finally, the outputs of the optimisation service are comprised of a set of optimal energy utilisation curves that reflect when and how much energy is to be stored, imported, exported and consumed. In accordance with these profiles, subsequent services depicted in the following sections will provide means of converting power consumption curves into concrete control actions depicting when to schedule appliance usage and how to set the references for heating, ventilation, and air conditioning (HVAC) devices.

16.2.4 Control services

After running the optimisation service, the system is provided with optimal energy utilisation curves as power values through time. However, this format of data usually cannot be considered to provide useful information without being processed first. This process entails the conversion of these curves into discrete "turn on"/"turn off" instructions for appliances and set-point values. These two processes will be further discussed in the following sections.

16.2.4.1 User recommender service and appliance controls

The first component of the control service mainly focuses on the electric domain and the conversion of optimal loading profiles into appliance use schedules. An approach for solving this problem, outlined in [15] and further explored in [16], makes use of a heuristic tabu search method. Namely, by discretizing the appliance usage schedule, the algorithm looks for the best arrangements of their activity in time such that this schedule, in conjunction with the fixed demand curve, results in a closest match to the demand curve that is deemed optimal. However, this approach only tackles the electric domain while applying it to the thermal one would be a much more challenging task as it would require the use of complex models. This issue is precisely what will be further discussed in the following section.

16.2.4.2 Building models and HP controls

To create a simulated building model, a site survey has been conducted to gather data related to construction characteristics, such as type of the walls, windows and roof. Furthermore, sensors and meters have been installed, measuring indoor temperature (°C), total electricity consumption (kWh) and heat pump electricity consumption (kWh). The collected data was used to develop a detailed and calibrated white-box model, using the Integrated Environmental Solution Virtual Environment (IESVE) software. Next, simulations have been carried out to identify the main parameters and heating transfer dynamics necessary to build a reduced order grey-box model. The parameters extracted from the white-box model were indoor air temperature increase and decrease rates for both, domestic hot water (DHW) and indoor temperatures, considering their behaviour during stationary conditions (system off) and when actions are performed (indoor heating or DHW on). Additional information about the house parameters and white-box model calibration, including the validation metrics applied,

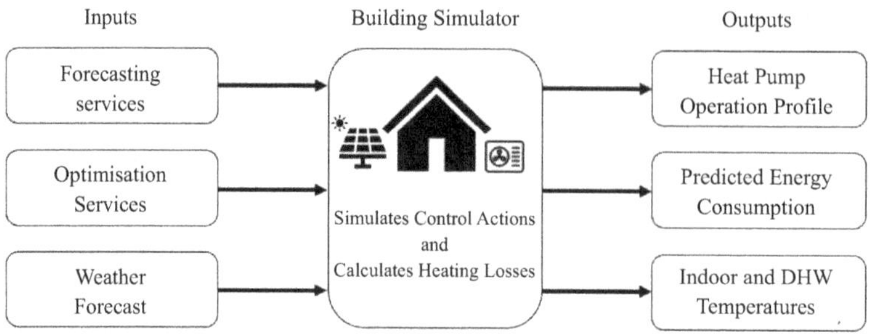

Figure 16.2 Building Simulator process

can be seen in [17]. Once developed, the building simulator is able to receive the optimisation and forecasting services as input and translate them to optimal heat pump control actions, as can be seen in Figure 16.2.

The Building Simulator reads the current environment state every 5 minutes and estimates the next DHW and indoor temperature values, calculated based on the rates established in the white-box model. This new environment allows for a more flexible framework where different control techniques can be tested. The forecasting and optimisation services, along with the weather forecast, are used as inputs, so they can be part of the DR control strategy. For instance, a control action can be scheduled to activate the heat pump when the PV production forecast will be higher in the next day. The output of the building simulator is a heat pump control operation profile, which indicates the control actions to be executed the next day. Furthermore, it calculates the predicted energy consumption and the expected indoor and DHW temperatures over the day.

16.3 Use case setup

The location of the experiments is Inishmore, the largest of the three Aran Islands, in Galway, Ireland. With a population of approximately 800 people, the island itself is very exposed to the weather elements, particularly during the winter months as it has very little shelter. The islands are connected to the mainland through a sub-sea cable, in which 1,855 MWh of electricity was imported in 2017 [18]. In 2016, a fault in the sub-sea cable resulted in a power outage on two of the three Aran islands, that lasted for four days and affected approximately 400 residents [19]. On that occasion, some islanders had to rely on local diesel-powered generators. This event showed the islands' vulnerability and dependency on the main island power generation leveraging the need for new reliable on-site solutions.

16.3.1 Pilot installations

On the Aran Islands, there is a potential to apply demand response in 450 dwellings that share similar characteristics in terms of construction materials. Moreover, as they are geographically close to each other, the external environment conditions do not vary considerably across the buildings, hence the daily heating losses dynamics tend to be similar. A total of nine houses have been selected to be part of the test cases. They already had individual PV production for self-consumption, a heat pump system for indoor heating and DHW, and appliances such as washing machines and tumble dryers. To allow DR capabilities and to take benefit of the services provided by RESPOND, a new set of devices from Energomonitor[†] were installed in each of the houses. The new architecture added smart capabilities for the legacy equipment, allowing for individual load measurement and control, besides as well as providing monitoring of room temperatures and CO_2 concentrations. The description of the deployed devices and their application can be found below:

- External meter interface (Energomonitor Optosense): measures electricity consumption or production by reading the optical impulse output of a digital electricity meter. Application: Electricity meter.
- Electricity wire sensor (Energomonitor Powersense): measures electricity consumption or production by induction coils installed on 1 or 3-phase wires leading to the main breaker cabinet/panel. Application: Heat pump, PV, and electric vehicle charger.
- Temperature sensor (Energomonitor Thermosense): is a thermometer for indoor or outdoor use. Application: Room temperature sensor.
- CO_2 and humidity sensor (Energomonitor Airsense): monitors complex air quality in the room – carbon dioxide (CO_2) concentration, temperature, humidity and noise level. Application: User comfort level measurements.
- Smart plugs (Energomonitor Plugsense): measures consumption over residential electrical appliances and can be used to switched them on and off remotely. Application: Individual load control (e.g., dishwasher, washing machine, tumble dryer).
- Gateway (Homebase): is the heart of the solution Energomonitor, wirelessly picking up data from up to 30 transmitters in the house through encrypted radio protocols. Application: To provide communication and data between the previous devices and services.

The final list of measurement points per house can be seen in Table 16.1, where each electrical appliance, equipment or sensor can be related to the aforementioned Energomonitor devices, following their specific application. The data gathered from the devices were utilised as input for the services described in sub-section 16.2 and also for user's verification and control, through a mobile application.

Finally, to assess the demand response capabilities and support the validation process, an average dwelling, that was built in the 1970s and has a total floor area of 110 m^2, was modelled following the process found in Section 16.2.4.2. In recent years, the dwelling has been upgraded, including additional external insulation to the walls and roof and installation of an 8.5 kW Mitsubishi air sourced heat pump along with a PV panel array consisting of eight panels, with a total nominal power of 2 kW_p.

[†] https://www.energomonitor.com/

Table 16.1 Number of measurement points per house

House number	01	02	03	04	05	06	08	10	12
Dishwasher	1		1				1	1	
Electricity meter	1	1	1	1	1	1	1	1	1
EV charger							1		
Heat pump	1	1	1	1	1	1	1	1	1
PV panel	1	1	1	1	1				1
Tumble dryer	1	1	1	1	1		1		1
Washing machine	1	1	1	1	1	1	1		1
Temperature sensor	3	5	5	4	5	5	5	5	5
Humidity sensor	2	5	2	1	2	2	2	2	2
CO_2 sensor	1	1	1		1	1	1	1	1

The heat pump connects to a 170 L hot water cylinder which is used to store hot water for both space heating and DHW.

16.3.2 User interface

In RESPOND, a mobile app has been deployed to increase the user's participation in the DR strategies. The app enables users to visualise energy-related consumption and generation, to check comfort matters and status of the devices. Moreover, information from the forecasting and optimisation services can also be visualised, helping user's to make informed control actions. Through the app, users can receive notifications asking to consume more or less energy according to the DR event. Some of the screens available in the user interface can be seen in Figure 16.3.

Detailed information about the mobile app can be found in the RESPOND report [20]. Starting with the main page screen, left screen of Figure 16.3, is shown once the RESPOND mobile app is loaded. This is the main screen where users can navigate and select other different screens to visualise the information available. For instance, the energy consumption screen shows the recent and historic energy consumption at a dwelling and neighbourhood levels. Furthermore, dwellers can visualise hourly, daily, weekly or monthly energy consumption information. The energy generation screen (centre screen of Figure 16.3) shows users the recent and historic values of energy being produced from the PV panels, also with hourly, daily, weekly or monthly resolution. This information may help users to realise the levels of PV production available, and combined with their energy consumption, to raise awareness of the potential reduction of energy coming from conventional non-renewable sources. The energy prices screen, right screen of Figure 16.3, shows hourly tariffs of the energy for the current and upcoming hours and days, so users can decide the best time to use or not some appliance.

Another important information found in the app is the comfort screen, which helps users to verify the indoor environment quality. It provides the mean temperature, humidity and CO_2 levels, which are considered the basic indicators of comfort.

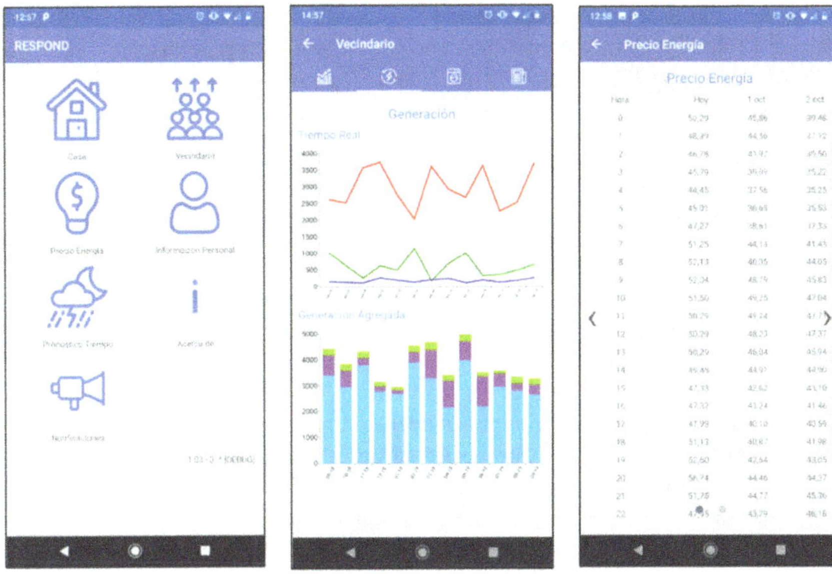

Figure 16.3 RESPOND App–general user interface

Moving to the device list screen, users can monitor and control devices within their houses. Users can then select each of these devices and take different actions. For example, for comfort devices, a user can check the temperature, humidity and/or CO_2 measurements. For appliances, users can check their current and historical energy consumption, as well as activating or deactivating them. The Weather Forecast screen is aimed at showing users the expected weather for the upcoming hours and days. This information, combined with other features such as the energy price, empowers users to strive towards more environmentally-friendly and energy-saving behaviour. For example, knowing that the next day is going to be hot and sunny, a user can decide to hang out their clothes instead of using the tumble dryer.

Finally, the notifications screen presents the notifications received. Users can receive different types of notifications such as recommendations or even alarms or warnings. These notifications will be received in the mobile app in the form of push messages instantly and also be available in the notifications screen.

16.4 Case studies and assessment

Four different test cases have been designed to assess different types of DR within the RESPOND project in the Aran Islands, considering implicit and explicit DR models. The use test cases aim to exploit as much as possible the benefits of the ICT architecture deployed in the pilot and the optimisation services available, presented in the previous sections.

The test cases results were calculated using the DEXMA Energy Intelligence Software [21]. The DEXMA platform enables real time energy management, with a Measurement and Verification (M&V) tool that contains an automatic baseline calculator [22] fully compatible with the International Performance Measurement and Verification Protocol (IPMVP) [23].

16.4.1 Test case #1

The objective of the first test case was to analyse the impact of the RESPOND smartphone app on user consumption behaviour. The main idea was to understand if, after having access to the information described in Section 16.3.2 (e.g. energy consumption per appliance, PV production, etc.), consumers changed their energy consumption pattern in a voluntary manner. For the assessment, comparison of the period before and after the RESPOND app release has been compared. This test case started to be applied on 13 April 2020 which was the day participants received their RESPOND app passwords. The final date of verification of voluntary behaviour change was 30 May. As this experiment aim is to verify users' willingness of changing their consumption, there was no notification or other kind of intervention asking them for some energy reduction or increase in the period of evaluation.

The baseline period used in this analysis was from 1 March 2020 until 12 April of the same year. Considering the accumulated values of all dwellings, the results of this test case presented a reduction of 20.28% in energy consumption in the period, compared to the baseline. Figure 16.4 shows the results of the M&V project created inside the DEXMA platform to calculate the energy savings key performance indicator (KPI) in the first test case. The blue line represents the expected consumption over

Figure 16.4 Example of M&V output from DEXMA platform (blue line – baseline, green bar – lower than baseline consumption, orange bar – close to baseline consumption, red bar – higher than baseline consumption)

the days (baseline), while the bars are the real consumption. Green bars are values where the real consumption is lower than the baseline, red is the opposite, and yellow means that they are close. The reduction of greenhouse gas emission is also estimated in the platform and in this test case, there was a total of 2.51 t CO_2e avoided in the referenced period.

The performance of the ICT communication infrastructure is a very important topic for guaranteeing the reliability of the test case. With this in mind, some houses were excluded from the validation process due to a lack of sufficient data. The criteria for exclusion in this test case was that houses should have at least 60% of data available in the discussed period, hence three houses that presented lower values were excluded.

16.4.2 Test case #2

The objective of the second test case is to maximise PV self-consumption during periods where there is a peak in energy production. On the day preceding the DR event, the forecasting services estimate the hourly PV production for the next day. With this information, a notification is sent to the participants informing them about the best time to consume the PV produced energy if a pre-defined threshold is achieved. Users can get energy savings and also help to reduce peak load in the grid. In the Aran Islands, if PV production is not consumed, it is injected directly into the grid, and users do not receive any payment from the energy provider.

This test case was applied when the prediction of energy production achieves a specific target. The first step was to define which houses were able to participate. Although there are nine houses in total, only houses 01, 02, 03, 04, 05 and 12 have PV production, therefore messages were sent only to this group. The Irish language, or Gaeilge, is unique to Ireland and it is, therefore, of crucial importance to the identity of the Irish people [24]. To better engage the participants in the actions, the notifications were sent to the participants in English and Gaelic:

"Tomorrow between HH:MM-HH:MM your PV panels are expected to have a period of high production. Try to use your appliances during this period to save money and energy."

"Amarach idir HH:MM-HH:MM meastar go mbeidh do phainéileacha fotavóltacha ag ginniúint roinnt mhaith leictreachas. Déan iarracht do chuid fearais tí a úsáid i rith an am sin chun airgead agus fuinneamh a shábháil."

The application of this test case is classified into one of two experimental periods. During the first experimental period, a message was sent to the customers if the predictions achieved the threshold of 900 W for at least an hour. The demand response events started sending the messages on 31 May. Since messages were not being sent due to the weather conditions in Ireland at that time (PV predictions rarely achieved 900 W), an analysis was performed to define a new threshold that could result in sending notifications 2–3 times per week on average. In August, the new value was then defined as 600 W for houses 01, 03, 04, 05 and 12 and 1100 W for house 02, which has a PV system with higher capacity, effectively defining the second experiment.

The KPI calculations showed a decrease of 6.11% in energy usage from the grid in the first experiment period, and 21.81% in the second period, considering the aggregated values of the participant houses. In the total test case duration, the final result is 17.89% of energy savings. The reduction of greenhouse gas emissions is the energy savings total converted to greenhouse gas emission equivalent and it is estimated that this test case avoided 49.66 t CO_2e over the entire period.

The total renewable energy consumption KPI shows the ratio of the total amount of renewable energy produced and the demand at the event period. Table 16.2 summarises the amount of PV energy production consumed in each day of first and second experiment. The analysis considers the aggregated consumption and production value of the participant houses. As a result, PV production was 72.7% consumed on average over the first experiment. The second experiment presented an even better performance, where 79.6% of PV energy produced during the event was consumed on average, and sometimes reached 100% of usage.

The baseline used for this use case was from 1 March 2020 to 4 April 2020. The main calculations were realised using the DEXMA platform. According to the IPMVP methodology, data backfilling is not allowed [25], and in line with this, periods with missing data were excluded during the calculation process. Outliers, such as accumulated values due to communication issues, have also been removed.

Table 16.2 Total energy production/consumption during DR events

Date	PV production (kWh)	Consumption (kWh)	% of PV usage
June 1st	7.55	5.66	75%
June 3rd	4.91	3.28	67%
June 5th	5.16	3.10	60%
June 8th	1.48	1.32	89%
Experiment one avg.	4.78	3.34	72.7%
August 6th	2.68	2.73	100%
August 7th	8.49	5.23	62%
August 8th	7.90	5.85	74%
August 9th	2.69	2.30	86%
August 10th	3.39	2.02	60%
August 11th	2.76	2.93	100%
August 17th	2.06	1.68	82%
August 19th	1.79	2.55	100%
August 20th	2.39	2.17	91%
August 22nd	7.61	4.24	56%
August 28th	3.84	1.77	46%
August 29th	0.91	2.02	100%
Experiment two avg.	3.87	2.96	79.6%

The rescheduled demand KPI aims to verify if the use case helped to move demand into the event period. For instance, if the PV production was higher from 15:00 to 17:00 and the user had received a message, it was expected that a greater consumption during this period and less activity before and after the event would be observed. After the period of the experiments, the real measurements were compared with the baseline. According to the baseline analysis, it was expected that 6.70% of the daily load would be in the event period for experiment one. However, after applying the demand response events, the real data showed 8.71% of the load in the period, which represents a demand increase of around 30% in the event hours. On the other hand, experiment two did not present the same performance, with a 1% of load decrease during the event, compared to the baseline.

The economic savings KPI compares the difference between the average baseline energy cost and the energy cost during the DR event. It considers the amount of energy consumed from the grid, so using appliances when the PV production is higher during the event reduces the final costs. As a result, imports from the grid represented only 4% of the total necessary energy during the event, which is 20% less than the expected baseline. It is important to note that the costs are much related to the way that end users distribute their load over the day. For instance, if the PV production is high and achieved 2 kW, the user has to be aware and try to avoid exceeding this value by controlling the amount of load that uses a certain time to optimise the usage. Otherwise, the amount of energy bought from grid can be greater than expected as all loads are concentrated during the same time interval.

16.4.3 Test case #3

The aim of the third use case is to maximise PV self-consumption by generating an optimal usage profile for heat pumps. This use case can be performed in different ways, such as through fully-automated operation of the heat pumps, which can achieve better energy savings and does not rely on the user behaviour, or manually operated, following a similar methodology as outlined in Test case 2 where users have to play their role and perform the actions when necessary. For remote operations, the installation of an additional device, that is not available in the selected houses, is needed. To assess the potential of this test case, tests have been performed in the simulated environment described in Section 16.2.4.2.

The overall methodology is similar as Test case 2, where at the night before the event, the PV production prediction is checked to identify the best period to perform actions (when production is expected to be high). Actions are simulated and the comfort parameters verified in the building simulator. As a result, an optimal heat pump operation profile is generated. In this test case, the focused was placed on optimising the DHW temperature regulation by changing the operation mode and boosting the tank temperature when PV production is higher. This will avoid DHW heating actions during peak consumption hours by creating a buffer of hot water. In the Mitsubishi heat pumps available in the houses, operation mode 1 is the standard mode, where the tank temperatures range from 40°C to 50°C. In mode 2, the maximum threshold is increased and temperatures can go up to 55°C.

As inputs for the building simulator, 10 days in August 2020 were selected where PV production prediction was considered good enough to aid the heat pump system operation. The forecasting services provided information about the peak hour of PV production for each of the days. The model was then simulated, prioritising DHW actions mode 2 in these peak generation periods. As a result, an optimal operation mode is set for each of the hours the energy from PV was generated. The operation modes can then be set manually by the user or autonomously, if technology for remote control is available. Figure 16.5 shows 2 days of simulations, where the blue line is the tank temperature and the dotted green line is the PV production. The grey bars are the control action for heating the tank. In the first chart, both actions are in periods of PV production, which occurs only once in the second chart. The reason is that the optimised model also takes into account the minimum setpoint to keep the user's comfort as the main premise, so regardless of the PV production, if the temperature drops to below 40°C, the system performs a DHW action.

If only operation mode 1 is considered, the heat pump would heat the 170 L tank to 50°C and then stay on hold until the temperature drops to 40°C, heating again to 50°C and so on. This mode can be costly, as it does not verify the best time to perform the actions, which may be when there is no PV production or when the energy demand peak is high. The optimised profile provided by the building simulator checks the PV production schedule and anticipates the actions needed to achieve the economic savings, without adversely impacting on users comfort.

The results from the building simulator were compared with the heat pump real consumption (baseline). Almost 30% of energy from the grid used for DHW heating could be saved if the actions had been performed as the optimal profile generated, mostly due to using the heat pump when the PV energy production was higher. This performance could drop to around 25% because the real world scenario can face more uncertainties related to users behaviour. For instance, the simulations consider an ideal profile of DHW usage, while the real user can suddenly decide to use all the water at once, thus making the control system activate more times over a day.

Another important metric is the rescheduled demand, which was calculated considering the total demand consumed inside the period of higher PV production incidence (10:00–18:00) and out of it, considering the average of the 10 analysed days. Looking at the collected data, only 37% of the heat pump consumption was

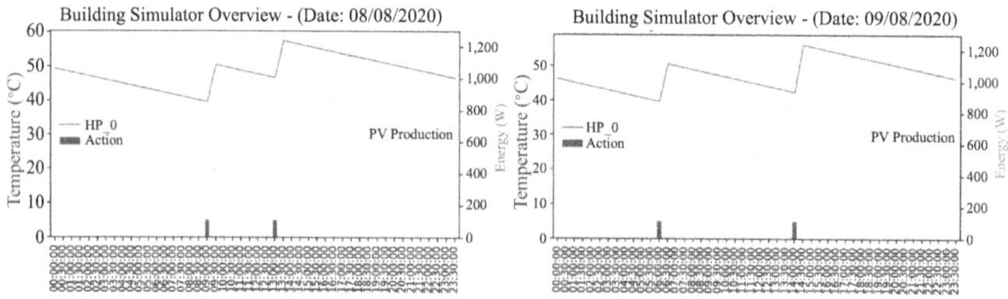

Figure 16.5 Building simulator dashboard

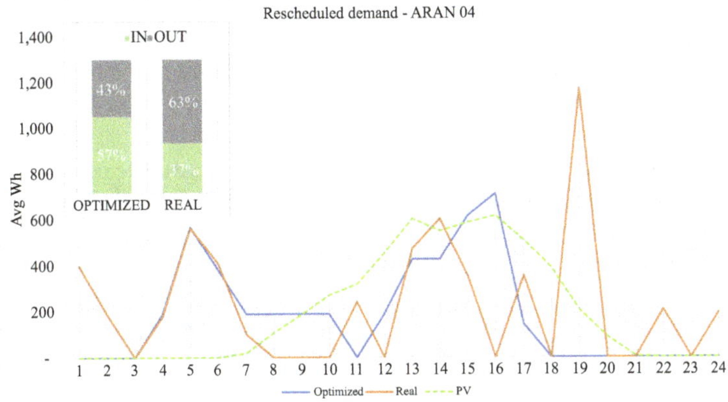

Figure 16.6 Rescheduled demand test case #3

inside the PV event range, while the optimised model increases this value to 57%, which effectively demonstrates an increase of 20%. Figure 16.6 presents the average consumption for both the optimised (blue line) and the real data (orange line), including also the average of PV production (dotted green line). Note that one of the benefits of the optimised model is the peak load reduction and load shifting, as the peak load that was originally between 18:00 and 20:00 in the real data was moved a few hours in advance in the optimised version.

Regarding the utilisation of PV production, the amount of real PV consumed and the optimal PV consumption was compared. For experimental purposes, that PV production was considered to be used exclusively for heat pump actions. For each of the days, it was calculated the amount of renewable energy used and, on average, the optimal profile model improved renewable energy usage by 39.14%, by concentrating DHW action in periods with higher PV, as previously presented in Figure 16.5.

16.4.4 Test case #4

The aim of the fourth test case is to verify the changes in customer behaviour after receiving a message asking to turn off some appliances at a specific hour of the day with the main objective to decrease carbon emissions. In these events, the idea was to ask the participants to not use electric energy, without offering any direct financial incentives. The message was sent one hour prior the event to all the participants of the pilot through an app notification in both English and Gaelic languages, as follows:

"Electricity consumption of Ireland peaks within the next few hours, which means higher CO_2 emissions. Turn off some of your appliances between HH:MM and HH:MM – and help us with saving the climate."

"Buaicfidh tomhaltas leictreachais na hÉireann sna cúpla uair amach romhainn, rud a chiallaíonn astaíochtaí CO_2 níos airde. Múch cuid de do chuid fearais idir HH:MM agus HH:MM – agus cuidigh linn an aeráid a shábháil."

As opposed to the other test cases where the focus was on individual parameters, such as demand or production information, in this case the timing of the event was based on electrical grid data provided by EirGrid [26]. The Aran Islands are connected to the electricity grid in Ireland, and during the test case period the highest demand for energy identified occurred between 17:00 and 18:00. This test case could also be applied using other inputs, such as periods of high consumption or lower production identified in the forecasting services. The baseline period for the test case was the first 2 weeks of July 2020 and the DR events happened in August 2020. Although the average peak in Ireland is between 17:00 and 18:00, before sending the message this range was confirmed on the demand system prediction provided by EirGrid [26] for each event day.

A reduction of 14.73% CO_2 was observed in aggregate during the days of the event compared to the baseline period. Considering the consumption of individual appliances, it was also verified that an additional 4.37 tCO_2e could have been avoided in the period if all the customers had carried out the proposed action. However, the hourly analysis of the users' consumption behaviour showed no significant reduction in the peak load during the event period compared to the baseline. Considering the communication performance, houses 02, 04 and 08 were excluded and not considered in the validation process due to the missing data over the baseline period.

Conclusion

In summary, this chapter presents a platform for integrated management of residential energy systems by incorporating demand response events into a measure-forecast-optimise workflow for automated and semi-automated control of appliances and heat pumps. Through smart use of a set of deployed sensors and actuators, as well as synergistic relation between different services within the platform, the proposed system takes into account both generation and demand-side constraints in order to provide the most cost-effective and energy-efficient scenario for energy management. Different methodologies that were utilised for different components of the system are outlined, followed by a set of four thoroughly analysed use cases focusing on the adaptation of electric loads and utilisation of heat pumps in relation to specifically generated DR events.

Various scenarios are depicted in the discussed test case results with them portraying, in line with the applicable time periods and baseline data, effective savings of 2.5 t CO_2 emissions by providing information to users regarding their energy use and slightly below 80% of renewable energy self-consumption achieved once adequate messages are sent to denote periods with high expected production levels. Furthermore, an estimated 25% reduction of grid-imported energy for DHW temperature regulation through heat pump usage optimisation is demonstrated, followed by a case showing a reduction in CO_2 emissions of over 14% when responding to DR messages intended to shift appliance activations away from peak times for the grid.

Finally, it should be noted that this chapter provides a quantitative analysis based on the data that was collected and processed through the presented platform.

The results presented are based on evaluations of the absolute available numerical data and are derived based on the selection of an appropriate baseline estimation methodology. Since energy use, especially for the residential case, is a complex multidisciplinary problem, there are also behaviour-related factors that influence the way in which energy is managed. Therefore, as a complementary addition to the presented results, related studies pertaining to the user experience domain [27] should also be considered.

Acknowledgement

The research presented in this chapter is partly financed by the European Union (H2020 RESPOND project, Grant agreement no. 768619 and SINERGY project, Grant agreement no. 952140) and the Ministry of Education, Science and Technological Development and the Science Fund of the Republic of Serbia (AI-ARTEMIS project, #6527051).

References

[1] Programme UNE, Construction GAfBa. 2020 Global Status Report for Buildings and Construction: Towards a Zero-emissions, Efficient and Resilient Buildings and Construction Sector – Executive Summary. Accepted: 2020-12-15. Available from: https://wedocs.unep.org/xmlui/handle/20.500.11822/34572.

[2] Matt Golden, Adam Scheer, Carmen Best. Decarbonization of electricity requires market-based demand flexibility. The Electricity Journal. 2019;32(7):106621. New York, NY: Elsevier. Available from: https://www.sciencedirect.com/science/article/pii/S1040619019302027.

[3] Barney A, Polatidis H, Jelić M, *et al*. Transition towards decarbonisation for islands: development of an integrated energy planning platform and application. Sustainable Energy Technologies and Assessments. 2021;47:101501. Available from: https://www.sciencedirect.com/science/article/pii/S2213138821005129.

[4] Farhangi H. The path of the smart grid. IEEE Power and Energy Magazine. 2010;8(1):18–28. Conference Name: IEEE Power and Energy Magazine.

[5] Lund PD, Lindgren J, Mikkola J, *et al*. Review of energy system flexibility measures to enable high levels of variable renewable electricity. Renewable and Sustainable Energy Reviews. 2015;45:785–807. Available from: https://www.sciencedirect.com/science/article/pii/S1364032115000672.

[6] Faruqui A, Sergici S. Household response to dynamic pricing of electricity: a survey of 15 experiments. Journal of Regulatory Economics. 2010;38(2):193–225. Available from: https://doi.org/10.1007/s11149-010-9127-y.

[7] Duffie JA and Beckman WA. *Solar Engineering of Thermal Processes*. John Wiley & Sons, Ltd; 2013. Available from: https://onlinelibrary.wiley.com/doi/book/10.1002/9781118671603.

[8] Gómez-Omella M, Esnaola-Gonzalez I, Ferreiro S. Short-Term Electric Demand Forecasting for the Residential Sector: Lessons Learned from the RESPOND H2020 Project. Proceedings. 2020;65(1). Available from: https://www.mdpi.com/2504-3900/65/1/24.
[9] Gomez-Omella M, Esnaola-Gonzalez I, Ferreiro S. Short-term forecasting methodology for energy demand in residential buildings and the impact of the COVID-19 pandemic on forecasts. In: Proceedings of 40th SGAI International Conference on Artificial Intelligence, vol. 12498; 2020. pp. 227–240.
[10] Favre-Perrod P. A vision of future energy networks. In: 2005 IEEE Power Engineering Society Inaugural Conference and Exposition in Africa; 2005. pp. 13–17.
[11] Mohammadi M, Noorollahi Y, Mohammadi-ivatloo B, *et al.* Energy hub: From a model to a concept – A review. Renewable and Sustainable Energy Reviews. 2017;80:1512–1527. Available from: https://www.sciencedirect.com/science/article/pii/S1364032117310985.
[12] Jelić M, Batić M, Tomašević N, *et al.* Towards self-sustainable island grids through optimal utilization of renewable energy potential and community engagement. Energies. 2020;13(13):3386. Number: 13 Publisher: Multidisciplinary Digital Publishing Institute. Available from: https://www.mdpi.com/1996-1073/13/13/3386.
[13] Batić M, Tomašević N, Beccuti G, *et al.* Combined energy hub optimisation and demand side management for buildings. Energy and Buildings. 2016;127:229–241.
[14] Achieving energy efficiency through behaviour change: what does it take? – European Environment Agency [Publication]. Available from: https://www.eea.europa.eu/publications/achieving-energy-efficiency-through-behaviour/.
[15] Esnaola-Gonzalez I, Jelić M, Pujić D, *et al.* An AI-powered system for residential demand response. Electronics. 2021;10(6):693. Number: 6 Publisher: Multidisciplinary Digital Publishing Institute. Available from: https://www.mdpi.com/2079-9292/10/6/693.
[16] Alfageme A, Esnaola-Gonzalez I, Díez FJ, *et al.* Metaheuristics for optimal scheduling of appliances in energy efficient neighbourhoods. In: Marreiros G, Melo FS, Lau N, *et al.*, editors. Progress in Artificial Intelligence. Lecture Notes in Computer Science. New York, NY: Springer International Publishing. pp. 151–162.
[17] Lissa P, Deane C, Schukat M, *et al.* Deep reinforcement learning for home energy management system control. Energy and AI. 2021;3:100043. Available from: https://www.sciencedirect.com/science/article/pii/S2666546820300434.
[18] Energy Master Plan 2018. Árainn and Inis Meáin [Publication]. Available from: https://156.234.107.34.bc.googleusercontent.com/community-energy/sustainable-energy-communities/tools-and-resources/energy-master-plan/Sample_Energy_Master_Plan.pdf.
[19] Clean Energy for EU Islands – Aran Islands (Ireland) [Publication]. Available from: https://www.euislands.eu/island-details/25.

[20] RESPOND D5.4 – Desktop Dashboard and Smart Mobile Client Demonstrator [Report]. Available from: http://project-respond.eu/repository/.

[21] DEXMA Platform – AI Powered Energy Savings Tool [WebPage]. Available from: https://www.dexma.com/what-is-dexma-platform/.

[22] DEXMA – Automatic Baseline Calculator [WebPage]. Available from: https://support.dexma.com/hc/en-gb/articles/360013577059-Apps-Market-Automatic-Baseline-Calculator-ABC-.

[23] International Performance Measurement and Verification Protocol (IPMVP) [WebPage]. Available from: https://evo-world.org/en/products-services-mainmenu-en/protocols/ipmvp.

[24] Ceallaigh TJO, Dhonnabhain AN. Reawakening the Irish language through the Irish education system: challenges and priorities International Electronic Journal of Elementary Education. 2015;8(2):179–198.

[25] Gallagher CV, Leahy K, O'Donovan P, *et al.* IntelliMaV: A cloud computing measurement and verification 2.0 application for automated, near real-time energy savings quantification and performance deviation detection. Energy and Buildings. 2019;185:26–38.

[26] Eir Group Webpage [WebPage]. Available from: http://www.eirgridgroup.com.

[27] Christensen TH, Knudsen HN, Yara A, *et al.* RESPOND Deliverable 6.3 – User Engagement Assessment. The RESPOND Consortium. Available from: http://project-respond.eu/repository/.

Chapter 17

Use case of artificial intelligence, and neural networks in energy consumption markets, and industrial demand response

Ashkan Safari[1] *and Amir Aminzadeh Ghavifekr*[2]

Despite all achievements, and advances in energy markets, microgrids, and smart grids within the world, issues such as power distribution, consumption, or optimization are among the important and significant areas within the industry and technology. As industrialization and technology improve, these subjects become more important. Most of the experts attempt to have far better control on power consumption/distribution, and technologies like combined heat, and power (CHP), or gas-electricity, or demand forecasting, especially in smart sustainable cities (SSCs). Using artificial intelligence (AI) and neural networks (NNs) can have an important role in performing, and optimization that will lead to lowering the issues in future power systems. An NN-long short-term memory (LSTM)-based model can help the experts to control, predict, and optimize the facility consumption, and power distribution. Conceptually, in industrial and SSCs, more they develop, more the quantity of data is going to be generated that a simple and practical tool to research about and analyze these big data is AI. Regarding an outsized amount of data, the training and predicting process of AI is going to be far more accurate, due to the low root mean square error (RMSE). Accordingly, the result is going to be near the actual and help the SSCs to possess controlled power consumption, distribution, and CHPs. Also, the combination of quantum technology with smart grids, and NNs are analyzed. Accordingly, the mentioned technologies cause preventing power loss and promoting a way to a smarter, technology-based, and sustainable world with high ability of demand response (DR).

17.1 AI in energy market

Recent advances in industry and technology have caused an exponential increment in energy through the last few years [1]. Especially in SSCs that energy, AI, and IoT can be considered its main bases. As well as industrialization, the growing need for electricity-consuming devices has been increased [2, 3], accordingly, most researches

[1,2]Faculty of Electrical and Computer Engineering, University of Tabriz, Tabriz, Iran

and developments are now in the fields of renewable energy, AI, and IoT in electricity [4]. Also, other optimization methods such as energy management in a virtual power plant (VPP) using robust Stackelberg game approach that is effective in Day-Ahead (DA), and real-time market transactions [5] are used. Even, bidding of VPPs in energy markets by a bi-level multi-objective approach [6] is in demand. Based on [7], in the markets with different self-interest subjects, consumers become prosumers with attribution of production and consumption by using distributed generators (DG). AI is one of the practical tools that can help energy consumption optimization in an easier way. In the big data-oriented fields to power consumption in control centers, AI tools should be used in energy sectors. Also, the companies if want to remain competitive should use it. The conventional power grid (PG) was not suitable to handle the combination of renewable energy sources (RESs), due to the fact that a change in the source can cause challenges in the microgrids in meeting variable loads [8]. Applying AI leads to:

- Efficient inverter of photovoltaic (PV) systems [9]
- Maximizes the ability of power point tracking [10]

Moreover, artificial Maximum Power Point Tracking (MPPT) approaches are efficient, and swarm optimization classes favor the approach for MPPT, since its fast and easy capability of consumption [11].

More technology and industrialization develop, the much Big Data (BD) is generated and the use of AI comes more sensible, especially in SSCs, so energy consumption can be reduced through better control, prediction, reliability, and automation. Nowadays, the AI technology advances in SSCs and smart buildings caused two concepts that are effective in the power consumption field [12]:

- Building management system (BMS)
- DR programs (DRP)

Ref. [13] says practical power system engineering and energy market problems require logical reasoning. Thus, since AI has developed, it can be a constructive tool.

Finally, in this chapter, a new and practical approach, and method in power consumption, dual fuels, and DR prediction are defined to have better control in distribution, and transmission that helps better control the energy, and D optimization, especially in SSCs. Moreover, it is in the combination of IoT, and so on that will cause a reduction in energy loss and waste. One of the fundamental methods is the NN-LSTM-based model that is entirely expressed in the further sections. Also, the quantum-based methods are defined. The structure of the chapter is as follows:

For Section 17.2, we have NNs description. Sections 17.3 and 17.4 represent the power consumption, and cogeneration & dual fuels, respectively. DR, power consumption prediction, and the proposed framework are expressed in sections 17.5, 17.6, and 17.7, respectively. Sections 17.8 and 17.9 describe the use case, overview, and benefits of the NN model, the difference between RNN, and LSTM, and their overview. Quantum technology, its applications in general, and smart grids are presented in Sections 17.10, 17.11, and 17.12. Sections 17.13 and 17.14 express the quantum techniques in forecasting variables of smart grids, and the conclusion, respectively. The brief structure of the chapter is illustrated in Figures 17.A1 and 17.A2.

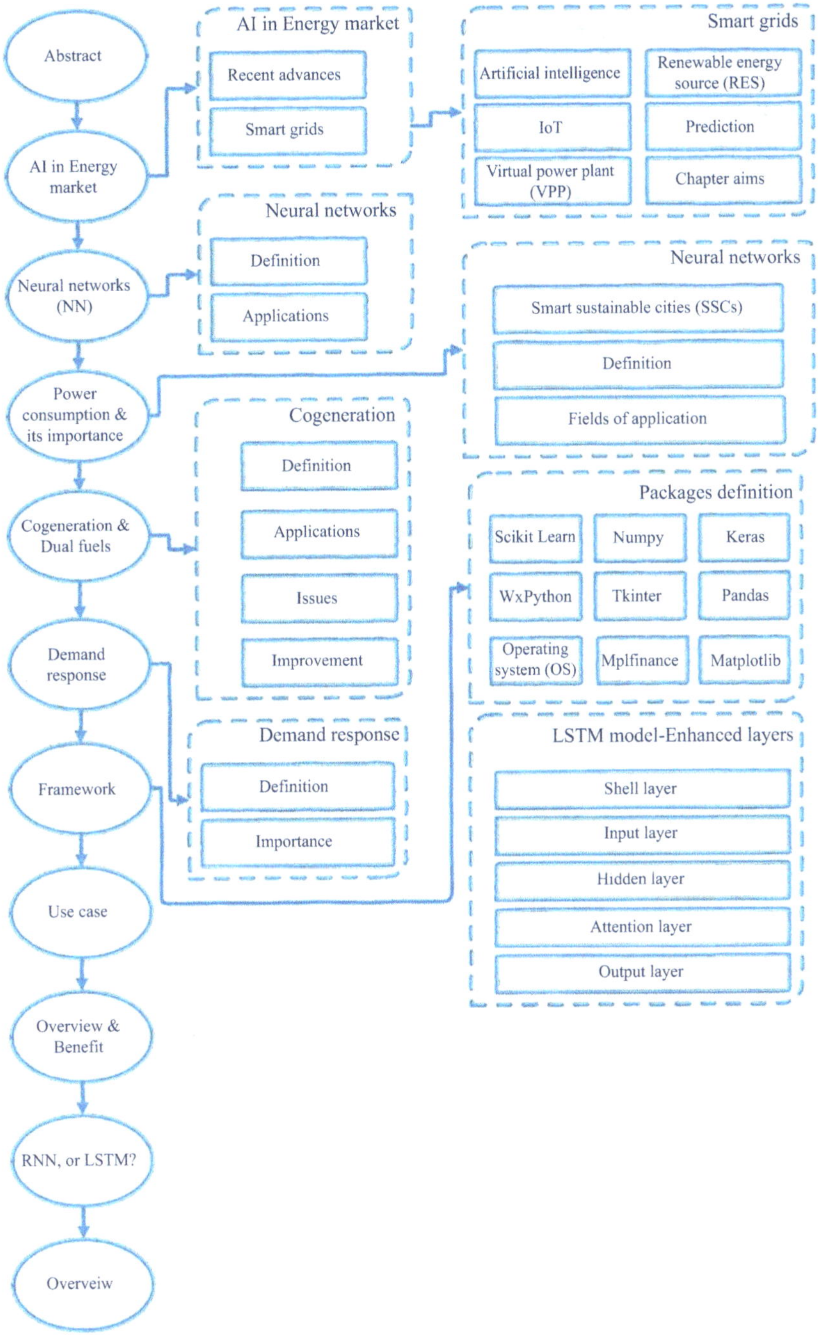

Figure 17.A1 The first part of chapter structure (regular AI and smart grids)

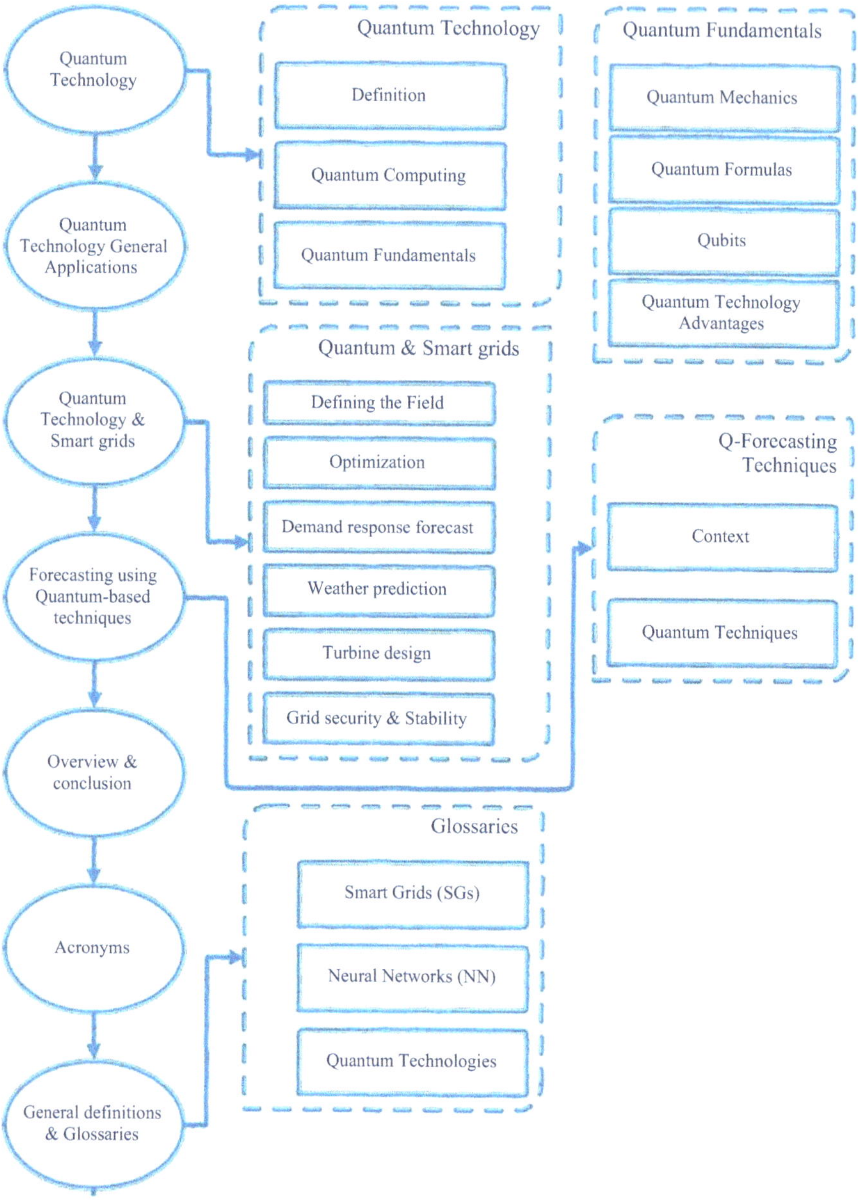

Figure 17.A2 The second part of chapter structure (quantum technology and smart grids)

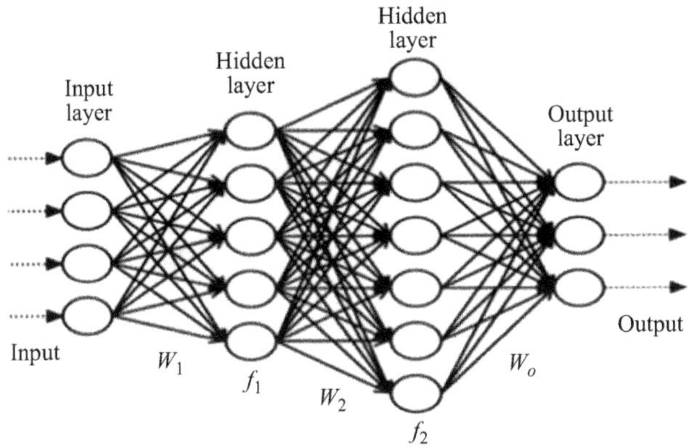

Figure 17.1 A 4 input–3 output NN model

17.2 NN

A NN is a group of algorithms that resemble the operations and processes of a human brain to recognize and distinguish correlations between large amounts of data and information. In a different form, an NN is based on a set of connected nodes called artificial neurons, which loosely model and trace the neurons in a biological brain. They are used in a variety of applications in financial services, forecasting and marketing research, power consumption prediction, micro and smart grids, sustainable cities, and Blockchain. Figure 17.1 depicts a 4 inputs–3 outputs NN.

17.3 Power consumption and importance of its prediction

Recent advances in technology have caused electrical energy and power to be used in a much larger amount than before. Accordingly, annual electricity consumption per capita is a crucial measure of electrical power development. Generally, electricity consumption develops faster as the industrialization and technology process quickly and performs down rapidly when industrialization is completed or near the achievement. The same is often remarked about annual electricity consumption per capita. The most important part of monitoring power consumption can be divided into three main factors:

1. Comprehend the capacity of the electric panel
2. Identify the energy costs and prevent the energy waste
3. Troubleshoot circuit breaker trips

One of the fields that experts are trying to optimize the power consumption is SSC. Based on UNECE and ITU definition, a smart sustainable city is an innovative city that uses digital technologies to improve the quality of different aspects such as

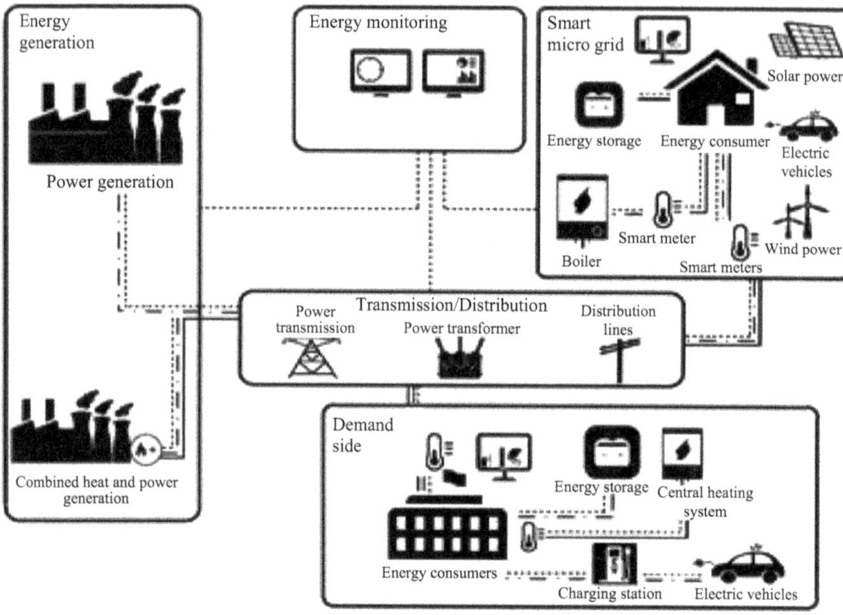

Figure 17.2 Energy system scheme with smart grid infrastructure

services while ensuring it fulfills present and future generations considering economic and social, as well as environmental, and cultural aspects.

In order to have optimized power consumption and energy transfer in SSCs, smart grids are being used. This subject is illustrated in Figure 17.2.

Figure 17.2 represents a schematic that indicates the process of energy generation to energy consumption and their communication line and transportation. Based on the model, the generated balk energy and the heat move to energy monitoring and control and the transmission/distribution system. Accordingly, the result is transmitted and distributed in microgrids and demand sides such as Energy storage, as well as EV charging.

Ref. [28] studies the history of smart grids in energy systems. Accordingly, there will be energy storage, high-temperature superconductor, and more distribution control centers in the future.

Finally, many smart grid technology areas span the entire grid from generation to transmission and distribution in order to supply energy for various types of electricity consumers. This subject is manifested in Figure 17.3.

According to Figure 17.3, the world will have Digital Grid AI, and smart building or factory based on, and completely controlled by AI [29]. Also, represents the promoted smart grids in the future, the collaboration with AI, the concept and growth of smart grids [14], and the differences of conservational grids with smart grids [15].

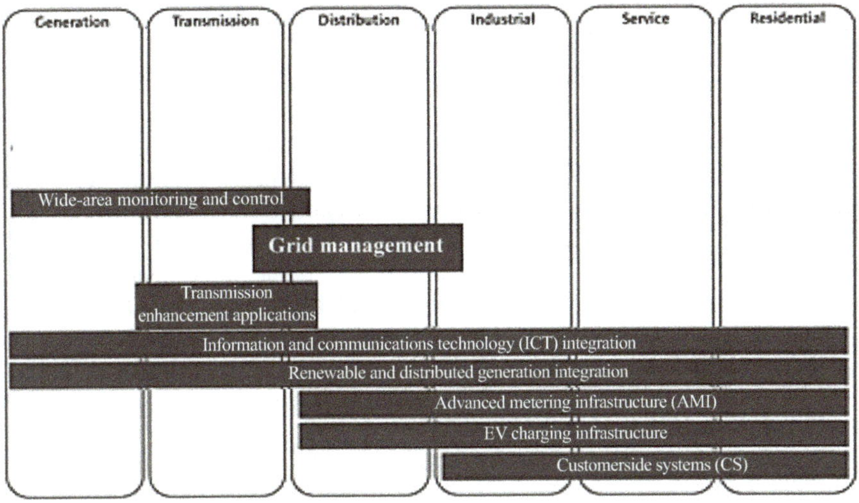

Figure 17.3 Smart grid technologies, consumption types, and the process

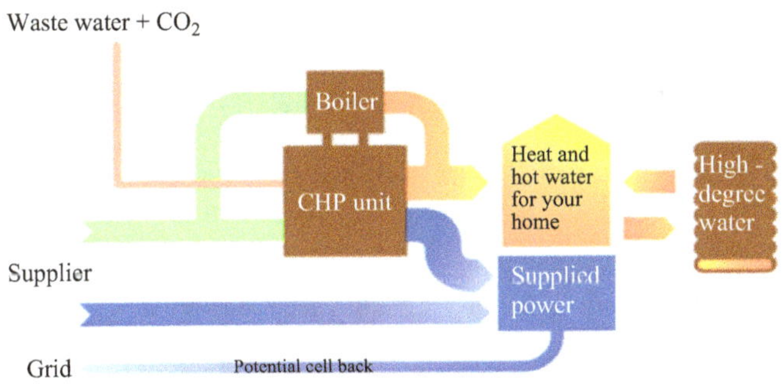

Figure 17.4 Fuel cell cogeneration

17.4 Cogeneration and dual fuels

Cogeneration also referred to as combined heat and power (CHP) is the synchronized production of two or more varieties of energy from one fuel source. Cogeneration power plants often operate at 50–70% higher efficiency rates than single-generation plants. This concept is illustrated in Figure 17.5.

Figure 17.4 represents heat-electricity cogeneration that is much more practical and economical than conventional approaches and methods. Figure 17.4 manifests that prosumers can sell excess electricity back to the grid by creating a decentralized energy system that leads to reduce carbon footprint and lower energy bills.

Conventional power channels produce electricity remotely from their users. This massive distance means it is not economically viable to move the wasted heat to homes to be used. Also, there is an extraordinary inconvenience related to the distance and it is that the facility generated in large power plants has to be transported to the client, with transmission losses occurring on the long route. However, cogeneration enables electricity to be produced directly where demanded. Consequently, it prevents network costs and transmission losses. Up to 90% of the fuel is manipulated [16], by the waste heat used for heating. Moreover, a residential fuel-cell-combined heat and power (FC-CHP) system is utilized as a promising low-carbon technology that will reduce residential energy consumption [17]. A CHP can also produce serviceable cooling when merge with a heat-driven absorption chiller. The process during this operation is defined as "trigeneration" or combined cooling heat and power (CCHP). Compared to the widespread production of energy, these systems contribute many benefits in terms of [18]:

- Primary energy savings.
- Economic and greenhouse emission (GHG) savings.

Dual fuel energy means you get your gas and electricity from an equivalent supplier. It could prevent money and be more convenient than using separate suppliers, leaving with one bill and one point of contact. Its only difference with the common ways is that the consumers both energies are supplied with the same supplier. However, there are challenges in the mentioned topics, smart grids, distributed intelligence for energy generation, and the role of distributed energy resources (DERs) in the future of power systems [19].

Using AI in this field will lead to creativity, efficiency, and environmental protection by saving energy and lower waste. The significant benefits can be as follows [19]:

- Analyze the consumers' generated data that are types of Big Data (BD), by AI to give better services and optimized energy.
- DERs can get access to all markets using VPP concept.
- Using VPP market intelligence to optimize the process in DERs.
- Joint (gas, water, heat) and electricity demand forecasting and improve the accuracy of the forecasting.

Figure 17.5 illustrates the energy generation capacity by RES.

Another application is modeling prediction process of required turbine power, or gas flow of an engine, that is depicted in *Use Cases* section.

17.5 DR and its importance

DR is the changes in electricity usage by end-use customers regarding their regular consumption patterns in response to change in the price of electricity over time, or to incentive payments in high commercial market prices [24]. DR circumscribes short-term impacts on the electricity markets, leading to financial benefits for both

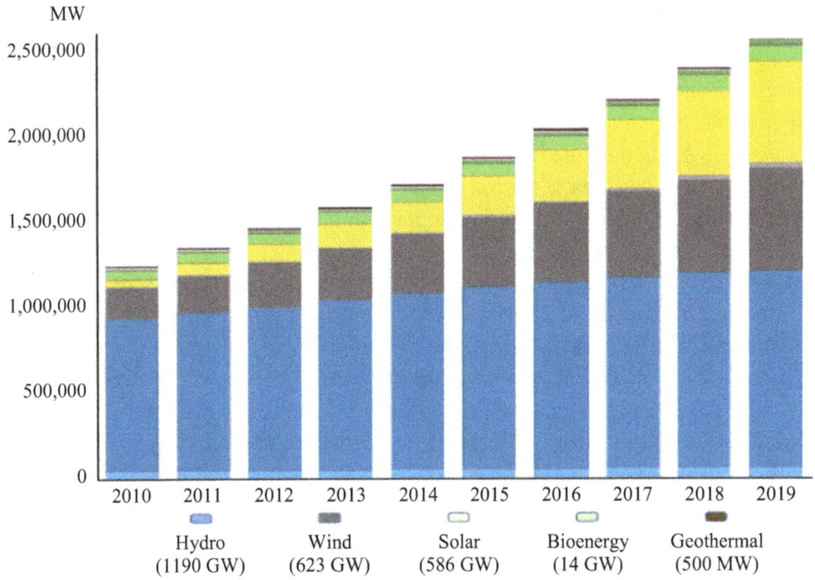

Figure 17.5 Energy generation capacity by RESs [19, 21]

the customers and the utility by promoting the interchange and responsiveness of the customers. Moreover, it improves the loyalty of the power system and, also, lowers peak demand reduces overall plant and capital cost investments, furthermore postpones the need for network upgrades [24].

17.6 Power consumption prediction using artificial NNs (ANNs)

As mentioned in the previous chapter, it is significant to have a prediction of power consumption, especially in a SSC that includes a wide range type of residential and industrial technologies such as IoT being used and consume energy and power. For having a high accuracy prediction, an ANN model can be used. This method is based on LSTM. Besides other methods such as real-time recurrent learning (RTRL), since the importance of accurate data, NN has become one of the efficient and sensible algorithms that are being utilized for modeling market behaviors [20]. ANN is a well-known approach that also consolidates technical analysis to make predictions in markets. One of the most effective methods in this area is LSTM [23]. LSTM is used in common since it predicts with high accuracy and low RMSE.

17.7 Framework of the NN-LSTM-based model

The program of the model runs with Python, which operates variety of applications such as numeric, and educational applications, IoT, internet development, and desktop graphical user interfaces (GUIs). In order to predict using the ANN model, the related AI and ANN packages are used:

Keras is an open-source implementing interface Python library for NNs. It operates as an interface for the TensorFlow library.

TensorFlow is a machine learning (ML) library. It can be used over a wide range of tasks, however, has a special focus on training and inference of deep neural networks (DNNs). It is an emblematic math library based on data flow and differentiable programming.

Numpy, a Python programming library, that enhances assistance for large, multidimensional arrays and matrices, concurrently with an outsized of high-level mathematical functions to run on the mentioned arrays.

Scikit-Learn is a ML-based Python library. It has diverse classification, regression, and clustering algorithms including support vector machines.

Pandas is a Python library for data administration and analysis. Remarkably, it extends data structures and operations for developing, modeling, training, and handling numerical tables and time series.

Tkinter is Python's standard graphical user interface (GUI) package and object-oriented (OO) interface.

WxPython is a type for the cross-platform GUI API (GUI Application Programming Interface). It is one of the alternatives to Tkinter GUI and performed as a Python extension module.

Matplotlib is a library for plotting. It implements an object-oriented (OO) API for implanting plots into platforms using GUI toolkits such as WxPython, Tkinter.

Mplfinance is a Matplotlib utility package for the visualization, and analysis of financial data.

OS (*Operating System*) module accommodates functions for interacting with the operating system in Python. It progresses under Python's standard utility modules. This module provides a portable way of using OS-dependent functionality. It incorporates many operators to connect with the file system and the intercommunication with the environment of Python.

In this chapter, the system code to be used in use cases is based on NN attention-based LSTM with a personalized security layer that the whole runs under Python. The model consists of five layers: shell & input layer, hidden layer, attention layer, as well as output layer.

The shell layer authenticates the user. The input data that meet the input requirements are read by the input layer. The hidden layer is associated with the linear network within the LSTM unit. The attention layer provides future substances based on the forecasts that are performed in the hidden layer. The final calculated and evaluated results are received by the output layer to manifest toward the user. The LSTM-based prediction is portrayed in Figure 17.6.

According to Figure 17.6, after the user imports the data, the system starts to create a data frame of the entered data. Then converts the data frame to a Numpy array to start performing algorithms in Python. In the next step, the system gets the number of rows to train the model. After scaling data, it creates a scaling training dataset. As the next

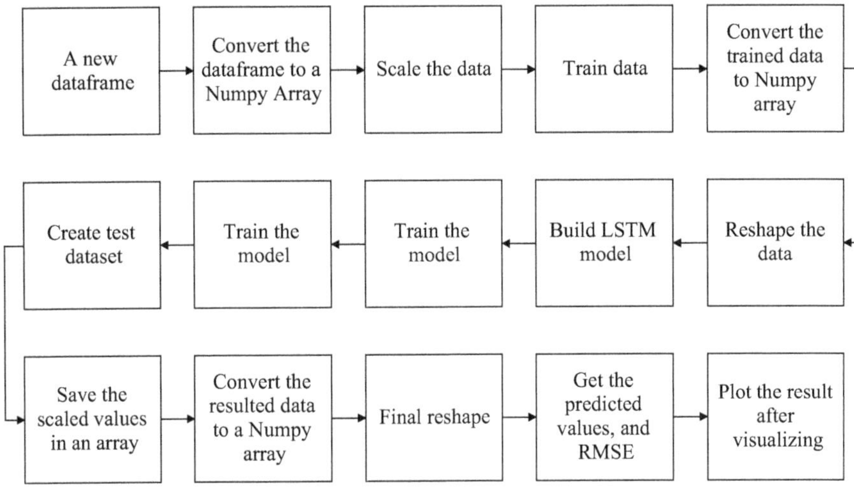

Figure 17.6 LSTM-based prediction diagram of the system

step, the dataset divides into two different datasets and the final ones are converted to the Numpy array. After reshaping data, the building and reshaping process of the model starts. After the converting process is done, reshaping and getting the predicted values begins. Finally, the result information is visualized and depicted to the user. This model can be used to predict power consumption, electricity costs, stock index, and other areas.

17.7.1 Shell layer

It is the first layer of the system in which the user will be authorized and authenticated. It works based on PIN or Voice that authenticate the user. Accordingly, if the user authorized, he/she will be allowed to move to the next layer. The algorithm of this layer is depicted in Figure 17.7.

17.7.2 Input layer

In this layer, user imports the data or if it is about electricity stock index, we can enter the stock name, and the Pandas package will get data and Numpy will perform the array-related tasks.

17.7.3 Hidden layer

It is the main layer of the system and the model. All predictions and LSTM method and RMSE are performed on the data in this layer. The block diagram of the layer can be exhibited in Figure 17.8.

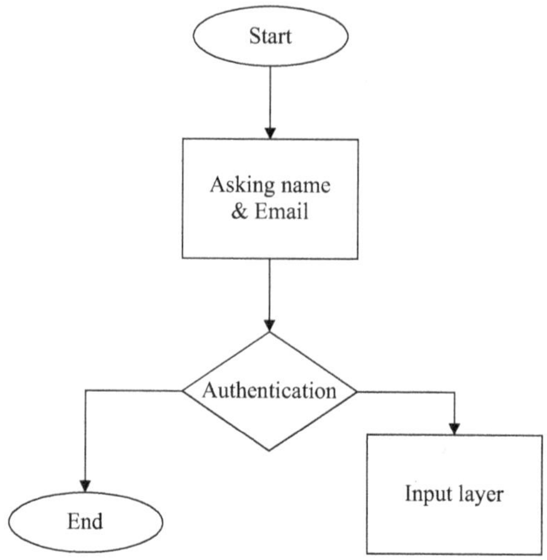

Figure 17.7 The algorithm of Shell layer in the system model

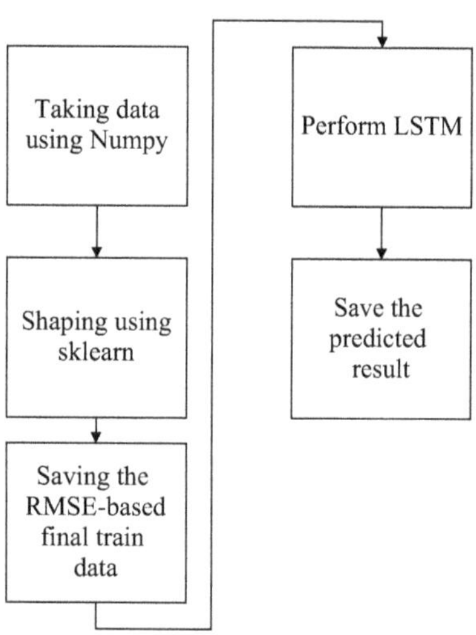

Figure 17.8 Block diagram of hidden layer

Depending on Figure 17.8, after the calculation finished, an email including Excel file of the result is sent to user's email that makes it more efficient. In this layer RMSE is adjusted on the final data based on the number of them. RMSE is defined as:

$$RMSE = \sqrt{\frac{\sum_{i=1}^{n} \left(Y_{i,\text{Predicted}} - Y_{i,Actual}\right)^2}{n}} \tag{17.1}$$

Based on (17.1), n is the total amount of imported data, $Y_{i,\text{Predicted}}$, and $Y_{i,Actual}$ ($Y_{i,\text{Predicted}}$, $Y_{i,Actual}$) are the predicted data, and the actual data, respectively. Accordingly, the more data imported, the less RMSE will be which manifests the error of the prediction is low and the predicted amount is near to the actual one.

17.7.4 Attention layer

After the time predicted data is ready in hidden layer, it moves to the Hidden Layer which learns the relation between data and predicts the future in a quantitative and qualitative way. When finished, it sends the result to the output layer.

17.7.5 Output layer

It is the final layer in which the results, sent from the previous layer (attention layer), are illustrated and shown.

17.8 Use case of NNs

AI and computational intelligence (CI) have a vast and vital range of applications in smart grids and power systems based on:

- ANNs.
- Neuro-Fuzzy systems.

As a use case that fits the subject of the chapter (prediction based on ANN), power prediction of steam turbines and gas flow prediction using ANN has been presented and shown below.

Based on Figure 17.9, the result is obtained, the RMSE value is 0.94% in the steam turbine model, and, in Figure 17.10, it is 2.75% in the engine, which is higher

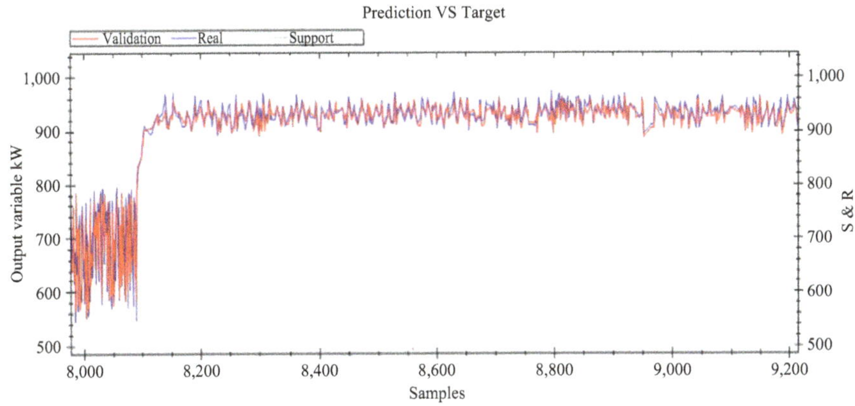

Figure 17.9 Real power and the power prediction model for a turbine using ANN [22]

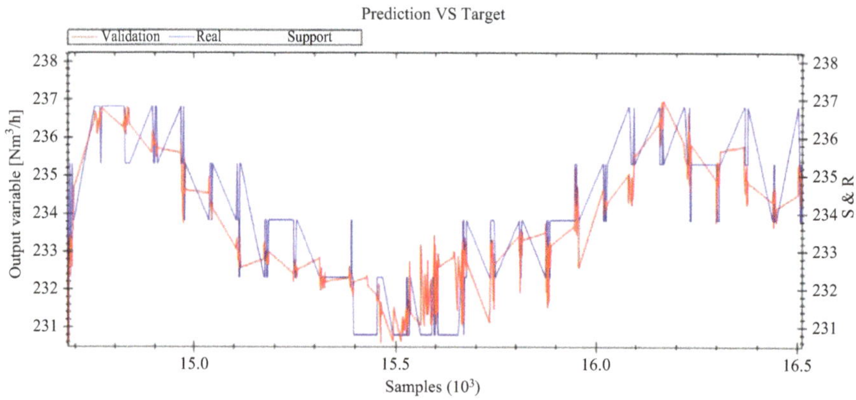

Figure 17.10 Real gas flow and gas flow prediction for an engine using ANN [22]

due to the load restrictions [22]. In both models, red lines are the prediction, and blue ones are the real values. Factors like this can help the experts to have better control of DR.

17.8.1 Overview and benefit

Based on the model and system presented, all kinds of fields that include pervious database can be predicted with high accuracy and near to the actual. The data can be the power consumption, or power consumption price of a city, company, or a factory. Accordingly, the future value will be predicted quantitatively and qualitatively. As a result, it can have a significant role in SSCs, and smart energy transportation, control, and optimization fields, and even smart grids.

17.9 RNN or LSTM: Which one is better for prediction?

When we want to evaluate the prediction ability of RNN, we can perform this in stock index prediction. The stock market is another channel for investment for individuals, even corporations. It can influence the electrical energy market and the reverse. Due to this fact, almost all famous companies have stock. However, success and proceeding in this market require experience and technical skills. Moreover, recent significant advances in the stock market such as cryptocurrencies and Blockchain technology have been started to be used in SSCs, makes them complex and unpredictable in a quantitative way. Also, recurrent NN (RNN) algorithms can be useful for stock market modeling. Whereas, a tuned AI-NN-based predictor program with high accuracy will play a significant role for investors to get the best result with the highest profit and the least disadvantage. The difference between RNN and LSTM is shown in Figure 17.11.

According to Figure 17.11, all RNNs perform feedback loops within the recurrent layer that lets them maintain information in memory over time. However, it is

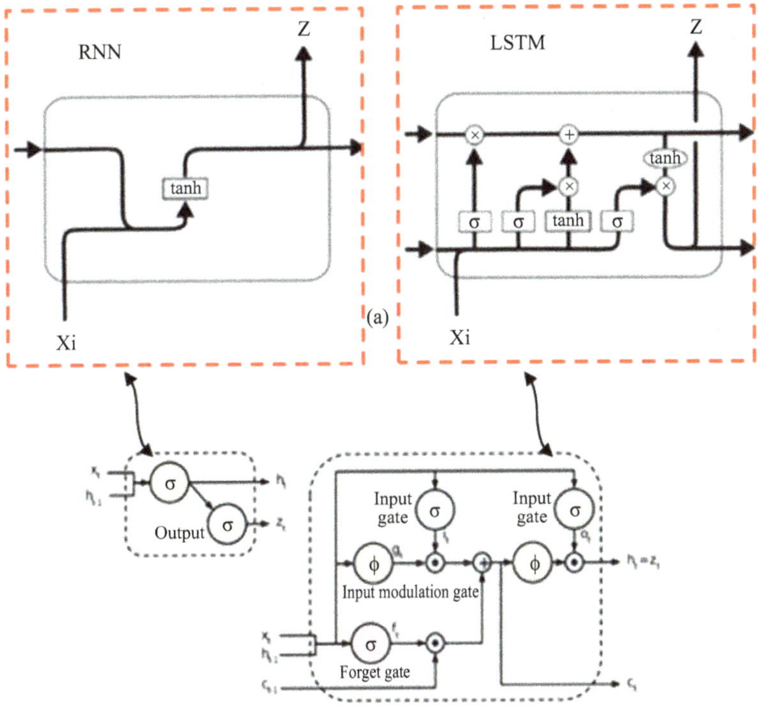

Figure 17.11 The difference between two models RNN and LSTM

challenging to train standard RNNs to solve problems that require learning long-term temporal dependencies. LSTM networks are a type of RNN that uses specific units besides standard ones. LSTM units include a 'memory cell' that can maintain information in memory for long periods. A set of gates is used to control when information enters the memory, its output, or when it is missed. This architecture lets them learn longer-term dependencies. This is the significant and main difference between LSTM and RNN.

17.9.1 Overview

Energy, its consumption, and effect on technology development, and the related companies and stocks are always important and play a significant role in this area. By ANN LSTM-based model, they can be predicted and lead to the best decision to improve and develop. Another application is to control the energy consumption/distribution in control centers of smart grids to prevent energy waste and loss. However, there is a challenge here. If the amount of Big Data that will be analyzed in control centers of smart grids is enormous, the processing will be challenging for supercomputers, and they can be inefficient. In this case, another stronger technology is known as *"Quantum Technology"* should be used.

17.10 Quantum technology

Quantum technology is a class of technology, and an emerging field of engineering and physics, which relies on the principles of quantum physics, that operates by utilizing the principles of quantum mechanics, including quantum superposition and quantum entanglement which leads to amazing technologies such as "Quantum Sensing," and high processing speed known as "Quantum Computing."

17.10.1 Quantum computing

Quantum computing is developed and processed by quantum mechanical phenomena that define the nature, energy at the level of fundamental subatomic particles. To get the aspired results, quantum computer controls the behavior of subatomic particles which permits it to possess enormously high processing power that can perform multiple functions simultaneously by using possible permutations [25].

17.10.2 Quantum fundamentals

Quantum mechanics is a science apportioning the performance of matter and light on, mostly, the subatomic scale. It attempts to represent the properties of molecules and atoms and their ingredients: electrons, protons, neutrons, and other more esoteric particles such as quarks and gluons. The main formulas of Quantum are written below:

$$\psi(r, s_z, t) = \frac{1}{\sqrt{2\pi h^3}} \int e^{\frac{ipx}{h}} \Phi(p, s_z, t) d^3 p_n \tag{17.2}$$

$$\Phi(p, s_z, t) = \frac{1}{\sqrt{2\pi h^3}} \int e^{\frac{-ipx}{h}} \psi(r, s_z, t) d^3 r \tag{17.3}$$

$$H(t)|\psi(t)\rangle = ih\frac{d}{dt}|\psi(t)\rangle \tag{17.4}$$

According to (17.2) and (17.3), Ψ is the position-space wave function, h, and Φ are Plank constant, and momentum–space wave function, respectively. Equation (17.4) is the most fundamental equation of quantum and quantum mechanics which is known as Schrodinger's equation. The most fundamental difference between quantum technology, and the regular ones is their particles. Regular technologies are based on bits (0, and 1), or Fuzzy logic. However, quantum technology is based on quantum bits known as *"qubits."* Regular bits can be 0, or 1. Whereas qubits can be across all available states between 0, and 1. Respectively, this feature makes them 1 million times faster than the regular computing technologies [25]. As well, it causes quantum technology to be used as an encryptor known as *"Quantum Cryptography"* in smart grids and SSCs. Qubits can be modeled by a unit sphere called *"Bloch Sphere"* [26] that is represented in Figure 17.12.

Depending on Figure 17.12, the North, and South poles, respectively, are $|0\rangle$, and $|1\rangle$. In qubits, the state can be anywhere between $|0\rangle$, and $|1\rangle$that's called superposition. On the other hand, particles can be in any states such as position, energy, or speed, or

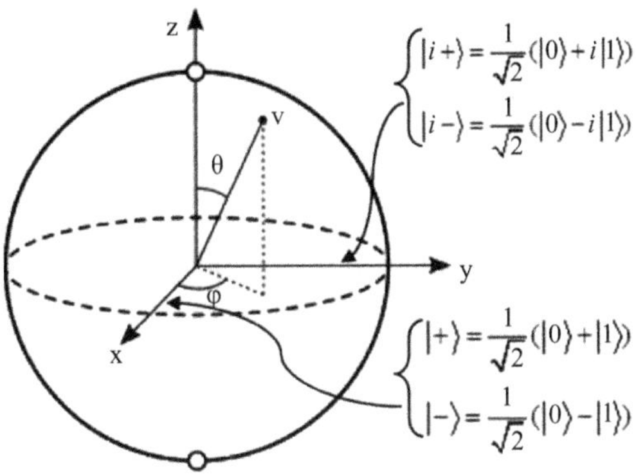

Figure 17.12 Visualization of a Qubit as a unit sphere known "The Bloch Sphere"

across all available states between 0, 1 that's clear in the figure. Based on the figure, the related equations are, also, presented for $\boldsymbol{x}$, and $\boldsymbol{y}$ axes, respectively.

17.11 Quantum technology general applications

As described in aforementioned sections, quantum technology is based on qubits which makes it be much faster than the regular technologies. Accordingly, it can be used in all fields such as:

- AI & ML
- Blockchain technology
- SSCs
- Smart grids (SGs)
- Security and cryptography

In the next section, quantum technology applications in smart grids will be expressed in detail.

17.12 Quantum technology and smart grids

Future grids of SSCs will face challenges that have not occurred before, accordingly, it will be challenging to perform them using the regular technologies. Quantum computing is a developing technology that has a great potential to last for a long time and be combined with other technologies such as AI, or Blockchain. As mentioned, in smart grids, AI, and NN are among the practical tools to analyze the gathered Big Data, however, they can perform by a limited amount of Big Data. When they face an enormous amount of Big Data, they will be inefficient. However, when they combine with quantum technology, QNN will be set up that runs under quantum theories and has an enormously fast processing speed that causes smart grids to have much more accurate, dependable result in a much less time. Hybrid systems that combine multiple RESs, such as CHPs or vector coupling technologies, are challenging to optimize. A grid that holds both wind and solar energy generation has the advantage of supplying less expensive energy as long as sunshine and wind blowing. Nevertheless, to satisfy customers' energy demands at night or during modest days, the grid needs to pull from stored power or ramp up energy generation from other resources. An IAS, Intelligent Automated System, that could trace demand, predict peaks in consumption, correspondent energy storage, and manage resources could remarkably boost efficiency, therefore, cover the way for a cheaper, and much more reliable, and abundant power [27]. Since quantum computing has substantial features, it can be used in significant aspects of smart grids that are as follows, but are not limited to:

- Optimization
- Demand forecasting
- Weather prediction
- Turbine design
- Grids stability & security

Based on the mentioned fields, we describe and focus on demand prediction, and DR, and management that are significant in SGs [24].

17.13 Forecasting in smart grids using quantum technology

Forecasting is always among important factors, and tools in smart grids even caused significant advances in solar or wind generation. It is performed by AI, and ML techniques such as ANN, RNN, etc. However, as mentioned in the previous sections, when gathered data of smart grids is the type of enormous big data, regular supercomputing methods are inefficient, and stronger technology is known as "Quantum Computing" should be used. Quantum computers have an enormously fast processing speed which makes the process and evaluate big data in much less time and perform the prediction process in a much accurate and fast way. Quantum technology has techniques such as QNN, Quantum behaved Particle Swarm Optimization (QPSO) to predict or forecast different variables. Quantum-based models and algorithms have more accurate, and faster outputs, and results than the regular models, and methods like Artificial Neural Networks (ANN) [25]. Based on (17.1), that is RMSE equation:

$$RMSE\alpha\frac{1}{\sqrt{n}} \tag{17.5}$$

In which, (17.5) expresses that when we face an enormous amount of Big Data, n increases accordingly, $RMSE$ decreases, and the prediction is near to the real value.

17.14 Final overview and conclusion

In this chapter, smart grids, their definition, and combination with AI, and quantum technology, prediction techniques of important variables like DR have been presented. Smart grids are among the significant and fundamental bases of an SSC which are responsible for electricity supply. One of the important factors in smart grids is DR. Experts try to have an enhanced prediction of it to control and optimize the grid. To predict these types of factors AI tools are being used with supercomputers. However, when a great number of Big Data is gathered, the prediction process with supercomputers will be inefficient. In order to cover this challenge, quantum computers, accordingly, quantum-combined AI methods such as QNN, QPSO should be used to have a more accurate prediction, in less time, in important variables like DR and weather prediction.

Acronyms

The acronyms and their definitions are presented in Table 17.1.

Table 17.1 Table of acronyms

Acronym	Meaning
SG	Smart grid
MG	Micro grid
QA	Quantum adiabatic
AI	Artificial intelligence
ML	Machine learning
NN	Neural network
IA	Intelligent automation
BD	Big Data
ANN	Artificial neural network
QNN	Quantum neural network
QKD	Quantum key distribution
SSC	Smart sustainable city
ITS	Information technology security
PMU	Phase measurement unit
DSM	Demand side management
IAS	Intelligent automation systems
CHP	Combined heat and power
RES	Renewable energy sources
QFNN	Quantum fuzzy neural network
RMSE	Root mean square error
ANFIS	Adaptive network fuzzy inference system
qRAM	Quantum random accessible memory
QPSO	Quantum particle swarm optimization
SOFC	Solid oxide fuel cell

References

[1] Himeur Y., Alsalemi A., Al-Kababji A., *et al.*: A survey of recommender systems for energy efficiency in buildings: principles, challenges and prospects. *Information Fusion*, 2021;72:1–21.

[2] Pyls, P., Lylykangas K., and Kurnitski J.: Buildings' energy efficiency measures effect on CO_2 emissions in combined heating, cooling and electricity production. *Renewable and Sustainable Energy Reviews*, 2020;134:110299.

[3] Himeur Y., Alsalemi A., Bensaali F., and Amira A.: Efficient multi-descriptor fusion for non-intrusive appliance recognition. In *2020 IEEE International Symposium on Circuits and Systems (ISCAS)*, IEEE, 2020.

[4] Refat K.H. and Sajjad R.N.: Prospect of achieving net-zero energy building with semi-transparent photovoltaics: a device to system level perspective. *Applied Energy*, 2020;279:115790.

[5] Shafiekhani M., Badri A., Shafie-khah M., and Catalão J.P.S.: Strategic bidding of virtual power plant in energy markets: a bi-level multi-objective approach. *International Journal of Electrical Power & Energy Systems*, 2019;113:208–19.

[6] Yin S., Ai Q., Li Z., Zhang Y., and Lu T.: Energy management for aggregate prosumers in a virtual power plant: a robust Stackelberg game approach. *International Journal of Electrical Power & Energy Systems*, 2020;117:105605.

[7] Ahmad T., Zhang D., Huang C., *et al.*: Artificial intelligence in sustainable energy industry: status quo, challenges and opportunities. *Journal of Cleaner Production*, 2021; 125834.

[8] Youssef A., El-Telbany M., and Zekry A.: The role of artificial intelligence in photo-voltaic systems design and control: a review. *Renewable and Sustainable Energy Reviews*, 2017;78:72–79.

[9] Seyedmahmoudian M., Rahmani R., Mekhilef S., *et al.*: Simulation and hardware implementation of new maximum power point tracking technique for partially shaded PV system using hybrid DEPSO method. *IEEE Transactions on Sustainable Energy*, 2015;6(3):850–862.

[10] Mao M., Cui L., Zhang Q., Guo K., Zhou L., and Huang H.: Classification and summarization of solar photovoltaic MPPT techniques: a review based on traditional and intelligent control strategies. *Energy Reports*, 2020;6:1312–27.

[11] Miyatake M., Veerachary M., Toriumi F., Fujii N., and Ko H.: Maximum power point tracking of multiple photovoltaic arrays: a PSO approach. *IEEE Transactions on Aerospace and Electronic Systems*, 2011;47(1):367–80.

[12] Farzaneh H., Malehmirchegini L., Bejan A., Afolabi T., Mulumba A., and Daka P.P.: Artificial intelligence evolution in smart buildings for energy efficiency. *Applied Sciences*, 2021;11(2):763.

[13] Ramos C. and Liu C.-C. AI in power systems and energy markets. *IEEE Intelligent Systems*, 2011;26(2):5–8.

[14] Chen Y., Niroumandi H., and Duan Y.: "Thermodynamic and economic analyses of a syngas-fueled high-temperature fuel cell with recycling

processes in novel electricity and freshwater cogeneration plant," *Energy*, 2021;235:121313.

[15] Isa N.M., Tan C.W., and Yatim A.: A comprehensive review of cogeneration system in a microgrid: a perspective from architecture and operating system. *Renewable and Sustainable Energy Reviews*, 2018;81: 2236–63.

[16] Arsalis A.: A comprehensive review of fuel cell-based micro-combined-heat-and-power systems. *Renewable and Sustainable Energy Reviews*, 2019;105: 391–414.

[17] Ozawa A. and Kudoh Y.: Performance of residential fuel-cell-combined heat and power systems for various household types in Japan. *International Journal of Hydrogen Energy*, 2018;43(32):15412–22.

[18] Papadimitriou A., Vassiliou V., Tataraki K., Giannini E., and Maroulis Z.: Economic assessment of cogeneration systems in operation. *Energies*, 2020;13(9):2206.

[19] Ali S.S. and Choi B.J.: state-of-the-art artificial intelligence techniques for distributed smart grids: a review. *Electronics*, 2020;**9**(6):1030.

[20] Mankar T., Hotchandani T., Madhwani M., Chidrawar A., and Lifna C.S.: Stock market prediction based on social sentiments using machine learning. In *2018 International Conference on Smart City and Emerging Technology (ICSCET)*, IEEE, 2018.

[21] Irena A.: Renewable capacity highlights. In *Proceedings of the International Renewable Energy Agency (IRENA)*; 2020.

[22] Seijo Fernandez S., Del Campo I., Echanobe J., Garcia Sedano J., Suso E., and Arbizu E.: Computational intelligence techniques for maximum energy efficiency of an internal combustion engine and a steam turbine of a cogeneration process. *International Journal of Energy and Environmental Engineering*, 2014;5(2):1–10.

[23] Safari A. and Ghavifekr A.A.: International stock index prediction using artificial neural network (ANN) and Python programming. In *2021 7th International Conference on Control, Instrumentation and Automation (ICCIA)*, IEEE, 2021.

[24] Siano P.: "Demand response and smart grids—a survey," *Renewable and Unstainable Energy Reviews*, 2014;30:461–78.

[25] Eskandarpour R., Ghosh K.J.B., Khodaei A., Paaso A., and Zhang L.: "Quantum-enhanced grid of the future: A primer," *IEEE Access*, 2020;8:188993–9002.

[26] Ketterer A.: "*Modular variables in quantum information*," Sorbonne Paris Cité, 2016.

[27] Shields A.: "Quantum cryptography: A new age in data security: Dr Andrew Shields highlights the importance of quantum cryptography in ensuring the future of data security," *Scientific Computing World*, 2018;163: 16–17.

[28] Graber G.: "Electric mobility: smart transportation in smart cities," PhD thesis, 2016.

[29] Moura P.S., López G.L., Moreno J.I., and De Almeida A.T.: "The role of smart grids to foster energy efficiency," *Energy Efficiency*, 2013;6(4): 621–39.

Index

www.ingramcontent.com/pod-product-compliance
Lightning Source LLC
LaVergne TN
LVHW011828200926
848484LV00002B/336